TIMBER

Its Structure & Properties

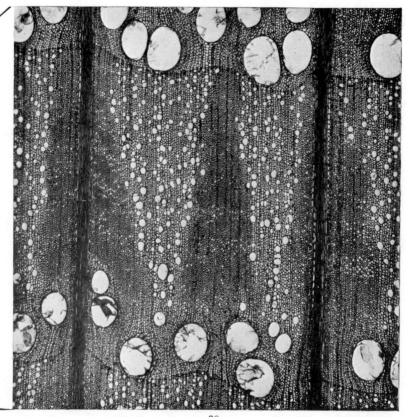

× 30

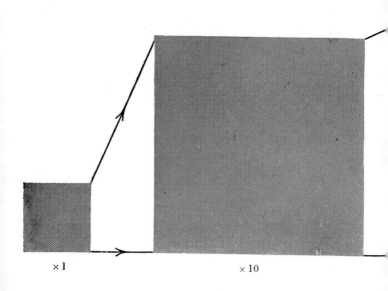

×1   ×10

Above: AREA MAGNIFICATION

Below: LINEAR MAGNIFICATION

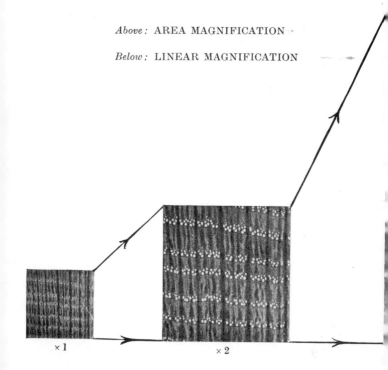

×1   ×2

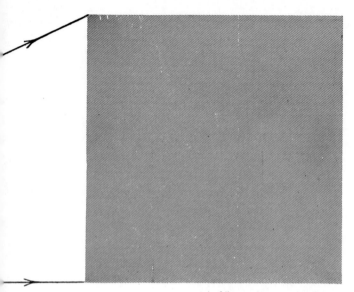

× 15

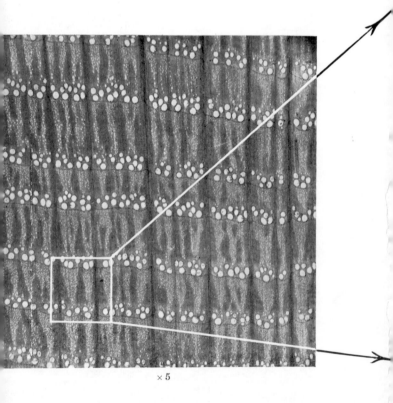

× 5

TIMBER

Its Structure and Properties

H. E. Desch
B.Sc., M.A., D.Phil.(Oxon), F.R.I.C.S.

Fifth Edition

Macmillan

First Edition 1938
Second Edition 1947
Reprinted (with additions and corrections) 1948
Third Edition 1953
Reprinted (with corrections) 1956
Reprinted (with corrections) 1962
Fourth Edition (revised and reset) 1968
Fifth Edition 1973

Published by
THE MACMILLAN PRESS LTD
London and Basingstoke
Associated companies in New York Melbourne
Dublin Johannesburg and Madras

SBN 333 14890 8

Reproduced and printed by photolithography and bound in
Great Britain at The Pitman Press, Bath

This Book is Dedicated to

S. H. CLARKE, ESQ., C.B.E., M.Sc.

*In Appreciation of His Assistance and Invaluable
Criticism During the Preparation of the Work*

Preface to the Fifth Edition

The need for a reprint of the fourth edition of *Timber: Its Structure and Properties* provided an opportunity for preparing an entirely new edition, which was fully justified because of the surprising amount of new material that has become available in the last six years. At the same time, I have expanded certain sections of the text to increase the practical usefulness of the book.

The first decision to be made today when producing a new edition of a technical book is whether the author should grasp the nettle, and adopt metric units. One difficulty immediately became apparent, namely that much original work published twenty, thirty, or even forty years ago, was as apposite today as when the papers were first written, illustrated by diagrams or graphs in imperial units. I consulted several colleagues faced with this problem, and all were agreed that it is not appropriate to tamper with original work merely to achieve absolute consistency. Those who have grown up with the more familiar imperial units will no doubt regret the change to the metric system – to them 12 in. is a dimension they can envisage, whereas 300 millimetres is not. That scales have recently been marketed to enable metric drawings to be interpreted in feet and inches is an indication that 'progress' is not acceptable to all. Moreover, many of those concerned with timber have so far rejected metrification.

Substantial additions have been made to Chapter 6 *Description of Some of the More Important Commercial Hardwoods*. As the primary aim of this chapter was to assist those interested in mastering timber identification, I decided to adopt a different yardstick for determining which timbers should be illustrated in this chapter. I decided to retain the twenty eight timbers illustrated in the previous edition, but to expand the number to ensure that all anatomical features used in lens identification of hardwoods were illustrated. I have, therefore, added seven new timbers, and have slightly revised the text under European oak, deleting references to the distinctive American red oak, which is now separately illustrated and described.

Chapter 9 – *The Strength Properties of Wood* – has been substantially revised consequent on the decision to adopt metric units. Appropriate conversion factors have been worked out for converting the earlier test data, which were in pounds per square inch or per square foot, to newtons per square millimetre. It has, however, been necessary to change the dimensions of test specimens, and the increments of loading, when working initially in metric units. The changes are fully discussed in *Department of Environment, Bulletin No. 50, 2nd Edition (Metric Issue), Strength properties of wood*.

New methods for drying wood have appeared from time to time but not all innovations can be said to have stayed the course – chemical seasoning, for example, is theoretically possible, but the comment made in a Madison publication in 1938, and quoted on page 239 remains a truism today. Similarly, the electrical drying of dimension timber, employing a high-frequency alternating current, remains a practical theory that has not been put to commercial use. There is, however, one development in the drying of timber, which originated as recently as 1964, which is by now an established alternative to kiln-drying. This is the application of dehumidifying equipment to the seasoning of timber. In 1970, Westair Dynamics evolved a specially designed plant for drying timber, which is now in commercial production.

I have taken the opportunity of rewriting the sections on knots and compression failures, and I have revised the description of *Coniophora* attack. Until recently, attack by this fungus was usually described as distinguishable from decay by *Merulius* because of the development of longitudinal fissures rather than cubical fractures. In technical Note No. 44, published in 1969, the Princes Risborough Laboratory included an illustration of cubical breakdown of timber attacked by *Merulius*, and, below it, indistinguishable cubical cracking following *Coniophora* attack. I am indebted to the Director of the Princes Risborough Laboratory for permission to reproduce the illustration in this book; it is Fig. 44.

I have removed the section on *Compressed Wood* from Chapter 15 *The Preservation of Wood* – to a new chapter on *Composite Wood Products*. Although not 'timber', these manufactured products make sizeable inroads into total world production of wood. In some cases the identity of the raw material is lost in the manufactured products, as with wood pulp for paper and some types of man-made fibres. With plywood the appearance of the timbers used for veneers is unchanged, but certain improvements in the properties of the woods used are secured. The properties of the original timbers are even more drastically changed in the various kinds of densified plywood, compressed wood, hydulignum or jabroc, wood plastic, and curifax, without masking the fact that the end product has been derived from wood. Finally, mention is made of the various kinds of particle board.

It has been necessary to rewrite the section on *Stress Grading*. The four grades laid down in C.P. 112: 1967 may have been theoretically sound, but they have never been commercially viable. A contributor to the *Timber Trades Journal* aptly summed up the position: 'To meet the requirements of C.P. 112 every inch of every foot of every face of every piece must be examined for the piece to pass no less than seven tests. . . . It was surely never intended to be a commercial proposition.' The obvious solution is to resort to mechanical stress grading of timber, the grading machines that are now available being no longer sophisticated laboratory equipment but plant suitable for commercial use.

I am indebted to the Director of the Princes Risborough Laboratory for permission to use certain new illustrations that are the subject of Crown Copyright. (I have to apologise for being inconsistent in referring to the name of the laboratory: when the new material was submitted to the publishers the Laboratory was still known as the Forest Products Research Laboratory. To have corrected all the references to the Laboratory after its name had been changed would have involved long delays in the production of the new edition. Moreover, the new name in full is rather cumbersome: Building Research Establishment, Princes Risborough Laboratory, Princes Risborough, Aylesbury, Buckinghamshire.) I am grateful to Messrs Charles Griffin & Co. Ltd for permission to reproduce figure 31, which first appeared in my book *Structural Surveying*, published by them. I am also indebted to Westair Dynamics Limited for the loan of original material used to illustrate one of the Company's leaflets, from which my new text figure, No. 41, has been made. I would also like to thank Mr J. S. Shaw of the Forest Department, Oxford, for preparing the additional drawings of transverse surfaces of the timbers added to Chapter 6.

In conclusion, I wish to record my appreciation of the assistance I have received from my secretary, Mrs Sheila Barralet, in preparing the text for this revised edition.

OXTED, SURREY
March 1973

H. E. DESCH

Contents

III THE PROPERTIES OF WOOD

IV CONSIDERATIONS INFLUENCING THE UTILIZATION OF WOOD

List of Plates

List of Text Figures

PART I
The Structure of Wood

CHAPTER I

Introduction

THE astonishing material progress in the 20th century not infrequently results in the consumer, seeking to satisfy a particular need, being completely bewildered because the choice is so wide. The quality of the many alternatives is, however, anything but equal. The progress is the result of much painstaking research, often not directed primarily to solving particular practical problems. Research into the properties of timber is one of the most pertinent factors in enabling wood to hold its own today, second to none, for so very wide a range of quite different end-uses. The practical significance of our new knowledge of wood has, however, yet to be fully appreciated, and generally applied. It is too often assumed that generations of practical experience have taught users all there is to know about a material in such general use.

There are numerous examples extant to prove how well the aesthetic qualities of wood have been appreciated in the past. On more critical examination, however, it will often be found that much of the earlier work in wood, although unquestionably beautiful, lacks some vital quality. For example, beautiful as are the many examples of the wood carver's art in our churches and ancient buildings, their critical study often reveals that the fullest use has not been made of the especial qualities of wood – as beautiful results could have been achieved in marble or stone. The best of the modern school of wood carvers do not make this mistake: working on a much more modest scale, they make the fullest use of the grain, texture, figure, and colours of different woods to enhance the beauty of their creative work, which could not be so successfully achieved in other materials. Similarly, the architects of past ages who designed the many elaborate forms of hammer-beam trusses or framed floors reveal that they did not fully appreciate the strength properties of the material with which they worked: unnecessarily large sections were used for some purposes and inadequate ones for others. The latter error was almost always made over the main beams of framed floors. In spite of the large size of such beams they are almost invariably the weakest members in the floor construction. When safe working loads are calculated, using appropriate engineering formulae that incorporate accepted factors of safety, and strength data for timber determined by tests carried out in accordance with standard modern practice, the values obtained for the main beams are usually found to fall far short of present-day bye-law requirements. This can be a matter of considerable importance when a change in user of a building arises, as, for example, when a mansion previously in ordinary domestic occupation is adapted for a different purpose. Plate 3 is an example of a framed floor where the apparently large beams were not sufficiently large for the span and loading conditions when the building was requisitioned for military purposes during the war. The main beams had a safe working load of only 957·6 N/m², whereas the joists could carry a load of 2872·8 N/m², and the flooring twice this figure. The main beams failed mechanically and the floors had to be shored up for safety. This particular building was a late Victorian one; in earlier buildings the main beams, even when 30 cm $\times$ 30 cm or even 40 cm $\times$ 40 cm in section, are often much weaker, being capable of supporting a calculated safe working load of only a few pounds. Such floors did not fail because the quality of the timber, or rather its strength properties, were substantially higher than the test figures obtained from tests on commercial qualities obtainable today. Nevertheless, with the passage of time, these framed floors are often found to be sagging appreciably, and they are not adequate when a change of user involves substantial increases in floor loads. There is the added factor of the effect of shrinkage of the large beams on the strength of the floor. It is not unusual to find that the subsidiary joists are no longer properly housed in the main beams, being dependent on the strength of their tenons. When the tenons are short, shrinkage of the main beams may result in some subsidiary joists having no bearing in the beams at all, being held up by the flooring! The modern architect, fortified with the research worker's data, can span much greater distances than his predecessors, with much less material, and, by using modern glues, can achieve shapes and forms no less beautiful than the earlier craftsmen's work. In effect, if the best and most economical use of wood is to be made, all the properties of the material require to be studied carefully, and used in the most appropriate manner (Plates 4 and 5).

It is always possible to recognize a piece of oak or mahogany, and to distinguish between these two timbers; it is not so readily apparent that no two pieces of the same timber are exactly alike. The great variation in structure is part of the charm of wood, making it suitable for widely different uses, but, in certain circumstances, it places timber at a disadvantage in competition with other more uniform materials. To use wood to the best advantage, it is necessary to understand its structure, and to know how and why that structure varies. This can best be done by seeking answers to such questions as: What is wood? How is it formed? And what purpose does it serve in the growing tree?

It is obvious that wood is produced by trees, not because of its usefulness to man, but because of its function, at one stage in its existence, as an integral part of a living plant. A study of the functions of wood in the life of trees is helpful in explaining the limitations and scope of timber as a useful material for so many different purposes, and will be found to justify the seemingly academic approach to practical problems adopted in the early chapters of this book.

THE TREE

Since Darwin first advanced his theories regarding evolution, it has become generally accepted that man has developed from more primitive ancestors, and that he represents the highest form of development in the animal kingdom. In the same way, advanced plants have evolved from earlier forms of plant life. Moreover, just as in the animal kingdom, we find primitive types existing side by side with the more advanced today.

To the botanist, trees are more primitive than herbs, because the tree habit, or form of growth, is less efficient for maintaining the existence of a species. In Nature there is a continuous struggle between individuals for survival, and in the long run it is only the more efficient that prevail. Fitness for ceaseless competition depends on rapid reproduction of individuals to make good the inevitable losses sustained in such competition. Most herbs grow from seed and develop and produce new seed in a single season, whereas trees require several seasons to mature, before reproducing their race. The production of a massive stem uses up much energy, which in herbs is devoted to the reproduction of the species. ⌈ In consequence, it may be inferred that, but for man's interference, the more effective 'economy', and rapid reproduction of herbs, would result in their effacing trees from the earth, although the process might well take millions of years. ⌋

Another aspect of the struggle for survival, of considerable practical importance to those concerned with the growing of timber, is the struggle between individual trees for the same area of ground, and the air space and light above. The forester makes use of the natural tendency of plants to compete against their neighbours by growing his trees just close enough to obtain the maximum volume of good-quality timber. The importance of the competition between individuals in producing clean, straight timber may readily be appreciated if the shape of a tree grown in park-land conditions be compared with one of the same species from high-forest: the former makes little height growth, and branches near the ground, whereas the latter is tall and straight, and the bole is clear of branches to a considerable height. From the economic standpoint, the forest-grown tree produces a greater volume of better quality timber than the park-land tree.

The tree habit, then, is a mode of growth assumed by certain plants to enable them to outwit their neighbours in the struggle for air and light, which is essential to the development of an individual, and its subsequent duty of reproducing its kind. It is not the most efficient mode of growth from the plant standpoint, but it results in the production of timber useful to man.

CLASSIFICATION OF TREES

Some trees belong to more primitive plant-types than others, giving rise to different classes of commercial timbers; all are the successors of still earlier forms of plant life, although the lines of development to the present-day representatives are rarely clear-cut, being more often suggestive of evolution along parallel lines from several common ancestors. Commercial timbers fall into two main groups, the **'softwoods'** and **'hardwoods'**, and the trees that produce these two different classes of timber are themselves quite distinct. The former are **gymosperms** – **conifers** or cone-bearing plants, characteristically with needle-shaped leaves and naked seeds; the latter are **dicotyledons** – broad-leaved plants, characteristically with broad leaves and seeds enclosed in a seed-case. Dicotyledons with **monocotyledons** (grasses and palms) constitute the **angiosperms.** Although the division into 'softwoods' and 'hardwoods' is a convenient one for differentiating two broad classes of timber, there are a few timbers, *e.g.*, pitch pine, among the softwoods that are actually harder than other timbers classed as hardwoods, *e.g.*, balsa, lime, willow. Further, the divisions are not always applied correctly, particularly in the tropics. For example, native softwoods in such regions are usually soft 'hardwoods', that is, they are broad-leaved species with soft wood, although they are frequently referred to as 'softwoods'; true 'softwoods' often do not occur in such localities.

Botanists early found the need for an orderly system of naming plants. They recognized that, although no two plants might be identical, minor variations between similar individuals did not alter the fact that several such individuals had many features in common, not shared by any other groups of plants, and these 'features in common' were reproduced in successive generations of such plants. This gave rise to the botanical concept that all plants could be separated into different **species.** It was also observed that several species shared certain 'features in common' – that is, they were more like one another than they were like other species. This gave rise to the second fundamental botanical concept of a **genus.** Recognizing the validity of these two concepts led botanists to adopt the binominal method for plant nomenclature, the first part of the botanical name indicating the genus to which the plant belongs, and the second part the species. Botanists have subsequently attempted to adopt a natural system of classification, based on evolutionary lines, arranging groups of similar genera in **families,** and bringing related families together into **orders.** The difficulty of reconciling all the complex factors that have to be considered has resulted in the systems of classification being arbitrary, rather than natural. Mistakes have occurred through placing an evolutionary significance on some feature that had no such significance, and through failing to recognize a significant feature or features as such.

Developments in the study of wood anatomy over the last thirty years, and the closer collaboration between wood anatomists and systematic botanists that has resulted, are helping to clarify difficulties. Parallel development from common ancestors, and the disappearance of some of the links in the evolutionary chain, however, give rise to real difficulties that may well be incapable of final and complete solution. Nevertheless, from the practical standpoint, the important fact emerges that every plant has one **botanical name**, made up of two parts, the first indicating the genus and the second the species. These names are, by general consent, in Latin.

Unfortunately, the position is not quite so simple as the foregoing paragraph suggests. A tendency of 19th-century botanists to name plants from inadequate material, the occurrence of actual errors of observation, and differences in interpretation of specific or generic characters, result in botanists not always being in agreement as to the correct botanical name of a plant: in consequence, some plants have been given more than one botanical name, and two or more different plants have been given the same name. Errors of observation arise through a failure to recognize the significance of some types of variation in morphological characters: 'immature' leaves, *e.g.*, the leaves of seedlings or even saplings, are often much larger, and very different in shape, from the leaves of a mature tree of the same species. Working with too little material, and seeing only sapling leaves, a botanist may make the mistake of thinking he is confronted with a 'new' species, which he proceeds to name and describe, whereas the species has already been named from mature leaves, or *vice versa*. Poor laboratory technique is responsible for actual errors of observation: the parts of a flower may be so broken or torn in dissecting that the 'evidence' is misinterpreted. Real differences of opinion as to the correct interpretation to be placed on observed variation in morphological characters are yet another source of confusion in plant nomenclature. Certain morphological features of plants with a relatively wide geographical range may exhibit considerable variation when specimens from the extreme limits of distribution are compared: in such circumstances it may well be almost a matter of personal opinion where to draw the dividing line between two very similar plants. This is perhaps a rather special case of variation, but it may help to underline the

PLATE 1

Mother and child by E. J. Clack illustrating the use of grain and figure of wood in a work of art

From a carving by E. J. Clack, by courtesy of the Editor of 'Wood'

PLATE 2

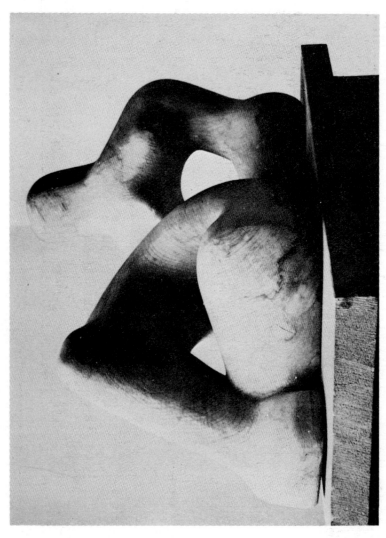

Reclining figure by Henry Moore illustrating the use of the grain and figure of wood in the modern sculptor's work

Henry Moore's 'Reclining Figure 1935–36' by courtesy of the sculptor

PLATE 3

A framed floor: the apparently substantial beams are suitable for a safe working load of only 957·6 N/m,² whereas the joists could carry a load of 2872·8 N/m³, and the floor boards twice this figure

PLATE 4

Fig. 1. Sheet laminated portal frames 12·7 m span

U.S. Forest Products Lab. Service Bulletin

Fig. 2. Church showing span of 13 m rise 12 m; spacing of glued laminate arches 4·2 m

U.S. Department of Agriculture

fact that differences in interpretation of morphological variation are likely to arise, and result in some botanists splitting what was previously considered a single species into several separate species, or, conversely, combining several formerly 'distinct' species into one species.

Rules for naming plants are open to less ambiguity, and are now regularized by accepted international procedure. The botanist receives dried specimens, called **herbarium material**, for study; it should consist of leaves, flowers, and fruit, but is seldom so complete. He has first to decide whether it is undescribed, that is, unnamed material, or whether it is additional material of an already described species. If the former, the botanist next has to satisfy himself whether the 'new' species can be regarded as another species of a known genus, or whether it is so distinctive as to necessitate establishing a new genus too. If it is a question only of a new species of an established genus, choice of a suitable specific name alone rests with the botanist, but when a new genus also has to be established, both generic and specific names are selected by the botanist 'describing' the plant. In selecting names, botanists have an entirely free choice, although it is usual either to adopt a word descriptive of some morphological character of the plant, or to commemorate the place, collector's name, or native name in the name chosen. The only obligation on the botanist is to Latinize the name or names he selects. To complete his task, and give the name validity, the botanist has to prepare a description of the new plant – in Latin – following long-established precedents as to the morphological data to be recorded, when the description is published in a recognized botanical journal. The material on which the description is based is recorded in this published description. Thereafter, provided the botanist has been correct in recognizing a new species, and has allocated it to its appropriate genus, the name he has selected becomes the valid botanical name of the species. In citing this name in future, the describing botanist's name or initials follow the selected name. This convention is a most important part of a botanical name, since it links the name with the authority for the name, and minimizes confusion later should the botanist have made a mistake.

Internationally accepted rules have been drawn up for dealing with the types of mistakes that do arise in the naming of plants: the essence of these rules is that when errors are detected the earliest name recorded in the literature must be revived and later names discarded, but, if the earlier name should refer the plant to the wrong genus, then only the earlier specific name is retained. Transference of a plant to another genus may necessitate the selection of a new specific name if the original specific name has already been used for another species in that genus. There are also rules regarding the use of capitals in specific names: place names should not be capitalized, only vernacular names and the names of people. Foresters have adopted the practice of de-capitalizing all specific names, but botanists have not accepted this departure from their rules. Critical workers today are constantly finding that two or three supposedly different plants are identical, or that some well-known botanical name must be dropped and an earlier, and obscure, name revived. This is apt to give the impression that botanical names are distinctly fluid, whereas the fault lies with botanists who have been too ready to describe and name plants from inadequate material, without searching the literature sufficiently thoroughly. It does indicate, however, that it is by no means always a simple matter to discover the correct botanical name of a species, and such names should be used with caution, and never without their authenticating authority.

NOMENCLATURE OF TIMBERS

It is not suggested that timber names present as complex problems as do botanical names, but there are nevertheless very real practical difficulties to be solved in selecting entirely satisfactory timber names. The precision essential in botanical work is seldom necessary in timber names, nor would it usually be practicable, because several botanically distinct species often provide a single commercial timber. Botanical names have the added disadvantages of being in a foreign language, often difficult to pronounce, and undesirably long. Moreover, it is sometimes more necessary to distinguish between the timber of one species from different localities, than it is to distinguish between the timbers of different species. An example of this is the timber of the two common species of oak, *Quercus robur* L. and *Q. petraea* Liebl., that occur both in the British Isles and in Europe. The timbers of the two species cannot be distinguished from one another on anatomical grounds, or by other identifying characters, but material from the richer soils of the south of England is suitable for very different purposes from that grown on poorer soils at higher altitudes in central Europe: for commercial purposes the two types are distinct timbers. In primitive communities, the problem is readily solved by the local inhabitants choosing words from their own tongue or dialects: these are **vernacular names**. Since the local inhabitant is not as critical an observer as the scientific botanist, vernacular names are rarely as precise as botanical ones; they often refer to more than one species, and they may on occasions be applied to quite different species because of some superficial similarity in form between distinct plants. These objections are not of serious practical importance so long as there is little movement of timber from district to district, but the position is very different in a market drawing its supplies from many different localities or even countries. Even in a small country dialects change from district to district, so that the vernacular name in one locality may be very different from that in another. This may lead to confusion in a distant, importing market. Nomenclature difficulties are further complicated by the deliberate hiding of the true identity of a timber under a 'trade' name, often that of some well-established timber, with the addition of a geographical or other qualifying, but sometimes misleading, adjective.

Some examples may assist the reader in clarifying the problem in his own mind. For example, the true oaks, beech, and sweet chestnut belong to one family, the *Fagaceae*; the oaks constitute one genus, *Quercus*, beeches a second, *Fagus*, and the true chestnuts a third, *Castanea*. The different kinds of true oak, *e.g.*, American red oak, American white oak, Turkey oak, are separate species of the genus *Quercus*. This is a simple case in which trade practice follows botanical classification closely, although different countries have different names for *Quercus* timber: in Britain it is oak, whereas in France it is chêne, and in Germany Eiche. When vernacular names are of the popular type absurdities may occur: the standard trade name adopted for *Eucalyptus regnans* F.v.M. is mountain ash in Australia and Tasmanian oak in the United Kingdom, but this timber has also been called swamp gum in Tasmania and Australian oak in Victoria; it is neither a true ash nor a true oak. Such anomalies are the outcome of European emigration: settlers name plants sometimes because of similarities of tree form or habit, sometimes because of colour similarities in the timbers, and sometimes because of other associated ideas. When the same species occurs over a wide area several names, based on these different concepts, may come into being.

The Malaysian Peninsula provides examples of confusion, arising from differences in meaning of vernacular names in different parts of the same country, which is accentuated by reason of a single trade timber being produced by several distinct botanical species, each with one, and sometimes more than one, vernacular tree name. A common source of *red meranti* in the Peninsula is the tree known to botanists as *Shorea leprosula* Miq. This tree, in the forest, is called *meranti tembaga*; its timber is sold in most parts of Malaysia as *meranti*, and in Singapore as *seriah* (pronounced *seraya*). *Seraya*, on the other hand, is the tree name of another species of *Shorea*; as a timber name in certain parts of the Peninsula it refers to the produce of several tree species that yield a grade of timber superior to common *meranti*. *Meranti* exported to the United Kingdom has to compete with commercially similar timbers from the Philippines called *lauan*, thereby confusing the importer and the layman. Both Malaysian *meranti* and Philippine *lauan* may be offered for sale as Philippine mahogany, when the layman is not only confused but deceived.

The practice of borrowing names of familiar timbers, and applying them to other and quite distinct woods, is at the root of much confusion in timber nomenclature. It frequently misleads the layman to the extent of causing him to use timbers for purposes for which they are unsuited. Alternatively, he may be induced to buy timbers that, were he more enlightened, he would not purchase, or, if he did consider them, he would not be prepared to pay the prices asked. It has been computed that the name 'mahogany' has been applied at some time or another to the timbers of more than two hundred distinct botanical species, and the name is in common use today for several distinct groups of timber. The original 'mahogany' of commerce was the so-called Spanish mahogany obtained from San Domingo and other West Indian islands then owned by Spain; Central American mahogany is a very close relative, being the timber of a different species of the same genus, *Swietenia*, but it is a very much milder timber to work than the original Spanish mahogany.

African mahogany, on the other hand, is produced by several species of two different, but related, genera (*Khaya* and *Entandrophragma*), belonging, however, to the same family as the American species of *Swietenia*. Philippine mahogany is produced by several species of more than one genus belonging to a different family altogether, the *Dipterocarpaceae*. In consequence, timbers bearing the name 'mahogany' vary appreciably in their appearance and properties, and in the opinion of the author some have no real claim to be considered in the same class as true mahogany (Plate 6). Similar confusion has arisen through the widespread use and misuse of such names as walnut, ash, oak, and teak, leading in some instances to lengthy and costly litigation between the sponsors of distinct timbers sold under the same trade name.

Attempts are being made to standardize trade names. The British Standards Institution, for example, has published a list of **standard names** for timbers known to the U.K. trade, *vide* BS 881 and 589. The Standards Association of Australia has issued a similar list for Australian timbers, *vide* AS 5.0.2–1940, and so has the Forest Research Institute, Dehra Dun, for Indian timbers. Publication of a list does not, of course, solve the problem of timber nomenclature, but it points the way. It has not yet been possible to compile a list based entirely on internationally acceptable rules; compromise has been necessary. The ideal solution would be the adoption of standard names on an international basis, whereby each trade timber would have a single unambiguous name. Persistence in the present confusion may well result in several potentially useful timbers being discredited, besides perpetuating deception of the laymen. It could give rise to much expensive and acrimonious litigation, such as marred relations for several years in the different sections of the American 'mahogany' trade. Every timber worth putting on the market is entitled to a name of its own; it does not need to masquerade as something else. Conversely, within the English-speaking world, the same 'commercial' timber does not require half a dozen names, merely because of different geographical sources of supply. In the absence of general agreement regarding timber names, the existence of a difficult problem must be recognized: ' standard names' are used throughout in the text, but to assist the reader the botanical equivalents are given in Appendix I.

DIVISIONS OF THE TREE

Almost all plants with which we are familiar have three main parts: **roots**, **stems**, and **leaves**. The characteristic that separates trees from other woody plants is that they have a single main **stem**, the **trunk** or **bole**.

Each of the three parts is specially adapted to a particular function: the roots anchor the plant in the ground, and take in water and mineral salts in dilute solutions from the soil: the stem conducts these solutions from the roots to the leaves, it stores food materials, and it has mechanical rigidity, supporting the leaves above competing vegetation: the leaves absorb gases from the atmosphere and, with the energy obtained from sunlight, manufacture complex substances required for carrying on life processes from the simple 'salts' obtained from the soil. (Fig. 1.)

The timber user is interested primarily in the trunk or bole. This bole has an outer covering, called the **bark**, which protects the wood from extremes of temperature, drought, and mechanical injury. The inner layers of the bark conduct the food manufactured in the leaves to regions of active growth, and into places where it can be conveniently stored. The bark, being a conductor of food materials, is often rich in chemical substances, such as tannin and dyes derived from plant metabolism.

Between the bark and the wood is a thin, delicate tissue, known as the **cambium**, which forms a complete, glove-like sheath covering the bole and branches. This tissue produces bark towards the outside and wood towards the inside of the tree, and the enlargement in girth of the trunk is brought about entirely by the activity of the cambial sheath. The production of wood and bark tissue occurs only when the cambium is growing: in temperate regions this is during the spring and summer months. In this period the bark may easily be peeled, because the cambial tissue is then less rigid and more easily torn than during the non-growing seasons, when it is tough and strongly attached to the bark and wood tissues.

DIVISIONS OF THE STEM

Under the bark is the cylinder of wood, in the centre of which is the **pith** (Plate 7), up to 1·25 cm. in diameter, but in many trees

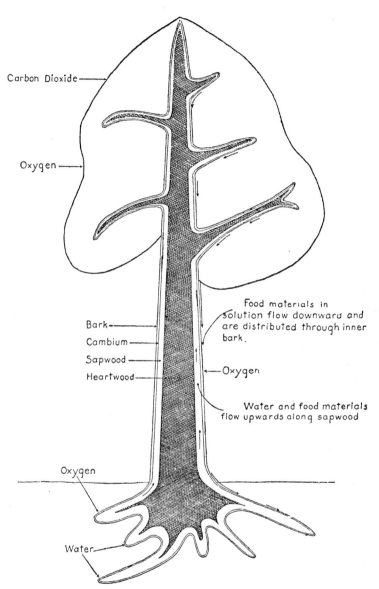

Carbon Dioxide

Oxygen

Bark
Cambium
Sapwood
Heartwood

Food materials in
solution flow downward and
are distributed through inner
bark.

Oxygen

Water and food materials
flow upwards along sapwood

Oxygen

Water

Fig. 1. Diagram showing main parts of a tree and how food
is manufactured and distributed

By courtesy of the Canadian Forest Service

barely visible. A cross section of a branch is similar to that of the main stem, but that of a root differs in having little or no pith and the anatomical structure is more variable.

If the end surface of the bole is planed, further details of the wood structure can be seen: the wood of trees grown under seasonal conditions consists of a series of concentric layers of tissue, called **growth rings** (Plate 7). Each **growth layer** comprises the wood produced by the cambium in a single growing season. The rings are actually layers of wood, extending the full height of the tree, a new layer being added each growing season, like a glove, over the whole tree. Thus the wood nearest the outside of the bole is the youngest. In temperate regions, and certain tropical countries, the alternation each year of a growing season, followed by a resting period, results in the growth rings being annual rings, thus providing a fairly accurate means of computing the age of a tree after it is felled. Double (or multiple) rings, consisting of two or more false rings, caused by serious interruptions to growth during the growing period, sometimes cause errors in such calculations. Where growing seasons are not well defined, as in many tropical regions, growth rings may be indistinct and they may not be annual, but, as in seasonal climates, new wood is formed in concentric layers.

Growth rings are apparent because the wood produced at the beginning of the growing season is different in character from that formed later in the season, and zones of **early wood** [1] and **late wood** [1] may be distinguished. Where this is the case the early wood is softer, coarser, or more porous, than the late wood.

The work of food storage and sap conduction is performed in most trees only by the outer, or youngest, growth layers; these are known as the **sapwood**. The sapwood forms a distinctive zone, typically 1·25 to 5 cm wide, but in a few species, particularly tropical species, the sapwood may be up to 20 cm or more wide, depending on the species and age of the tree, and the mode of growth of individual trees. Trees of the same age and species have a wider zone of sapwood when grown in the open than when grown under forest conditions in close competition with other trees. The central part of the tree is concerned with providing mechanical rigidity to the stem and support for the crown; it is known as **heartwood.**

---

[1] The early wood is sometimes called **springwood**, and the late wood, **summer-** or **autumn-wood.**

The sapwood is usually lighter in colour than the heartwood and less durable, and, when green, contains much more moisture. The line of demarcation between the two zones may be sharply defined or indefinite, and in some species there is no colour differentiation between the two: such trees are popularly spoken of as 'all-sapwood' trees, although this is not an accurate description. In many trees the conducting channels are blocked in various ways when the wood becomes heartwood, and any remains of stored food material become changed to tannins and other substances; it is to these changes that the durability of the heartwood may be ascribed. In the absence of colour differences such changes are the only indication of transition to heartwood.

The Units Composing Wood

In common with all living tissue, plant or animal, wood is built up of individual units called **cells**. These units are either tube-like, with blunt or pointed ends, or brick-shaped. They may be empty or they may contain various kinds of solid or semi-solid substances. Cells differ considerably in size and shape, and each is adapted to one or more of the three primary functions of the stem. The majority are invisible to the naked eye, varying from $0 \cdot 025$ to $0 \cdot 5$ mm. in their largest dimension.

The formation of cells is a 'vital' or 'living' process, which we describe as **'growth'**, the increase in size of plants being brought about by the formation of additional cells much more than by the enlargement of existing ones. Growth in plants is restricted to regions where cell-forming tissue occurs. The main stem and branches of a tree increase in length solely at their tips,[1] and growth in thickness occurs in the sheath of cambial tissue, one or more cells thick, situated between the bark and wood. Growth in thickness continues after height growth has practically ceased, and up to the time the tree dies.

New cells arise as a result of repeated division of the cambial cells.

[1] The growing tip of the main stem and branches is composed of (1) **meristematic** cells of the apex and (2) the **pro-cambial strands** derived from normal meristematic cells and situated immediately below the growing tips. Growth in length is brought about by activity in the region of the pro-cambial strands. These strands give rise later to isolated patches of cambial cells that eventually link up to form the cambial sheath.

Before division occurs these cells swell, and certain changes take place in their contents. Partition walls are then formed, either in the longitudinal, oblique, or horizontal planes, dividing each cell into two. The longitudinal division provides the cells of the bark and wood, and the oblique or horizontal division adds cells to the cambial sheath, necessitated by the increase in circumference of the bole as growth proceeds. The two cells that result from longitudinal division are identical at first, but their subsequent development is different. One, after enlargement to the size of the original cambial cell, resumes the function of these cells, and divides again, while the other develops into either a unit of the **secondary xylem** (wood tissue) or a unit of the **phloem** (bark tissue). The two cells resulting from oblique or horizontal division of a cambial cell merely increase to the size of normal cambial cells, and then undergo longitudinal division in the normal manner to form bark or wood cells as described above.

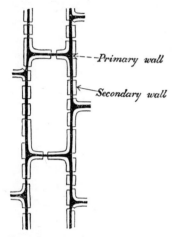

Fig. 2. A cell showing the primary and secondary walls (much enlarged)

—Primary wall

—Secondary wall

The secondary xylem is the timber of commerce, and our study will be confined to the cells that compose it. These cells develop rapidly after formation, completing the process in a few weeks, after which the majority die and undergo no further change in size or shape. Those that remain alive assist in growth, *e.g.*, by storing food, and when no longer required for this purpose they also die. In other words, the bulk of the stem and branches of a living tree is composed of dead cells.

When first formed, the young cell is in a plastic condition, and capable of considerable increase in size and change in shape, rather like a partially inflated balloon. Increase in size and change in shape are rapid, and when the final size is attained the walls are thickened by the addition of further layers of wall substance laid down from the inside of the cell. The original unthickened wall is called the **primary wall**, and the layers added afterwards constitute the **secondary wall** (Figs. 2 and 3).

Unthickened areas, called **pits**, are left in the primary wall during the formation of the inner layers, and these serve as means of communication between cells: liquids moving in the tree pass mainly through the pits. The pits of different types of cells show modifications in their structure, which sometimes increase their efficiency in controlling the movement of liquids into and out of the cells (Fig. 3, II). As might be expected, conducting cells have more pits than

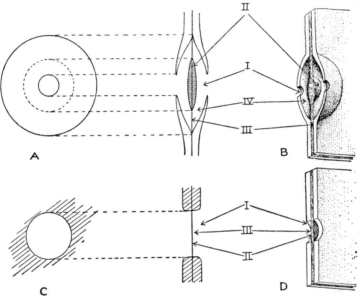

FIG. 3. A, surface view and section through pits in conducting cells; B, solid view of two pits cut in half (after a woodcut by Dr L. Chalk): I, pit opening; II, torus; III, primary wall; IV, pit cavity (much enlarged); C, surface view and section through pits in storage and strengthening tissue; D, solid view of two pits cut in half: I, pit opening; II, primary wall; III, pit cavity

those concerned merely with the provision of mechanical rigidity, and the pits are specially adapted for controlling the movement of liquids in these cells.

Unlike most of the other structural features discussed, pits are a 'laboratory feature', visible only with the aid of a microscope in thin sections of wood, or specially prepared slides.

With the formation of the secondary walls, chemical changes occur that increase the rigidity of the walls. Among the substances formed is **lignin**, which gives its name to the process: **lignification.**

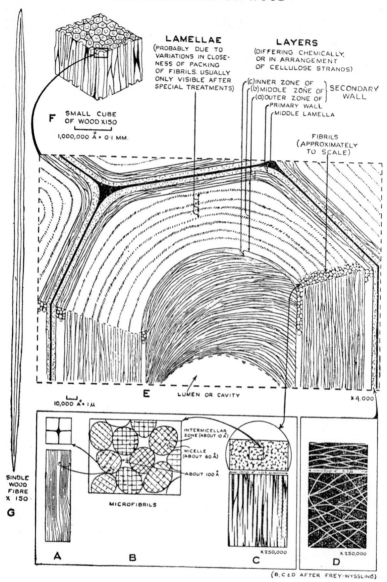

FIG. 4. Diagrammatic representation of cell wall structure
(for legend see facing page)

*Drawn by Miss M. S. Smith, and reproduced by
courtesy of the Director, F.P.R.L., and the Editors of 'Nature'*

FIG. 4. Diagrammatic representation of cell wall structure

A. Cellulose chain-molecules, showing here and there zones of regular and parallel arrangement (the micelles indicated by X-rays).

B. Group of microfibrils, showing approximate relative sizes of micelles and spaces revealed by study of material impregnated with silver.

C. Cellulosic and non-cellulosic systems in the secondary wall. Cellulosic white, non-cellulosic materials black; transverse section above, longitudinal section below. Large circle indicates approximate size of fibril at same magnification. (Note that the dimensions given by the author of Figs. B and C do not correspond exactly, but they serve to indicate the approximate sizes. The linking arrows have been inserted by the author of the paper from which this figure is taken.)

D. Cellulosic and non-cellulosic systems in the primary wall. Cellulosic white, non-cellulosic materials black; transverse section above, longitudinal section below.

E. Small piece of wood, showing the relative sizes and dispositions of the cell wall constituents as revealed by microscopic examination.

F. Small cube of wood fibres magnified 150 times.

G. Single wood fibre magnified 150 times.

Fig. 4 is from a paper entitled 'Fine structure of the plant cell wall', by S. H. Clarke, published in *Nature*, No. 3603, pp. 899-904, November 1938, and subsequently reprinted as F.P.R.L. Special Report, No. 5, 1939: *Recent work on growth structure, and properties of wood*. This paper is a most authoritative survey, with a comprehensive bibliography, which the reader wanting to pursue the study of cell wall structure should not fail to read. The account given in the third paragraph on page 22 reduces an extremely complex subject to a skeleton outline essential to the understanding of certain phenomena to be discussed later.

This concept of cell wall structure is based on direct microscopic examination, aided by staining reagents, to the limits of resolution of the compound microscope (up to × 1000 linear magnification), inference between × 1000 and × 15,000, and X-ray studies above × 15,000. Microscopic examination reveals:

1. The **middle lamella**, which is shared by adjacent cells, and differs chemically from the other cell wall layers.

2. The **primary wall**, which is laid down during extension growth – that is, after division of the cambial initial, when the new xylem cell is increasing in size and changing its shape.

3. The **secondary wall**, which is the thickening material added after extension growth is finished; it may be differentiated into at least two zones; the layer labelled 'inner' in Fig. 4 is sometimes not distinguishable.

The cell wall is composed of cellulose and other constituents, the relative proportions of which vary in the different layers and lamellae. The lamellae can be dissected into fibrils – threads of cellulose, the direction of which, with reference to the main axis of the cell, may vary in the different layers. The fibrils can be dissected chemically into fusiform bodies.

The fibrils are composed of microfibrils, which are aggregates of **micelles**. Micelles are aggregates or bundles of cellulose chain-molecules. The long chain-like molecules form the framework of the cell wall, and have been compared with the steel rods of reinforced concrete. They are mainly parallel with each other, but between them are many ultramicroscopic spaces – intermicellar zones – which are capable of holding water and other substances. Some theories attribute shrinkage and swelling to the movement of water from and to these intermicellar spaces.

Lignification should not be confused with the changes that take place in the transition from sapwood to heartwood. Lignification is dependent on the cell being alive, whereas changes occurring in the transition from sapwood to heartwood can take place in dead cells, since they are confined to changes in the contents of the **cell cavity or lumen** (the space enclosed by the cell walls), and to the addition of infiltrates to the cell wall that do not alter the chemical composition of wall substance itself.

We have learned that the stems of trees perform three functions, and that these different functions are carried out by different types of cells. A group of similar cells performing the same function is called a **tissue**; thus we may speak of the storage tissue, the conducting tissue, and the mechanical or strengthening tissue.

The Composition of Cell Walls

Cell wall structure is decidedly complex, but as a result of the combined efforts of physicists, chemists, and botanists in the 1930s, we now possess a reasonably clear picture of the fine structure of the plant cell wall. A complete picture is still impossible: visible and ultra-violet light permit of accurate and direct observation of particles larger than 2500 Å,[1] and X-rays make it possible to study only details of much smaller dimensions; between these limits is a gap that can only be filled by inference and conjecture (Fig. 4). Reduced to simple terms, our present knowledge may be summarized as follows: the cell walls of all plants may be visualized as series of thin, concentric layers or sleeves. The individual layers are composed of spiral, thread-like strands, known as **fibrils** (Fig. 5), which may be likened to valve springs making one or more turns in the length of a cell. The fibrils in turn are composed of minute, spindle-like units, **fusiform bodies (microfibrils)**, which may be dissected into small **spherical units**. The ultimate composition of the spherical units are the molecules of **cellulose**, a substance composed of carbon, hydrogen, and oxygen.

The purest form of cellulose in plant tissue is cotton; actually the hairs of the cotton seed, which are individual cells. In the cells of wood, cellulose is associated with other substances, the most important of which is lignin. To this latter substance wood owes its

[1] $1000 \text{ Å} = 0\cdot1 \ \mu\text{m}; \ 1 \ \mu = 0\cdot001 \ \text{mm}.$

stiffness. Nevertheless, the cellulose of wood is chemically identical with that of cotton, and various methods are employed commercially to remove the other constituents of the cell wall, leaving a fairly pure form of cellulose. The resultant substance is the raw material of the chemical paper pulp and artificial silk industries.

In addition to cellulose and lignin, which are the main constituents of the cell walls of *all* woods, other substances, spoken of as **infiltrates**, are present in the cell walls and cell cavities of *some* woods. These infiltrates have an important bearing on problems of utilization. For example, tannin renders the heartwood of oak durable, and black-coloured infiltrates are responsible for the decorative appearance of ebony. On the other hand, absence of such infiltrates is of great importance to the manufacturers of paperpulp and artificial silk, and the presence of gums and resins may adversely affect the working and painting qualities of timber. Even in extreme cases, however, infiltrates rarely exceed 10 per cent. of the dry weight of wood, and more usually account for only 2 to 3 per cent.

The chemical composition of cell walls influences strength properties, working qualities, and utilization of timber, and the physical

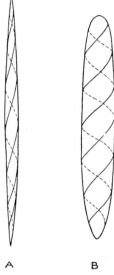

A   B

FIG. 5. Diagrammatic drawing showing the spiral alignment of fibrils in the cell wall. Note the difference in pitch of the spirals in A and B

By courtesy of A. Koehler, Esq.

structure explains certain other properties of wood, *e.g.*, electrical conductivity and insulating properties, and its behaviour in relation to changes in atmospheric conditions.

CELL CONTENTS

In addition to the 'living' content or **protoplasm** of all living cells, **crystals** of calcium oxalate (Fig. 6, *a* and *b*), deposits of **silica** (Fig. 6, *c*), and plant food materials may occur in the storage tissue of both sapwood and heartwood, and gums and other solid deposits in the vessels of the heartwood. Plant food materials are of particular importance, because in some forms they are the food of certain insects and fungi that attack wood, and in other forms they are repellent to such foes. For example, starch, which only occurs in any quantity in the sapwood, is an essential food of powder-post

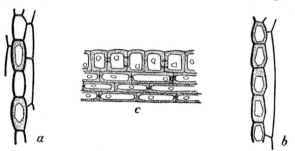

FIG. 6. *a* and *b*, crystals in wood-parenchyma cells; *c*, deposits of silica in ray cells

(*Lyctus*) beetles and sap-stain fungi, and the aromatic oils that occur in the heartwood of some timbers are toxic to fungi and insects.

The foregoing pages have dealt with problems of nomenclature and the structure of wood in outline; the next two chapters deal in greater detail with the structure of the different kinds of cells of which wood is composed. As the tissues of softwoods are simpler in many respects than those of hardwoods they are described first.

PLATE 6

TYPES OF COMMERCIAL MAHOGANY

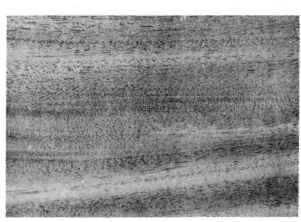

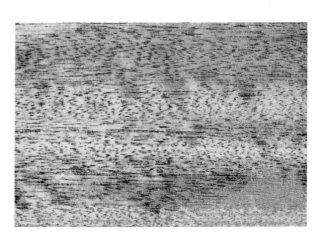

Red meranti – any one of at least 20 species of *Shorea*. Wt per cu. ft 30 to 45 lb. (480 to 720 kg/m³). Shrinkage from green to 12 per cent. M/C = Tangential 4·5; radial 1·9

Central American mahogany – *Swietenia macrophylla*. Wt per cu. ft 35 lb. (560 kg/m³). Shrinkage from green to 12 per cent. M/C = Tangential 2·6; radial 1·9

African mahogany – principally *Khaya ivorensis*. Wt per cu. ft 34 lb. (530 kg/m³). Shrinkage from green to 12 per cent. M/C = Tangential 2·5; radial 1·8

PLATE 7

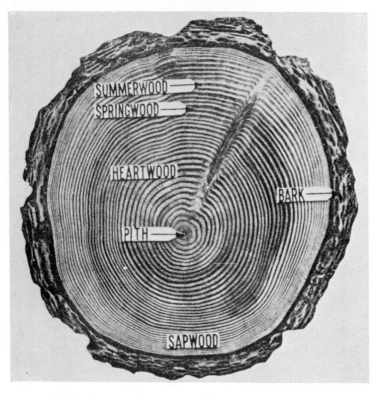

Cross section of softwood log showing bark, wood, and pith

By courtesy of the Canadian Forest Service

PLATE 8

FIG. 1. A cube of softwood magnified

FIG. 2. A cube of hardwood highly magnified

Photos by F.P.R.L., Princes Risborough

PLATE 9

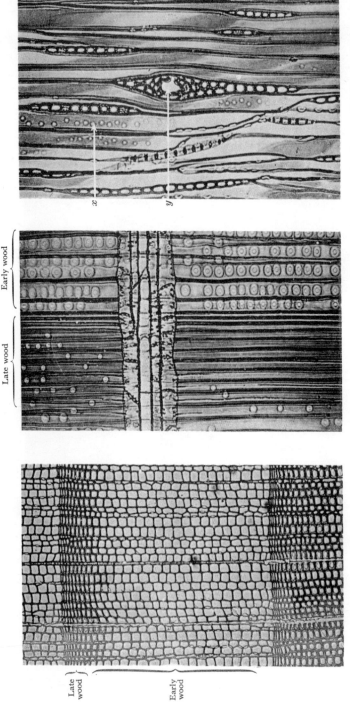

Late wood   Early wood

x

y

Late wood   Early wood

Fig. 1. Transverse section of Scots pine (×75) showing one complete growth ring. Note the thin walls of the early wood tracheids and the thick walls of the late wood, and the abrupt change from early to late wood

Fig. 2. Radial-longitudinal section of Scots pine (×150). Note the numerous and large pits in the early wood, and the few small pits in the late wood

Fig. 3. Tangential-longitudinal section of Scots pine (×150). Note the small pits in the late wood (x) and the resin canal in the ray (y)

Photos by L. A. Clinkard

PLATE 10

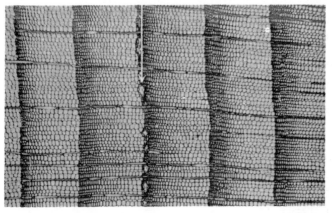

FIG. 1. Transverse section of Scots pine (about ×300) showing a vertical resin canal

FIG. 2. Transverse section of larch (×10) showing normal distribution of vertical resin canals

FIG. 3. Transverse section of cedar (×10) showing a tangential series of traumatic resin canals

Photos by L. A. Clinkard

Photo by F.P.R.L., Princes Risborough

PLATE 11

Scalariform perforation plates in the vessel members of American whitewood (×40), as seen on a radial face of the wood

Photo by F.P.R.L., Princes Risborough

PLATE 12

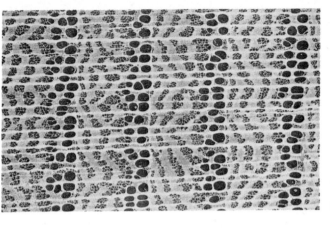

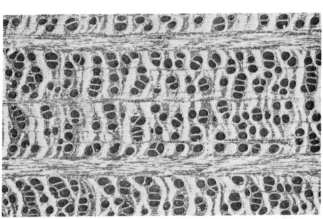

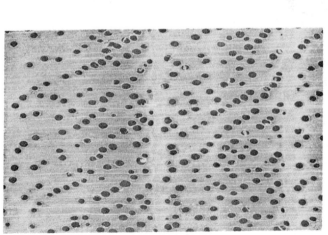

Fig. 1. Transverse section of Tasmanian oak (mountain ash) (×10) showing solitary vessels

Fig. 2. Transverse section of Australian silky oak (×10) showing tangential groups of vessels

Fig. 3. Transverse section of wych elm (×10) showing clustered arrangement of vessels in late wood

(For radial arrangement of vessels, see makoré, Plate 16b, Fig. 6)

Photos by F.P.R.L., Princes Risborough

PLATE 13

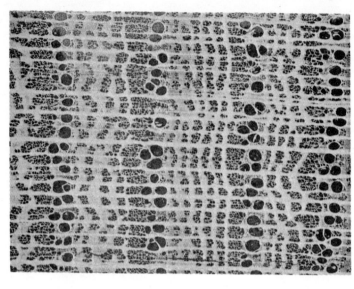

FIG. 2. Transverse section of a ring-porous wood – elm (× 10)

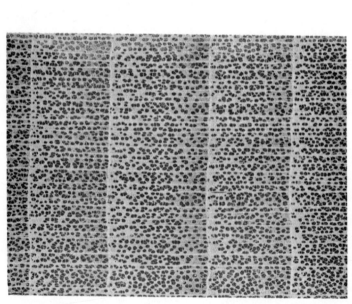

FIG. 1. Transverse section of a diffuse-porous wood – American whitewood (× 10)

Photos by L. A. Clinkard

CHAPTER 2

Softwood Tissues

GENERAL CONSIDERATIONS

CELLS have been described as being tube-like, with blunt or pointed ends, or brick-shaped, but when examining wood with a lens or under the microscope it is seldom possible to see the cells in more than one plane at a time. It is, however, important to keep in mind when examining a piece of wood that cells have three dimensions, and that their appearance in the three different planes may be different (see Plate 8). In other words, in a given section of wood, any one cell will be viewed in one of its three planes, and its appearance will depend on whether that plane presents a cross section of the cell or a longitudinal view; the longitudinal view may be radial-longitudinal or tangential-longitudinal, the cell itself being seen either in section or in elevation. There are therefore five different, alternative views: one in cross section or plan, one in radial-longitudinal elevation, one in radial-longitudinal section, one in tangential-longitudinal elevation, and one in tangential-longitudinal section. Differences between longitudinal sections and elevations of individual cells are, however, rarely discernible with the naked eye or an ordinary hand lens. At the outset of a study of wood anatomy, it is often difficult to appreciate that a given cell may present different appearances on the cross, radial, and tangential sections of a piece of wood, and until this is realized it may be difficult to understand illustrations. In the ensuing pages the different types of cells are described as individual units, but for the most part they are illustrated in one plane at a time, and in association with other cells of the same or of a different type, as they actually occur in wood. Individual cells can be separated out by **maceration**: thin shavings of wood are boiled in a solution of potassium chlorate in nitric acid (commercial acid diluted in 50 per cent. of water), which allows of the individual cells being teased apart with mounting needles. The separate cells, suitably stained, can then be mounted on slides for microscope examination.

The Conducting and Strengthening Tissues

In softwoods the conducting and mechanical functions are performed by a single type of cell, each unit of which is known as a **tracheid**. These cells are hollow, needle-shaped units attaining as much as 10 mm. in length, but more usually varying from 2·5 mm. to 5·0 mm. They are packed closely together so that a cross section through them resembles a honeycomb (Plate 9, fig. 1).

Examination of Plate 9, fig. 1 reveals differences in thickness of cell walls and in the size of cell cavities. It will readily be appreciated that the larger the cavity the better is the cell fitted for conduction, and, conversely, the thicker the walls and the smaller the cavity, the less suitable is that cell for this purpose, but the better is it fitted for strengthening purposes. In the tree the thin-walled tracheids with large cavities are primarily concerned with the conduction of sap, and the thick-walled ones with maintaining mechanical rigidity, although the latter may also play some part in conduction. Radial sections of softwoods show a second modification of structure between the thin- and thick-walled tracheids (Plate 9, fig. 2). In this figure it will be seen that the pits in the thinner-walled, conducting tracheids are larger and more numerous than those in the thicker-walled tracheids that function as strengthening tissue.

If we turn to Plate 9, fig. 1 again, a further and conspicuous feature of the wood structure of softwoods may be noticed. The distribution of thin- and thick-walled tracheids is not haphazard: the thin-walled conducting tracheids are laid down at the beginning of a growing season, when the water requirements of the leaves are at a maximum, whereas the thick-walled strengthening tracheids are formed later, giving rise to alternating zones of thin- and thick-walled cells. This arrangement, incidentally, renders growth rings in softwoods conspicuous to the naked eye; the early wood, containing a smaller proportion of wall substance, appears lighter in colour than the denser late wood. In some species the transition from thin- to thick-walled tracheids is abrupt, *e.g.*, larch, Douglas fir, European redwood, but in others *e.g.*, white pine and true firs, it is gradual.

The quality of a softwood depends largely on the proportions of thin- to thick-walled tracheids, and on the contrast between the wood of these two zones. The higher the percentage of late wood the stronger is the timber; moreover, marked differences in thickness of the walls of the early and late wood cells may cause the two zones to behave differently under tools and in service, and may give rise to painting problems, *e.g.*, as in Douglas fir.

In popular language tracheids are often called 'fibres', particularly in connection with wood-pulp in the paper industry, but this is incorrect, as true fibres occur only in hardwoods.

THE STORAGE TISSUE

The storage tissue is known collectively as parenchyma; it consists of two kinds of cells that are essentially similar in details of structure, but that differ in their manner of distribution in the wood. These cells are brick-shaped, with the longer axis horizontal in the **ray-parenchyma** cells, and vertical in the **wood-parenchyma** cells. The cells have relatively thin walls with numerous pits. They differ from tracheids in remaining alive for some years after their development is completed. This is because plant food is usually stored in some form other than that required by the growing cambium, and its conversion to a suitable state can only occur in a living cell. When no longer required for storage the parenchyma cells die like any other cells of the secondary xylem.

The ray tissue occurs in narrow, horizontal bands or plates called **rays** (medullary rays), which radiate outwards from the centre of the tree to the bark, although on a small area from near the outside of a large tree they may appear as a series of parallel layers between several rows of tracheids as in Plate 9, fig. 1. These plates of tissue are continuous outwards because the cambial cells from which they arise produce only ray cells at each division, and never wood-parenchyma cells or tracheids. As a tree increases in girth, additional groups of specialized cambial cells are formed that produce only ray cells. In this way the number of plates of ray tissue per unit length of circumference of a stem remains approximately the same, irrespective of the age of the tree. The number of rays per unit of circumference, however, varies appreciably in different species: from less than one to more than ten per millimetre of circumference. The use of the term 'medullary ray' should be avoided: **medulla** is an alternative term for pith. Hence, only those rays that originate in the first year's growth, or pith, are strictly medullary rays; away from the pith, such rays cannot be distinguished from those that originate later in the life of a tree and, therefore, are not true 'medullary rays'.

The rays are usually just visible to the naked eye on radial surfaces, where they appear as narrow, horizontal ribbons 0·05 to 0·5 mm. wide (Plate 9, fig. 2). They may appear discontinuous because the cut surface is rarely truly radial and the rays may run out of the section.

On end and tangential surfaces the rays can usually be seen with the aid of a low-power lens. On end surfaces they appear as narrow lines radiating outwards, crossing the growth rings at right angles, and on tangential surfaces (Plate 9, fig. 3), where the rays themselves are seen in section, they appear as short, vertical, boat-shaped lines.

The wood-parenchyma cells are derived from normal cambial cells that also produce tracheids. After longitudinal division of a cambial cell, the cell that is to become a unit of the storage tissue divides transversely, one or more times, to give a vertical series of cells. The individual cells are wood-parenchyma cells, and the series is known as a **wood parenchyma strand**. Several strands may be united end to end.

In softwoods the wood-parenchyma tissue is sparse in amount, and usually visible only under the microscope. The strands are scattered through the wood, or restricted to definite zones, or in a layer at the end of a season's growth. In the last-mentioned case, the layer is usually visible to the naked eye on end surface as a narrow line, lighter in colour than the surrounding tissue.

Resin Canals and 'Pitch Pockets'

A characteristic feature of many softwood timbers is their resinous nature, which is often sufficient to give them a pronounced odour, and may cause freshly sawn timber to be 'tacky'. The resin is formed in parenchyma cells, and in some species occurs in special channels called **resin canals** or **resin ducts**. These canals are not cells, but cavities in the wood, lined with an **'epithelium'** of parenchyma cells. The epithelial cells secrete resin into the canals.

Resin canals run vertically in the stem and horizontally in the rays; they are just large enough to be seen with the naked eye. They are a useful feature for distinguishing some timbers, since they are always present in some species, *e.g.*, the larches, Douglas fir, the true pines (Plate 10, fig. 1), and spruces, but they are normally absent in others, *e.g.*, the true firs, sequoia, and yew.

Vertical resin canals may develop, as the result of injury to the tree, in timbers from which they are normally absent, as well as in those in which they are a normal characteristic. Such canals are said to be **traumatic**; they differ from normal canals in that they occur in short or long rows parallel to the growth rings (as seen on transverse section), whereas the normal canals are scattered in distribution (compare Plate 10, figs. 2 and 3).

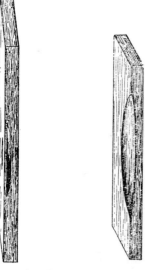

Serious injury to the cambial cells may result in the formation of **'pitch pockets'** (Fig. 7). These vary in size from about 3 mm. to several millimetres wide tangentially, and up to 300 millimetres or more longitudinally. They are saucer-shaped, with the concave face towards the pith, and up to 2·5 cm. or more at their greatest depth radially. They contain liquid resin,

Fig. 7. Pitch pockets in Douglas fir (considerably reduced)

which flows out readily when the pockets are sawn through. Openings of the wood along the growth rings may also become more or less filled with pitch in a liquid or granulated state: these are known as **pitch seams** or **pitch shakes.**

MICROSCOPIC FEATURES

Early and late wood tracheids, and wood- and ray-parenchyma cells, are the only types of *cells* that occur in softwoods, but certain distinctive types of cell-wall thickening, and the shape and distribution of pits, characterize the anatomical structure of the woods of different genera, providing the only positve means of identification. Hence, although these features have no bearing on the practical

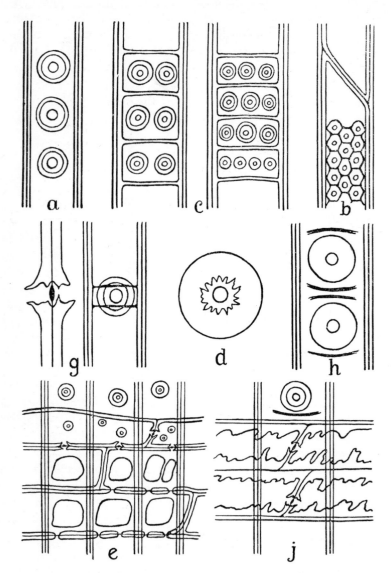

FIG. 8. Microscopic features used in the identification of softwoods
(for explanations see pages 32 and 33 of text)

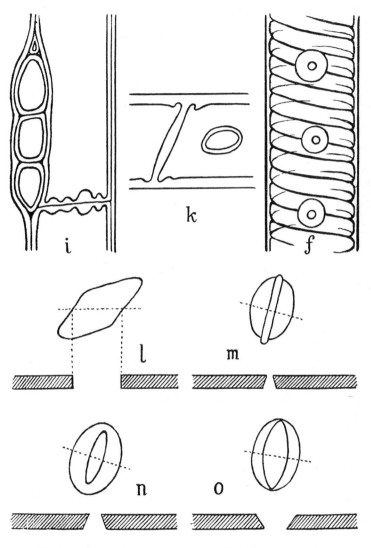

FIG. 8. Microscopic features used in the identification of softwoods
(for explanations see pages 32 and 33 of text)

utilization of the timbers concerned, nor can they be used diag-
nostically except in the laboratory, they are of some importance even
to the strictly practical user, and require defining here.

Tracheid pitting. The bordered pits of tracheids typically occur in
one or two rows, as seen on radial sections (fig. 2, Plate 9 and Text
fig. 8, *a*), but, in the *Araucariaceae* particularly, the pits are distinctly
angular in outline, and the pits in one row alternate with those in
the rows above and below, *i.e.*, **alternate pitting**, instead of being
arranged in parallel lines, one above the other (Fig. 8, *b*). Yet
another variation in distribution of pits in tracheids is their occurrence, mainly in rows of three, the pits in one row being immediately
above and below those in the adjacent rows, *i.e.*, **multiseriate** and
opposite pitting (Fig. 8, *c*).

In *Cedrus* the margin of the torus, as seen in radial section, is
regularly scalloped (**scalloped tori**), providing a reliable diagnostic
feature for distinguishing the genus (Fig. 8, *d*).

Ray tracheids. The ray cells of some softwoods are of two kinds –
ray parenchyma and **ray tracheids**. The latter are not parenchyma
cells, but mechanical tissue and physiologically inactive; they are
equipped with bordered pits, which can usually be seen in section
on the radial face; pitting between ray tracheids and ray-parenchyma cells is half-bordered (Fig. 8, *e*). Ray tracheids are normally
confined to the margins of rays, but in some species of *Pinus*, ray
tracheids may also occur in the middle portions of a ray, and the
low rays of the hard pines may consist wholly of ray tracheids.

Wall thickening. (a) *Spiral thickening* occurs as a characteristic
feature in Douglas fir and yew, and is present in some other species
of no commercial importance. The spirals are inclined in one
direction, but, because of the depth of focus of the microscope, the
spirals in the wall of the cell below may also be seen, producing a
reticulate pattern on the wall of the tracheid above (Fig. 8, *f*). In
some cases, *e.g.*, *Taxus baccata*, the thickness or depth of the cell
exceeds the depth of focus of the objective, and no reticulate pattern
is seen. The spirals are actual bands of thickening in the secondary
wall. Care must be exercised to distinguish checking in the tracheid
walls from spiral thickening.

(b) *Callitroid thickenings*: pairs of thickening bars across the pit
border occur in a few species, particularly in the genus *Callitris*
(Fig. 8, *g*).

PLATE 14

Heartwood of robinia showing tyloses (×10)

Photo by F.P.R.L., Princes Risborough

PLATE 15

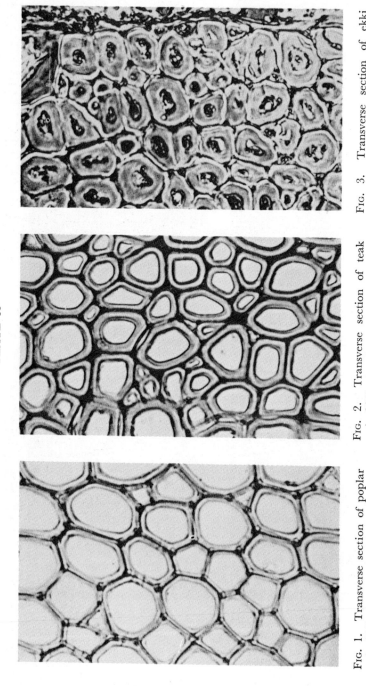

Photos by L. A. Clinkard

FIG. 1. Transverse section of poplar (×800) showing thin fibre walls

FIG. 2. Transverse section of teak (×800) showing moderately thick fibre walls

FIG. 3. Transverse section of ekki (×800) showing thick fibre walls

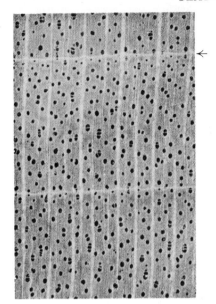

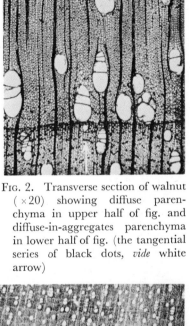

FIG. 1. Transverse section of syca-more (×15) showing conspicuous bands of terminal parenchyma, *vide* arrow

FIG. 2. Transverse section of walnut (×20) showing diffuse paren-chyma in upper half of fig. and diffuse-in-aggregates parenchyma in lower half of fig. (the tangential series of black dots, *vide* white arrow)

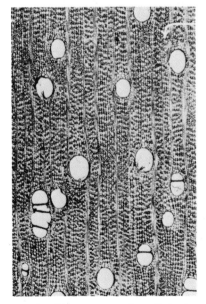

FIG. 3. Transverse section of obeche (×15) showing diffuse-in-aggre-gates parenchyma, *vide* white arrow

FIG. 4. Transverse section of *Enantia chlorantha* (×10) showing fine lines of metatracheal parenchyma, which is also reticulate (see area within white square)

Photos, Fig. 1 *by L. A. Clinkard.* Figs. 2 and 3, *Crown copyright reserved.* Fig. 4 *by H. F. Woodward.*

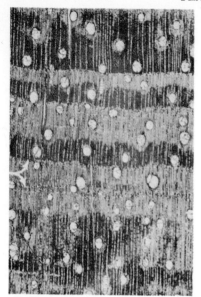

FIG. 5. Transverse section of rengas
(×10) showing fine lines of metatracheal parenchyma irregularly
spaced (banded) *vide* the white
lines indicated by arrows

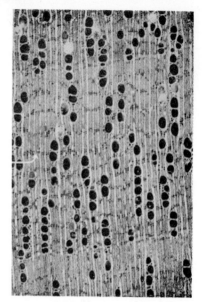

FIG. 6. Transverse section of makoré
(×15) showing broad bands of
metatracheal parenchyma (the
dark-coloured lines running across
the rays). Note the radial arrangement of the vessels

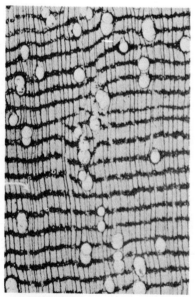

FIG. 7. Transverse section of ekki
(×10) showing broad conspicuous
bands of metatracheal parenchyma
(the dark lines indicated by white
arrow)

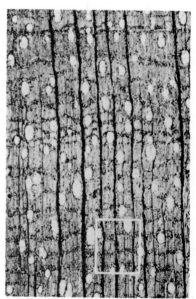

FIG. 8. Transverse section of *Tarrietia
argyrodendron* (×10) showing broad
bands of metatracheal parenchyma,
tending to be reticulate (area
within white square)

Photos, Figs. 5, 7 and 8 *by H. F. Woodward*. Fig. 6 *by L. A. Clinkard*

(c) *Crassulae* (formerly **'Bars of Sanio'**): concentrations of inter-cellular substance, appearing as horizontal bars, occur in the radial walls of all tracheids (except in the *Araucariaceae*) above and below the rows of pits (Fig. 8, *h*).

(d) *Nodular walls*: the transverse or end walls of parenchyma cells of some softwoods are nodular or bead-like in appearance; in a few species the end walls of the ray cells may be similarly thickened (Fig. 8, *i*).

(e) *Dentate thickening*: the lateral walls of ray tracheids of the 'hard pines', and, to a lesser degree, in *Picea*, are thickened in an irregular manner, giving the walls the appearance of rows of ir-regular teeth (Fig. 8, *j*).

(f) *Indentures*: pit-like hollows in the horizontal walls of rays, in which the ends of the vertical walls stand, have been described as **indentures** (Fig. 8, *k*). This feature has been observed in all families of softwoods, except the *Araucariaceae*, but is strongly developed only in some genera.

Cross-field pitting: the area of wall contact between a ray cell and a vertical tracheid is referred to as a **cross field**; the pitting occurring in a cross field takes one or other of five more or less distinct forms. In *Pinus*, the cross field is occupied by one to three large simple, or nearly simple, pits, or one to six small simple, or nearly simple, pits (**pinoid** type) (Fig. 8, *l*). The **piceoid** type refers to early-wood pits with narrow apertures, sometimes extending beyond the margins of the pits (Fig. 8, *m*). In the **cupressoid** type the apertures are included, and rather narrower than the border (Fig. 8, *n*), and in the **taxodioid** type the apertures, which are included, are ovoid to circular, and wider than the border (Fig. 8, *o*). The distinction between the last two types calls for careful observation of sections in proper focus.

CHAPTER 3

Hardwood Tissues

TYPES OF CELLS IN HARDWOODS

WHEREAS in softwoods both conducting and strengthening functions are undertaken by a single type of cell, in hardwoods there is a more distinct division of labour, and the conducting cells, called **vessels** or **pores**, are quite different from the **fibres** that provide mechanical support. The presence of specialized conducting tissue provides a simple means of differentiating hardwoods from softwoods.

The cambial cells of hardwoods are shorter than those of softwoods, and so are the mature cells that arise from the division of these cambial cells. The maximum length rarely exceeds 2 mm., as compared with 10 mm. attained by some softwood tracheids. This difference in length, between fibres and tracheids, is one of the reasons why paper-pulp manufactured from hardwoods is almost invariably inferior to that manufactured from softwoods. In hardwoods, as in softwoods, the same cambial cell may give rise successively to conducting, mechanical, or storage cells, or bark tissue; as in softwoods, a special type of cambial cell gives rise only to ray cells.

THE CONDUCTING TISSUE

The counterpart in hardwoods of the thin-walled, conducting tracheids of softwoods are the vessels or pores, illustrated in Fig. 9. This figure shows a vertical series of three fully-developed conducting cells, each of which is known as a **vessel member**. These members are always produced in vertical series, which may extend for a considerable distance in the tree. In Fig. 9 it will be seen that the vessel members have no 'end' or transverse walls, but are open top and bottom. When first formed these cells have end walls like other cells, but early in their development the cells swell and the end walls split and are absorbed, forming rims at either end of each vessel member, so that the members form a continuous tube, like a drain-pipe, in the tree. The individual members are frequently visible on longitudinal surfaces of wood as fine to coarse scratches – the **vessel lines.**

In some species, *e.g.*, birch, alder, American whitewood, the end walls of vessels do not disappear completely; instead, grid-like partitions, **scalariform perforation plates**, are left. The open ends of vessels are called **simple perforation plates.** Scalariform plates are always oblique, and in the radial plane; they can be seen with a hand lens on split radial surfaces as in Plate 11, if the vessels are not too small in cross section.

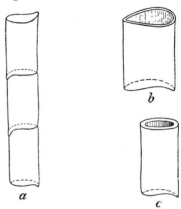

It will be clear that the conducting tissue of hardwoods is more effective in providing the water requirements of leaves than are tracheids in softwoods, and this is necessitated by reason of the larger leaf area of broad-leaved species compared with that of conifer needles. Moreover, in addition to the open ends of the vessel members, pits occur in the longitudinal walls, but these pits are smaller than those in the walls of softwood tracheids.

Fig. 9. Vessel members. *a*, vertical series of three vessel members; *b*, thin-walled early-wood vessel member; *c*, thick-walled late-wood vessel member. (Highly magnified)

Some hardwoods, of which oak and sweet chestnut are examples, have tracheids in addition to vessels to assist in conduction. These **vascular tracheids** are similar in appearance to softwood tracheids, but they are shorter, and the pits resemble those of vessel members, and are not restricted to the radial walls.

Vessels are distributed singly, or in radial or tangential groups, or in clusters throughout the wood (Plates 12 and 16b, fig. 6). As a general rule, those formed at the beginning of the growing season are wider and thinner-walled than those formed afterwards. In some species the decrease in size is gradual throughout the ring; these are the **diffuse-porous** woods, *e.g.*, beech, birch, poplar,

sycamore, and especially tropical hardwoods (Plate 13, fig. 1). In
a few species the vessels of the early wood are comparatively large,

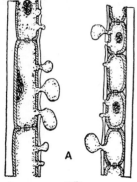

and there is an abrupt change in size to the
small and thicker-walled vessels of the late
wood; these are the **ring-porous** woods, so
called because the early-wood vessels form a
distinct ring that can be seen with the naked
eye on end surface, *e.g.*, oak, ash, elm (Plate
13, fig. 2).

The vessels of the heartwood do not conduct, but are often blocked with what appear
to be foam-like structures, known as **tyloses**,
or they may contain solid deposits of a
gummy type. Tyloses, as may be seen in
Fig. 10 and Plate 14, are ingrowing,
bladder-like structures from adjoining ray-
or wood-parenchyma cells; on a clean-cut
end surface of a piece of wood tyloses
appear as foam-like structures in the vessels
that often glisten because of differences in
light-reflection from the vessel wall and the
membraneous walls of the tyloses. In bright
light the walls of the tyloses produce rainbow
effects, similar to those in soap bubbles.

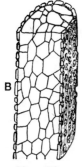

Fig. 10. Diagrammatic
drawing of the develop-
ment of tyloses: A, stage
1, development from
storage cell through the
pits into a conducting
cell. B, stage 2, tyloses
completely blocking a
vessel

By courtesy of Prof. Forsaith

When tyloses are themselves filled with gum-
like infiltrates, as in meranti, they rarely
glisten, and are then liable to be mistaken
for solid deposits filling the vessels. All the
vessels of the heartwood of some timbers, *e.g.*,
robinia (Plate 14), chengal, tembusu, are
blocked with tyloses, rendering the line of

demarcation between sapwood and heartwood distinct, even in
decayed or discoloured wood, where colour differences between
sapwood and heartwood are obscured. The formation of tyloses is
brought about by differences in pressure between the parenchyma
cells and adjacent vessels: when the vessels are actively conducting,
the pressures inside the parenchyma cells and vessels are more or
less equal, but when the vessels cease to conduct the pressure inside
the parenchyma cells is greater than the pressure in the vessels. In

consequence, the thin primary walls of the parenchyma pits become distended, being blown out like a child's balloon, to fill the vessel cavity.

Tyloses are important from the utilization standpoint: their presence may be beneficial or otherwise. They are beneficial in that they hinder the spread of fungal **hyphae** – the vegetative parts of fungi – that bring about the decay of wood, but tyloses are undesirable in timbers to be treated with wood preservatives, because they impede the absorption of preservatives which, in hardwoods, travel mainly through the vessels and only very sparingly transversely through the pits in the cell walls (in softwoods, where tyloses naturally do not occur, the movement of wood preservatives is largely through the large bordered pits in the tracheid walls).

The presence or absence of tyloses is also a useful character for distinguishing between certain woods. For example, they are absent from the true mahoganies (species of the genus *Swietenia*) and present in meranti (lauan or 'Philippine mahogany') (species of *Shorea* and *Pentacme*, family *Dipterocarpaceae*).

THE STRENGTHENING TISSUE

The mechanical tissue of hardwoods consists of **wood fibres** or **libriform fibres**. These are narrow, spindle-shaped cells, not unlike the late-wood tracheids of softwoods, but they usually have more pointed ends and are shorter. The walls of these cells may be comparatively thin, or so thick that the cell cavity is reduced almost to vanishing point (Plate 15). In discussing the factors that affect the quality of softwoods, the importance of the proportions of thin- to thick-walled tracheids, the thickness of the walls, and the distribution of the different tissues were mentioned. In hardwoods the thickness of the fibre walls and their physico-chemical nature are in many cases the most important factors in determining the strength, shrinkage, and working properties of a timber.

Pits in fibre walls are fewer and smaller compared with those in other kinds of cells. They are not confined to a particular wall, but tend to be more numerous in the radial walls. Pits may be simple or bordered. Fibres with bordered pits are called **fibre-tracheids.**

In some timbers, *e.g.*, teak and gabon, the cavities of the fibres are divided into small compartments by thin horizontal partitions;

such fibres are called **septate fibres**. The reason for the partition-ing is not known, but such fibres are more common in species with little parenchyma.

Fibres are sometimes arranged in very regular rows (as seen on cross section); such arrangement is a characteristic and constant feature of several timbers. When the individual fibres are sufficiently large in cross section, their radial arrangement is discernible with the aid of a hand lens (× 10 or × 20 area magnification), *e.g.*, in gabon.

The Storage Tissue

In hardwoods the storage tissue is essentially similar to that of softwoods, but it is frequently more abundantly developed, and it displays greater variety in distribution and arrangement. In conse-quence, wood parenchyma and rays are among the most useful features for distinguishing between different hardwoods.

Wood parenchyma. Two distinct types of distribution may be differentiated: **apotracheal parenchyma**, which is parenchyma independent of the vessels, and **paratracheal parenchyma**, which is parenchyma associated with the vessels. Both types perform the same function in the tree – that is, they are storage tissue, composed of wood-parenchyma strands. The two types can be further sub-divided; apotracheal into **terminal**, **diffuse**, and **metatracheal**, and paratracheal into **vasicentric**, **aliform**, and **confluent**. The foregoing divisions are convenient for purposes of identification, but they are by no means clear-cut. Further subdivisions are possible, which, however, grade into one or other of the divisions enumerated: *vide* Fig. 11. The appearance of the different types of wood paren-chyma, as seen on end surface, is described below.

1. *Terminal parenchyma* is the name for the narrow layers of parenchyma cells occurring at the close of a season's growth. If wide enough, the layers are visible to the naked eye as light-coloured lines, marking the boundaries of the rings, as in sycamore (Plate 16a, fig. 1). **Initial parenchyma** is used to differentiate 'terminal' parenchyma formed at the beginning of the next season's growth instead of at the close of the previous season.

2. *Diffuse parenchyma* consists of single strands distributed irregu-larly among the fibres, as in pear and box; this type is, as a

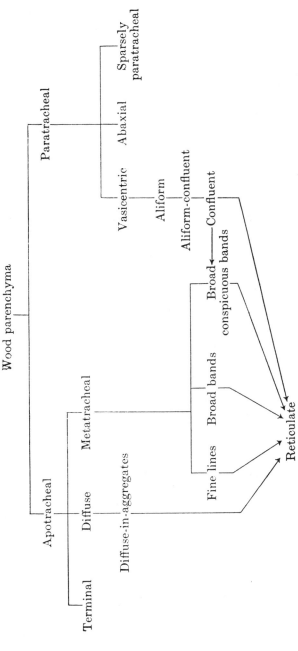

Fig. 11. Diagram showing different types of parenchyma. The arrows pointing to 'reticulate' indicate that the parenchyma may also be reticulate. The arrow between 'confluent' and 'broad conspicuous bands' is to indicate that the former may not always be easily differentiated from the latter

rule, distinct only under the microscope. When the individual strands are of sufficiently large cross section, they are discernible with a hand lens as indistinct, light-coloured dots. In woods with numerous parenchyma strands of rather large cross section the end surface may appear characteristically speckled, as in punah.

Diffuse strands tend to aggregate to form fine lines from ray to ray, a condition seen in European walnut, but often only in some parts of the growth ring (Plate 16a, fig. 2). In a few woods the fine lines from ray to ray are very regular, and alternate with rows of fibres to produce a very characteristic pattern, as in obeche (Plate 16a, fig. 3). The term **diffuse-in-aggregates** has been suggested for diffuse parenchyma forming fine lines from ray to ray.

3.   *Metatracheal parenchyma* occurs in tangential layers that are independent of the vessels. These layers appear as lighter-coloured bands, concentric with the growth ring boundaries. They may be very narrow and only visible with a lens – **fine lines** – as in *Enantia chlorantha* Oliv. (Plate 16a, fig. 4), or the layers may be visible to the naked eye – **broad bands** – as in makoré (Plate 16b, fig. 6). Fine lines or broad bands differ from the diffuse-in-aggregates type in that they appear to cross several rays, instead of extending only between one pair of rays; they are often more or less continuous concentric layers. Fine lines or broad bands may be more or less regularly spaced and close together as in ebony and makoré, or widely and irregularly spaced, as in rengas (Plate 16b, fig. 5), when they are not readily differentiated from layers of terminal parenchyma. The individual layers vary in width radially; they may be one to three cells wide, as in makoré, or several cells wide as in ekki (Plate 16b, fig. 7), when they are conveniently referred to as **broad conspicuous bands.**

When the layers of parenchyma and the rays are of about the same width, and the distance between the layers is similar to the distance between the rays, a net-like effect is produced on cross section: the parenchyma is then said to be **reticulate** (Plate 16a, fig. 4 and 16b, fig. 8).

4.   *Vasicentric parenchyma.* Where the tissue is sufficiently abundant to form complete sheaths or borders around the vessels, as in ash, it is said to be vasicentric (Plate 17a, fig. 4). In many timbers the borders or sheaths are not complete and therefore the parenchyma is not strictly vasicentric. To meet this objection, the term **sparsely**

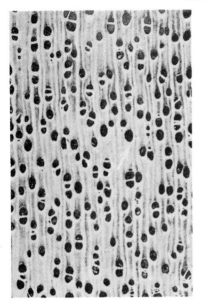

FIG. 1. Transverse section of African mahogany (×10) showing sparsely paratracheal parenchyma, *vide* arrow

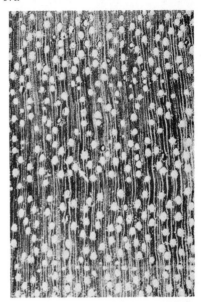

FIG. 2. Transverse section of *Goupia glabra* (×10) showing ab-axial parenchyma, *vide* arrow

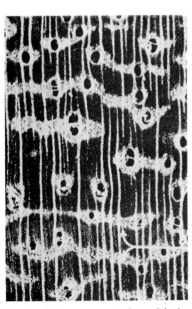

FIG. 3. Transverse section of iroko (×10) showing the aliform-confluent type of parenchyma, *vide* arrow

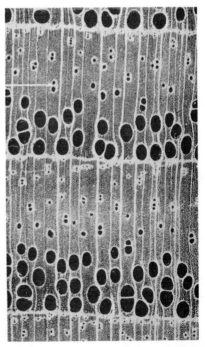

FIG. 4. Transverse section of ash (×15) showing vasicentric parenchyma (the white borders to the vessels)

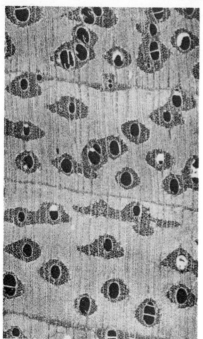

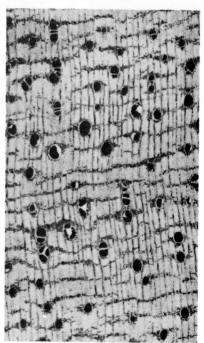

FIG. 5. Transverse section of merbau (×15) showing aliform parenchyma (the dark-coloured tissue surrounding the vessels)

FIG. 6. Transverse section of Indian rosewood (×15) showing confluent parenchyma (the dark-coloured bands joining up the vessels) that is progressively more continuous outwards in each growth ring

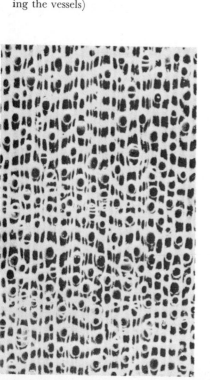

FIG. 7. Transverse section of *Symphoria globulifera* (×10) showing broad conspicuous bands of confluent parenchyma (the light-coloured bands joining the vessels); compare with ekki (Plate 16b, fig. 7)

Photos, Figs. 5 and 6 *by L. A. Clinkard.*
Fig. 7, *Crown copyright reserved*

PLATE 18

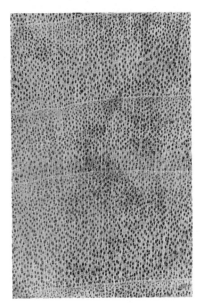

FIG. 1. Transverse section of willow (× 10) showing fine rays

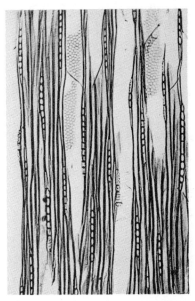

FIG. 2. Tangential section of willow (× 100) showing uniseriate rays

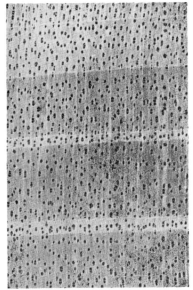

FIG. 3. Transverse section of birch (× 10) showing moderately fine rays

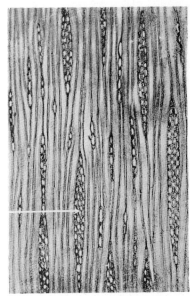

FIG. 4. Tangential section of birch (× 100) showing the rather small multiseriate rays

Photos by L. A. Clinkard

PLATE 19

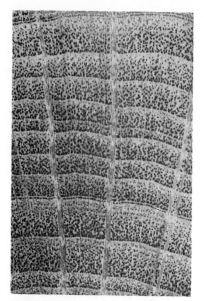

Fig. 1. Transverse section of beech (×10) showing broad rays

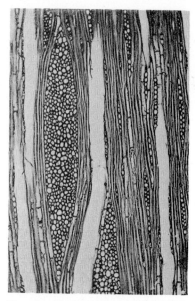

Fig. 2. Tangential section of beech (×75) showing broad, multiseriate rays

Fig. 3. Transverse section of oak (×10) showing rays of two distinct sizes (the fine uniseriate rays are indistinct)

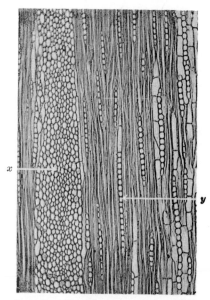

Fig. 4. Tangential section of oak (×75) showing a part of a broad ray (x) and several uniseriate rays (y): note rays of intermediate size do not occur

Photos by L. A. Clinkard

PLATE 20

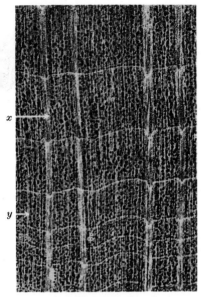

FIG. 1. Transverse section of alder ($\times 10$) showing aggregate rays (x) and fine rays (y)

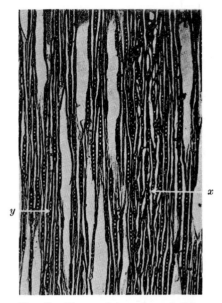

FIG. 2. Tangential section of alder ($\times 50$) showing an aggregate ray (x) and several uniseriate rays (y)

Photos by L. A. Clinkard

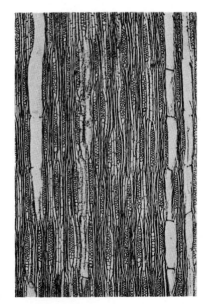

FIG. 3. 'Ripple marks' as seen on a tangential section of mansonia ($\times 35$)

FIG. 4. Ripple marks on a flat-sawn face of mansonia ($\times 12$)

Photos, Fig. 3, *by L. A. Clinkard.*
Fig. 4, *by F.P.R.L., Princes Risborough*

PLATE 21

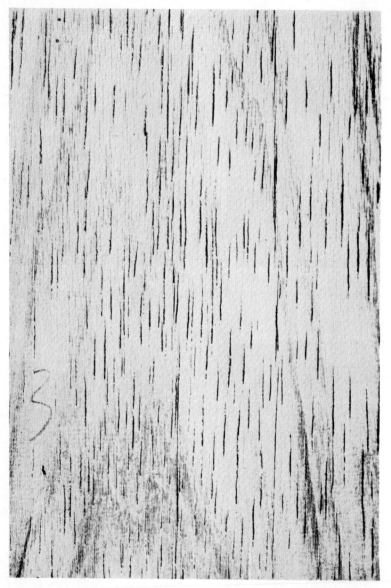

Tangential surface (three times natural size) of *Tarrietia argyrodendron* showing storeyed tissue other than rays.

PLATE 22

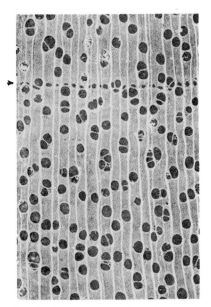

FIG. 1. Transverse section of meranti (×10) showing a tangential line of vertical 'resin' canals. Note the canals appear black in the figure because the 'resin' is dissolved in the process of mounting the section

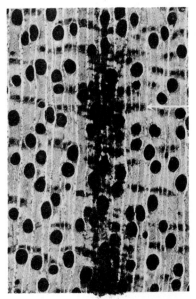

FIG. 2. Transverse section of keruing (×10) showing the short tangential series of vertical 'resin' canals

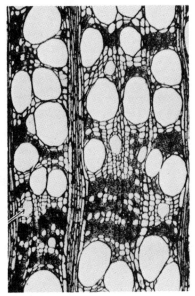

FIG. 3. Transverse section of mersawa (×75) showing scattered distribution of the vertical resin canals

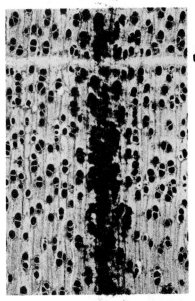

FIG. 4. Transverse section of African walnut (×10) showing traumatic 'gum ducts'

Photos, Figs. 1, 2, and 4 *by L. A. Clinkard*. Fig. 3 *by F.P.R.L., Princes Risborough*

PLATE 23

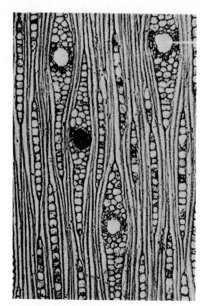

FIG. 1. Tangential section of rengas
(×75) showing radial canals

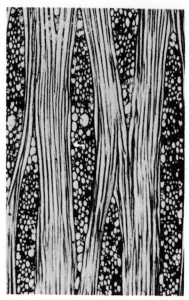

FIG. 2. Tangential section of *Shorea
Faguetiana* showing radial canals
(×75). Compare the small canals
with the larger canals in rengas

Photos, Fig. 1 *by L. A. Clinkard.* Fig. 2 *by F.P.R.L., Princes Risborough*

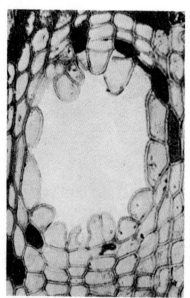

FIG. 3. Transverse section of a
vertical resin canal in *Shorea guiso*
(×300)

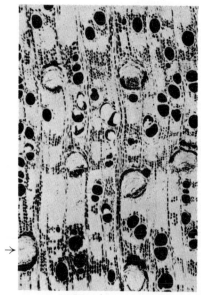

FIG. 4. Transverse section of *Strychnos non-blanda* showing islands of
included phloem (×30)

Photos by F.P.R.L., Princes Risborough

paratracheal has been proposed for paratracheal parenchyma around the vessels, but not forming a complete sheath (Plate 17a, fig. 1). In a few woods the paratracheal parenchyma is confined to one side of the vessel, the tangential face, as in *Goupia glabra* Aubl. (Plate 17a, fig. 2); this type of distribution has been called **abaxial.**

5. *Aliform parenchyma.* In many timbers the borders extend tangentially in wing-like arrangement, and appear in cross section as diamond- or lozenge-shaped masses containing the vessels, as in merbau. This is aliform parenchyma (Plate 17b, fig. 5).

6. *Confluent parenchyma.* When the tangential projections extend and link up with those of neighbouring vessels, the parenchyma is said to be confluent. Further distinctions in form of confluent parenchyma in different species can conveniently be made. Exceptionally, confluent parenchyma may occur in more or less continuous concentric layers, as in *Millettia* and *Symphoria* (Plate 17b, fig. 7), the layers being as broad and as regular as the broad layers of meta-tracheal parenchyma in ekki (Plate 16b, fig. 7). More frequently, the layers of confluent parenchyma are interrupted. In some timbers, where the parenchyma is often no more than vasicentric or aliform at the beginning of a ring, the interrupted layers become more and more continuous outwards in a ring, *e.g.*, Indian rosewood (Plate 17b, fig. 6), and other species of *Dalbergia*, and the genus *Pterocarpus*. In other timbers there is no definite distribution of confluent and aliform parenchyma in different parts of the growth ring, the two types intermingling as in iroko (Plate 17a, fig. 3), and several timbers of the genus *Terminalia*, typically in afara; Dr Chalk has adopted the appropriate term *aliform-confluent* for this type of parenchyma. All the foregoing variations in distribution of confluent parenchyma are distinct from the typically aliform type, which inevitably produces short tangential layers where two or more vessels are close together – contrast the short layers towards the end of the bottom growth ring in Plate 17b, fig. 5, with the inter-rupted layers marked with a white arrow in Plate 17a, fig. 3; the real nature of the parenchyma in Plate 17b, fig. 5, is quite clear over the rest of the cross section illustrated. It will be found useful in identification work to recognize the distinction between aliform, aliform-confluent, and confluent parenchyma as typified in the examples illustrated.

Broad layers of confluent parenchyma are sometimes not easily differentiated with a lens from broad layers of metatracheal parenchyma, when the comprehensive term **conspicuous broad bands** can be used to embrace both types. With careful examination, it is usually possible to establish whether the broad layers are in fact paratracheal parenchyma or apotracheal parenchyma: if the parenchyma is paratracheal occasional vessels will be observed with parenchyma reduced to the aliform type, whereas reduction in apotracheal parenchyma results in occasional vessels independent of any parenchyma.

The arrangement of the parenchymatous tissue is a useful aid in identifying many timbers: in some timbers only one type is present, but in others two or more types occur. In some, however, the arrangement is too variable to be of any diagnostic value, and other features have to be depended upon for purposes of identification.

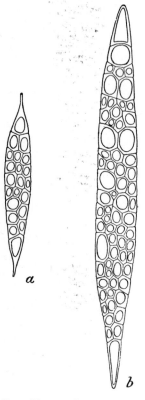

Fig. 12.   *a*, homogeneous ray; *b*, heterogeneous ray (much enlarged)

Rays. In softwoods ray tissue is sparsely developed and typically only one cell wide in the tangential direction, *i.e.*, the rays are **uniseriate**, but in hardwoods there is a considerable variation in both size and number of the rays.[1] Some hardwoods have only uniseriate rays, *e.g.*, poplar and willow (Plate 18, figs 1 and 2), but in the majority the rays are **multiseriate**, *i.e.*, more than one cell wide. In some timbers the rays are comparatively uniform in size; they may be relatively small, and not easily visible to the naked eye, as in birch (Plate 18, figs 3 and 4), or they may be broad and high, and conspicuous to the naked eye, as in beech (Plate 19, figs 1 and 2). In other woods rays of two distinct sizes occur: very large rays in association with uniseriate ones, as in oak (Plate 19, figs 3 and 4). In a few species groups of small rays occur in aggregations

[1] In softwoods ray tissue accounts for about 6 per cent. of the total volume of the wood; in hardwoods the figure is 18 per cent. and upwards.

that appear to the unaided eye, or at low magnifications, as single large rays; these are known as **aggregate** rays. The apparently broad rays of hornbeam, hazel, and alder are of this type (Plate 20, figs 1 and 2).

Very broad rays give rise to the handsome 'silver figure'[1] of quarter-sawn timber[2] of the true oaks and Australian silky oak. The presence of broad rays is also an indication that the timbers will split readily in a radial direction, an important property for certain specialized purposes, *e.g.*, the best quality 'tight' barrel staves.

Rays are sometimes arranged in regular storeys or tiers that appear on tangential surfaces as wavy, parallel, horizontal lines, known as **ripple marks.** If tangential surfaces of such woods are examined with a hand lens the individual rays are seen: they will be observed as series of storeys or tiers of short vertical lines, the top and bottom ends of each line or ray terminating at the same level as the rays on either side (Plate 20, figs 3 and 4), and the wavy horizontal lines, visible to the naked eye or hand lens, are the zones without rays that appear lighter in colour; these 'lines' are an optical effect, caused by differences in light reflection from the rays and non-ray tissue, and, in consequence, are not seen in sections of wood examined at higher magnifications, *vide* Plate 20, fig. 4. Ripple marks are a useful feature for distinguishing some timbers, *e.g.*, mansonia (Plate 20, figs 3 and 4) and the true Central American and Cuban mahoganies, although in the true mahoganies ripple marks are not always present.

In most timbers with storeyed rays the wood parenchyma tissue, and sometimes the fibres, are also storeyed. In a few, however, the wood parenchyma or fibres, or both, are storeyed but not the rays. In these latter circumstances, although the storeys are visible on both radial and tangential surfaces, it is not possible to determine from the radial surface whether or not the rays are storeyed, and the feature has to be confirmed from the tangential surface. As with storeyed rays, horizontal lines are seen on the tangential surface, but usually much less well defined than with storeyed rays, and the rays cross these lines instead of being bounded by them (Plate 21). Examination of sections under the microscope is necessary to deter-

---

[1] 'Silver figure' is more commonly called 'silver grain', but this is a mis-use of the term 'grain' (see section on Grain, texture, and figure, page 59).

[2] Quarter-sawn is defined on page 60.

mine which elements are storeyed. There is no one technical term to describe this feature, which is referred to as '**tissue** (or **elements**) **other than rays storeyed**'.

The individual ray cells may be either more or less similar in size and shape (in which case the rays are said to be **homogeneous**, Fig. 12, *a*), or distinctly variable (in which case the rays are **hetero-geneous**, Fig. 12, *b*). The example of a heterogeneous ray illustrated is a rather special case, where the size of ray cells varies throughout the ray. More usually, the term is used to define **marginal** or **sheath cells**. Marginal cells are typically upright cells, of greater height than width, that form margins or tails (according to the view-point radially or tangentially). Sheath cells are upright cells enclos-ing normal ray cells, as seen on tangential surface. A special type of ray cell, called **tile cells**, is common in certain of the *Tiliales* and *Malvales*: these are defined in the 'Glossary of terms used in describing woods' (Tropical Woods, No. 107, pp. 1–36) as a 'special type of apparently empty upright or square cells of approximately the same height as the procumbent cells and occurring in indeter-minate horizontal series usually interspersed among the procumbent cells'. The presence of heterogeneous rays and tile cells cannot usually be detected with an ordinary hand lens. The type of ray tissue is a great help in identification, but, without considerable experience in working with a low-power lens, it is not possible in most woods to determine whether the rays are homogeneous or heterogeneous.

CRYSTALS AND DEPOSITS OF SILICA IN WOOD

The storage tissue of many timbers contains crystals, usually of calcium oxalate (Fig. 6, *a* and *b*). These may be confined, in different species, to the wood parenchyma or rays, or they may occur in both tissues. More rarely, these cells contain deposits of silica, *e.g.*, the ray cells of white meranti, Queensland walnut, and apitong (keruing, gurjun, and yang) (Fig. 6, *c*). Crystals and deposits of silica, particularly the latter, may have an important bearing on the working qualities of timbers: an appreciable amount of silica in a wood renders ordinary machine tools and feed speeds uneconomic in the conversion of logs to sawn timber. For example, the standard type of circular saw in use in Malaysia for cutting red meranti, when freshly sharpened, will not make a single cut through a 4 metres long log of *meranti temak*, a form of white meranti with exceptionally high silica content (up to 3 per cent. of the dry weight of the wood). The particles of silica in the sawdust have an abrasive effect on the saw-teeth, producing rapid blunting of cutting edges and heating of the saw. Experiments at Princes Risborough indicate that saws of thicker gauge and wider gullet than normal, and teeth tipped with carborundum, are suitable for the conversion of timbers with high silica content. The presence of silica deposits in wood is of much less importance in the rotary peeling of logs in plywood and match manufacture: this may be explained partly by the logs being peeled after boiling, and partly by the different mechanical effect on the cutting edge of a knife held against a rotating log, compared with the spinning of saw-teeth in an abrasive mixture of sawdust and silica. In the former case, the knife edge tends to push the silica particles to one side, whereas in the latter the silica is ground around the saw-teeth. Timbers containing silica should, whenever practicable, be converted green, as they are then appreciably easier to work. A few timbers without crystals or deposits of silica are equally difficult to saw because of deeply interlocked fibres,[1] *e.g.*, keledang.

Resin Canals or Gum Ducts

Normal 'resin' canals or '**gum ducts**' are comparatively infrequent in hardwoods, although they are a constant feature of certain families, of which the most important commercially is the *Dipterocarpaceae*, *e.g.*, meranti (lauan), the apitong group, and mersawa. They may occur either as vertical canals in the wood, or horizontally in the rays, or, more rarely, both vertically in the wood and horizontally in the rays, in the same species. The vertical canals may occur in tangential series, producing the appearance of growth-ring boundaries (Plate 22, fig. 1), or they may be distributed in short tangential series throughout the wood (Plate 22, fig. 2) or scattered singly through the wood as in mersawa (*Anisoptera* spp.) (Plate 22, fig. 3), resak (*Vatica* spp.), and agba (*Gossweilerodendron balsamiferum* Harms). The contents of the canals of the *Dipterocarpaceae* usually consist of white or yellow, solid, dammar deposits, but in

[1] A definition of interlocked fibres is given on page 62.

Dipterocarpus spp. the deposits are viscous oleo-resins, which tend to ooze over sawn surfaces even after the wood is thoroughly seasoned, causing difficulties in painting and other finishing processes. The white deposits, particularly when the canals are in more or less continuous tangential series, are often conspicuous to the naked eye on all surfaces, appearing as prominent white lines, erroneously called 'mineral streaks'[1] in the trade.

Radial canals are illustrated in Plate 23, figs. 1 and 2.

Intercellular canals or gum ducts are produced as a result of wounding in many hardwoods. Such canals are said to be traumatic; they may be distinguished from the normal type because they are invariably in tangential series and they usually contain dark-coloured, more or less viscous, gum-like deposits (Plate 22, fig. 4). Further, traumatic canals are usually larger than the vessels, and typically widest tangentially. In some species traumatic canals are sufficiently frequent in occurrence to be regarded almost as a characteristic feature of the timber, *e.g.*, African walnut.

LATEX CANALS

Special cells or tubes, concerned with the storage of latex, occur in the ray tissue of certain timbers. They are usually invisible to the naked eye, but where they can be detected they are a helpful feature in identification. In a few timbers, *e.g.*, jelutong and mujua, specialized parenchymatous tissue, containing numerous **latex canals**, develops from leaf-traces and continues outwards during the subsequent growth of the bole; such canals, as seen on tangential surfaces, are up to 12·5 mm. high and lens-shaped in section (Plate 24). As the leaf-traces occur in whorls the latex tissue is found in tangential series at intervals of 0·6 to 0·8 metres, disfiguring long lengths of timber, and rendering it unsuitable for many purposes. Long splits often develop from the latex passages during seasoning.

INCLUDED PHLOEM

A few timbers contain strands or layers of phloem tissue included in the secondary xylem, as a result of abnormal development of the cambium. This phloem tissue is known as **included phloem.** The

---

[1] A definition of mineral streaks is given on page 246.

zones extend up and down the tree, but they may be quite small in cross section (Plate 23, fig. 4), or several millimetres wide tangentially and up to 12·5 mm. radially (Plate 25). Included phloem, being of different structure from normal wood, may affect the working qualities and seasoning properties of timbers. In some timbers the included phloem is softer than normal wood, and is inclined to pull or tear out when longitudinal surfaces are worked with machine or hand tools, but in other timbers, *e.g.*, kempas (Plate 25), the included phloem consists of harder tissue than normal wood, and behaves differently in seasoning, giving rise to serious splits; this abnormal wood also impedes the penetration of wood preservatives, even when the timber is subjected to 'full-cell' pressure processes.

In the foregoing pages the units composing woody structure have been discussed only in the detail necessary for a proper understanding of the properties and identification of wood. Readers who wish to go further into the subject are referred to such works as Eames and MacDaniels's *An introduction to plant anatomy*, and standard text-books on botany.

PLATE 24

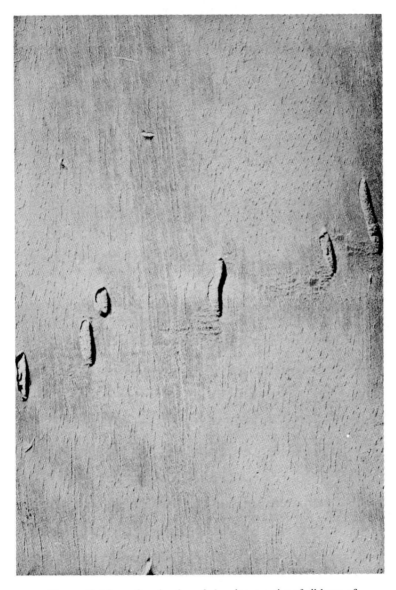

Tangential face of mujua board showing a series of ribbons of
parenchymatous tissue containing latex canals (natural size)

Photo by F.P.R.L., Princes Risborough

PLATE 25

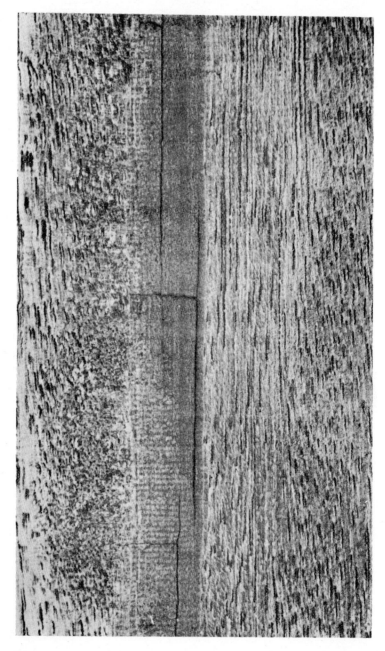

Quarter-sawn board of kempas with included phloem. Note the splits that have developed in seasoning

Photo by F.R.I., Kepong, Malaysia

PLATE 26

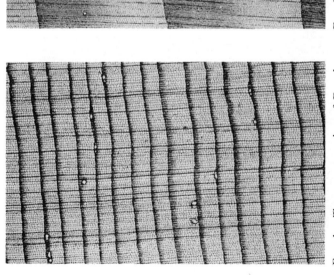

FIG. 1. Transverse section of Douglas fir (×12). Slow grown (54 rings per 25 mm)

FIG. 2. Transverse section of Douglas fir (×12). Medium-slow grown (8 rings per 25 mm)

FIG. 3. Transverse section of Douglas fir (×12). Fast grown (under 3 rings per 25 mm)

Photos by L. A. Clinkard

PLATE 27

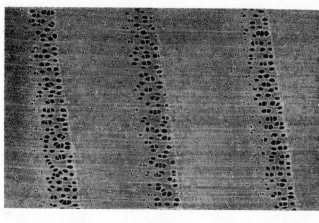

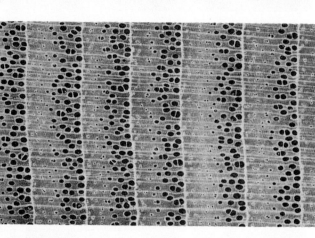

Fig. 1. Transverse section of ash (×7). Slow grown (40 rings per 25 mm)

Fig. 2. Transverse section of ash (×7). Medium-slow grown (19 rings per 25 mm)

Fig. 3. Transverse section of ash (×7). Fast grown (9 rings per 25 mm)

Photos by L. A. Clinkard

PART II

The Gross Features of Wood

CHAPTER 4

General Characters

In Part I it has been shown how the structure of wood is the outcome of the requirements of the living tree. It remains to be seen how this structure determines the usefulness of timber to man. It has been stated that the kinds of cells, and their arrangement, chemical composition, and physical structure, determine the properties of wood. These factors also govern the details of the gross features such as colour, sapwood, heartwood, and growth rings, which are readily seen with the naked eye, and enable the quality of timber to be assessed.

Sapwood and Heartwood

A striking feature of the majority of woods is the differentiation into sapwood and heartwood. Generally speaking, sapwood is lighter in colour and less durable; with softwoods, and in the log, the sapwood is wetter than the heartwood.

It is usual to regard sapwood as inferior to heartwood, so that the first consideration is to examine how far the specification of timber free from sap is justified. That this is a point of considerable importance will be realized when it is seen how high is the percentage of sapwood in an average log. For example, a 5 cm ring of sapwood is not exceptional, but it represents 55·6 per cent. of the total volume in a log 30 cm in diameter, 30 per cent. for 60 cm diameter, 21 per cent. for 90 cm diameter, and 16 per cent. for 120 cm diameter. A 2·5 cm ring in a 30 cm diameter log represents 30·5 per cent. of sapwood. Some tropical timbers have as much as 30 cm of sapwood in 75- to 90-cm diameter logs. In general, it may be said that average-size commercial logs contain between 25 and 30 per cent. of sapwood, which, if discarded, represents appreciable waste. The properties common to sapwood and heartwood, and those that differ, may be summarized as follows:

Colour. In some timbers there is no colour distinction between sapwood and heartwood, but in the majority the heartwood is more deeply coloured.

Weight. There is usually no significant difference between the weight of sound sapwood and sound heartwood of the same moisture content. Exceptions to this statement are timbers with high infiltrate content, *i.e.*, 5 per cent. or more, when the heartwood is appreciably heavier than the sapwood. In green timber the moisture content of the sapwood is usually higher than that of the heartwood, offsetting to some extent weight differences resulting from infiltrates in the heartwood.

Strength properties. Mechanical tests indicate that sapwood of the same moisture content and density as heartwood, and free of defects, is approximately equivalent in strength properties. The figures for sapwood are a little lower in some cases, but the differences are not of practical significance.

Durability. Sapwood is rich in plant food material that is attractive to certain wood-rotting fungi and insects. Different timbers vary appreciably in this respect, and fungi and insects are selective in their hosts. For example, *Lyctus* powder-post beetles must have starch but they cannot attack softwoods[1] or small-pored hardwoods because of the lack of facilities for egg-laying. Further, starch in itself is not sufficient to attract powder-post beetles, the presence of traces of other substances appears to be essential to render timbers liable to attack, and these substances are removed by prolonged soaking of timber in water. The infiltrates of the heartwood, on the other hand, are frequently positively toxic to fungi and insects. In positions where wood is exposed to the risk of decay, or to insect attack, sapwood is usually much more readily attacked than heartwood of the same species. Moreover, the presence of large quantities of sapwood, resulting, in favourable circumstances, in vigorous growth of wood-rotting fungi, or heavy infestation

---

[1] This statement, although generally true, appears to require qualification; the infestation of the sapwood of *Pinus canariensis* C. Sm. by *Lyctus* powder-post beetles has been recorded in South Africa. Infested sapwood of susceptible hardwoods immediately in contact with softwood sapwood may result in the latter becoming attacked by the feeding larvae, but this, of course, is a different matter from infestation originating in such timber.

of an insect pest, may lead to spread of such attack to adjacent heartwood.

Special considerations arise when repairs to wood-work are necessitated by decay resulting from fungal activity: in these circumstances it is rarely possible to ensure that all traces of fungal hyphae are eradicated, so that, as a precautionary measure, only timber free from sapwood should be used for the repairs, unless pressure-treated timber is employed to replace the defective wood.

Permeability. The conducting tissue of wood usually undergoes modifications at the time of heartwood formation so that the free movement of liquids is interrupted. Further, various substances are deposited on the walls of most cells during transition to heartwood, which renders them more or less impermeable to moisture movements. In consequence, heartwood is not so easily impregnated with preservatives or dyes as is sapwood. This may be of less practical importance than is apparent on the surface: heartwood possesses more natural resistance to fungal and insect attack than sapwood, and the reduced absorption of wood preservatives may still be sufficient to ensure that the more lightly treated heartwood will outlast the mechanical life of the treated timber. Where service conditions impose no limits on the growth of fungi, it is not unusual for the heartwood to decay, while the outer, heavily impregnated sapwood remains quite sound.

The case for using or rejecting sapwood. Where colour is of primary importance it is usually necessary to exclude sapwood, but the difficulty may sometimes be overcome by judicious staining. For some purposes, however, absence of colour is considered desirable, and in these circumstances the light colour of the sapwood is an advantage. Where colour is unimportant, durability may be the controlling factor, and durability depends on the conditions under which the timber is to be used. For outdoor uses generally, *e.g.*, fencing, posts, gates, and railway sleepers, sapwood should be excluded unless the wood is to be given adequate preservative treatment, in which case it may safely be retained. In well-ventilated, dry, internal situations, such as carcassing and joinery work generally, there is no objection to sapwood (1) if the timber is thoroughly seasoned *before* it is installed, (2) if the site conditions cannot reasonably be expected to alter adversely *after* the timber is installed, and (3) if the timber is not one prone to powder-post beetle and borer attack, or if the risk of such attack is remote. In effect, in temperate regions the sapwood of softwoods can be retained in most indoor situations if the likelihood of some furniture beetle infestation in the future in such timbers as European redwood is acceptable. For outdoor use, European redwood should be adequately treated as it is not possible to obtain timber free from sapwood, which is particularly perishable. Western hemlock and spruce should be avoided for external joinery because these timbers do not take preservatives well. The small additional cost of pressure or diffusion treatments is a good argument for using such treated timber in new work, both internally and externally, when unlimited sapwood would be permissible. The sapwood of all *Lyctus* susceptible hardwoods, *e.g.*, oak, agba, iroko, afzelia, should be excluded. If the distinction between sapwood and heartwood is not readily apparent, as in agba, there is a risk of *Lyctus* infestation occurring on an appreciable scale, which makes the use of this timber in joinery distinctly hazardous. In the tropics the risk of wood-borer infestation of several types is so much greater that even the sapwood of timbers immune to powder-post beetle attack should be severely restricted in timber to be used untreated for semi-permanent or permanent work, and its total exclusion from furniture and high-class flooring and joinery is advisable.

Although sapwood may be used in certain circumstances, it must not be overlooked that its presence is a potential source of danger should the site conditions change at any time, and this fact must be borne in mind when drawing up a specification. For example, there is a greater risk that built-in wall plates and ground-floor, basement, and roofing timbers may be exposed to damp than there is that first-floor and ceiling joists will be, and damp conditions render wood liable to fungal attack. More recently cases of widespread 'wet rot' in external joinery have come to light in post-war buildings, attributable in the main to the large quantities of sapwood likely to be present today in joinery grades of European redwood (See Building Research Station Digest 73 (Second Series): *Prevention of Decay in Window Joinery*). It would, therefore, be reasonable to allow sapwood in positions where the risk of attack is small, and to exclude it, where practicable, if the likelihood of infection is considerable. It will rarely be possible in practice to obtain softwood timber entirely free from sapwood, so that the problem resolves itself into paying proper attention to 'site' conditions, or insisting on the timber being adequately treated with preservatives. Damp-proof courses and air bricks provide the means for maintaining good internal conditions, and it is imperative that they should be given proper attention. It is not unusual, for example, to find ground-line air bricks blocked-up, or for flowerbeds to be raised above the damp-proof course: in such circumstances the protective measures are rendered ineffective. Other causes of damp interiors, likely to lead to decay of timber in a building, are neglected pointing, inadequate or ineffective rainwater disposal arrangements, or plumbing leaks. Stopped-up rainwater heads and gutters, cracked down pipes, and neglect of flashings are probably the commonest causes of 'dry rot' in houses, and are of greater significance than the amount of sapwood in the timbers of such houses. Since, however, neglect of maintenance of rainwater disposal arrangements is so prevalent a failing, the increasing use of timber containing large quantities of sapwood tends to make the situation still worse: temporary neglect provides the opportunity for fungi of the *Merulius* type to become established, and recurring neglect results in serious decay. The extended use of impervious floor coverings, laid tight up to skirtings, in rooms subject to condensation, or where floors are frequently washed, also gives rise to conditions favourable to fungal infection, and the large quantities of sapwood ordinarily used today tend to increase the hazards of decay originating in this way.

As sapwood is more readily impregnated with preservatives than heartwood it should be retained whenever the material is to be properly impregnated, particularly if the timber is in the round, or roughly squared, and is completely encircled by sapwood. If the application of preservatives is confined to brush coating the ends of beams, joists, or posts, sapwood pieces should not, of course, be selected in preference to timber free from sapwood.

In certain other circumstances, *e.g.*, sports goods, tool handles, shuttles, spools, and bobbins, sapwood is sometimes preferred to heartwood, but in most cases there is either no justification for the preference, or the heartwood of the timbers used for such purposes does not differ in colour from the sapwood. For example, there was a preference in America for the sapwood of hickory to the exclusion of the heartwood, but exhaustive tests by the Forest Products Laboratory, Madison, showed that the heartwood was equally suitable for all purposes for which the sapwood was preferred. Again, the favourite timbers for shuttles, spools, and bobbins (persimmon, Turkish cornel, and European box) have little or no heartwood, or the heartwood is the same colour as the sapwood. In one or two tropical timbers the sapwood is distinctly lighter in weight than the heartwood, and for this reason is preferred for tool handles, because the strength properties of the lighter timber are more than adequate for the purpose. Keranji is a case in point: the heartwood of keranji contains a high percentage of infiltrates, which increase the specific gravity of the wood appreciably without increasing the strength properties; this added weight is a disadvantage in an already heavy timber for the purpose. Sapwood is, of course, freer from such defects as knots and shakes, but this advantage is minimized in comparison with the outer heartwood of large-sized trees.

GROWTH RINGS OR LAYERS

The second gross feature to be considered is the presence or absence of growth rings or layers. These, it has been explained, occur in timber grown in regions with distinct seasonal climates, in which a growing period alternates with a resting state, and where the wood laid down at one period of the growing season differs from that produced later in the season. The character of the growth ring is sometimes useful in identifying a timber, and is often of value in assessing the quality of a piece of wood.

The width of rings, or the number of rings per cm of radius, is a measure of the rapidity of growth, and is some indication of the strength properties of wood. In softwoods, and the ring-porous hardwoods, variations in ring width are associated with variations in the proportion of late to early wood (Plates 26 and 27). In the diffuse-porous woods, in which the wood produced in a single growing season is not differentiated into early and late wood, variations in ring width are associated with variations in porosity. In all three types of wood extremely narrow and extremely broad rings are an indication of exceptionally weak timber; probably in all species there is an optimum rate of growth for the production of the strongest timber, but the rate differs with the species. In softwoods this optimum is about 7 to 20 rings per 2·5 cm, and within these

FIG. 1. Cross section of spruce log showing compression wood (the dark, wide-ringed portion shown in the lower part of the section).

FIG. 2. Cross section of beech log showing tension wood on the uphill side of leaning trees, and compression wood on the lower side of branches and the downhill side of leaning trees

Note: tension wood is formed on the upper side of branches and on the uphill side of leaning trees, and compression wood on the lower side of branches and the downhill side of leaning trees

Photo by U.S. Forest Products Laboratory

Photo by F.P.R.L., Princes Risborough

PLATE 29

FIG. 1. Flat-sawn European redwood

By courtesy of E. H. B. Boulton, Esq.

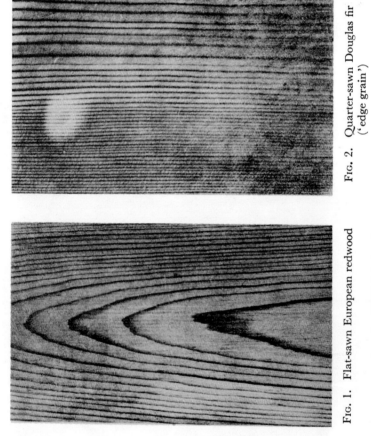

FIG. 2. Quarter-sawn Douglas fir ('edge grain')

By courtesy of the Editor of 'Wood'

FIG. 3. Rotary-cut veneer of Douglas fir

Photo by F.P.R.L., Risborough; sample lent by the Editor of 'Wood'

PLATE 30

Fig. 1. Silver figure in quarter-sawn oak

Fig. 2. Flat-sawn oak

Photos by F.P.R.L., Princes Risborough

PLATE 31

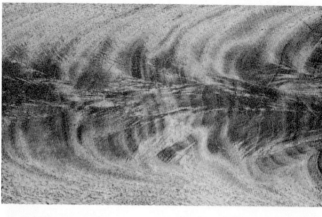

Fig. 3. Curl figure in Spanish mahogany

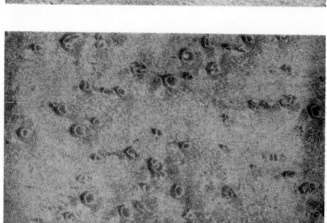

Fig. 2. Bird's-eye figure in maple

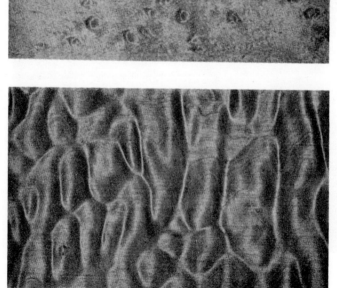

Fig. 1. Blister figure in Pacific maple

Photos by F.P.R.L., Princes Risborough; veneers lent by Messrs. John Wright (Veneers) & Sons

limits the *narrower* the ring the *narrower* is the layer of early wood, and, consequently, the *higher* is the proportion of late wood. In ring-porous woods the optimum is roughly 6 to 10 rings per 2·5 cm, and within these limits the *wider* the ring the *wider* the layer of late wood, and, consequently, the *higher* is the proportion of late wood. These limits are too strict for most practical purposes: work on ash at the Forest Products Research Laboratory, Princes Risborough, has shown that wood within a range as wide as 4 and 14 rings to 2·5 cm is likely to be stronger than that of faster or slower growth. It may be recalled that the late wood is composed largely of strengthening material and, therefore, the higher the proportion of late wood the stronger the timber. It follows that within the optimum limits for the species, the narrower the rings of softwoods and the wider the rings of ring-porous hardwoods the stronger the timber.

The weakness of the very slowly grown softwoods and ring-porous hardwoods is explained by the very narrow layers of late wood that their rings contain. The weakness of rapidly grown softwoods, on the other hand, is explained by the very wide layers of early wood. In hardwoods grown faster than the optimum limits, however, although there is an increase in the proportion of late wood, the individual fibres are abnormally thin walled, and the timber is, in consequence, weaker than less rapidly grown material in which the late wood fibres have thicker walls.

Other factors than ring width are more important in determining strength properties of individual pieces of wood, but ring width is a useful rough guide, so long as its limitations are recognized, and it can be used in the timber yard or workshop. For example, the writer classified a stock of ash tool handles into grades on a ring-width basis and, by imposing an arbitrary maximum number of rings per 2·5 cm of radius, eliminated practically all the inferior timber. On the other hand, mechanical tests showed that the arbitrary classification excluded a considerable proportion of good timber. Had density been utilized as a second factor a more accurate estimate of quality would have been realized. The value of the rings-per-2·5 cm classification depends on whether, in practice, the acceptance of a certain amount of inferior timber would be more economical than paying the higher price resulting from the imposition of a stricter specification, and this is a point that must be settled separately for each case.

No specification covering supplies of timber on a commercial scale should rest on a ring-width classification alone: a more important factor is the percentage of late wood. An average of 50 per cent. of late wood is recommended, for example, with some pines for exacting constructional purposes. Other factors, to be discussed later, *e.g.*, density, irregular grain, and defects, also require to be taken into consideration.

Strength properties are not the only factors that determine the merits of a timber: for some purposes working qualities are of equal or greater importance. In such circumstances, timber produced under other than the optimum growth conditions for strength may be superior to that produced under the optimum conditions. For example, mildness in working is associated with narrow-ringed material. The mildest softwood timber is that from northern Europe and the higher altitudes of central Europe, and the rings frequently exceed 20 to 2·5 cm; such timber is unsurpassed for joinery purposes. In the same way, the milder, slower-grown, and consequently narrower-ringed 'Austrian' oak is preferred for flooring, panelling, etc., to the faster grown, wider-ringed, and stronger English oak. The English oak is, however, better for constructional work.

Compression Wood

In softwoods a special type of tissue, known as **compression wood**, is developed on the under (compression) side of branches and the lower sides of leaning stems. The outstanding feature of this type of wood is its abnormally high longitudinal shrinkage. Whereas normal wood shrinks 0·1 to 0·2 per cent. longitudinally in drying from the green to the oven-dry condition, compression wood may shrink as much as 5·78 per cent., and is commonly 0·3 to 1·0 per cent.[1] In consequence, boards and planks containing compression wood are liable to bow in seasoning. The abnormal wood is exceptionally dense, but the extra weight is not accompanied by proportional increase in strength; in particular, compression wood has relatively low bending strength and lacks toughness. The changed properties may be attributed to abnormally high lignin content, and for this reason compression wood is not suitable for chemical

---

[1] Figures from 'The longitudinal shrinkage of wood', by A. Koehler, in *Trans. Am. Soc. Mech. Engineers*, Jan.–April 1931, vol. 53, No. 5.

paper-pulp, and its lack of toughness is equally objectionable in mechanical pulp.

In most species compression wood may be recognized by its relatively dark red-brown colour, and by the lack of contrast in colour between the early and late wood (Plate 28, fig. 1). In boards and planks the abnormal wood frequently occurs in streaks running the length of the timber.

TENSION WOOD

In hardwoods, **tension wood** may be formed on the upper sides of branches and the upper sides of leaning stems (Plate 28, fig. 2). In effect, hardwoods are likely to produce tension wood in circumstances where softwoods produce compression wood, but on the upper sides of branches and on the uphill sides of leaning trees. Tension wood is paler than normal wood, and more lustrous when viewed by obliquely reflected light. It differs from normal wood of equal density in being exceptionally weak in compression parallel to the grain. It is, however, slightly stronger in tension and toughness than normal wood of the same density. As with compression wood, tension wood has abnormally high longitudinal shrinkage; the radial shrinkage is normal, and tangential shrinkage rather greater than normal. Gelatinous fibres are characteristic of the tension wood of many species, *e.g.*, of beech and walnut but not of ash. The lignin content of the cell walls is deficient compared with normal wood. S. H. Clarke has summarized the working qualities of tension wood as follows: 'in the lathe, turnings came from tension-wood cylinders in long, pliable pieces, but those from normal wood were more brittle and broke into small chips. When surfaced on a rotary planer the tension wood was inclined to be woolly where the cutting went against the grain.'

GRAIN, TEXTURE, AND FIGURE

Grain and texture should be used to refer to two quite distinct characters of wood, but more often than not they are confused in everyday use. An attempt has been made by timber research laboratories to standardize the use of the terms, restricting each to a single feature. It is proposed that **grain** shall refer to the direction

of the fibres, relative to the axis of the tree or the longitudinal edges
of individual pieces of timber, and that **texture** shall apply to the
relative size, and the amount of variation in size, of the cells.

Figure refers to the pattern produced on longitudinal surfaces of
wood, as a result of the arrangement of the different tissues, and the
nature of the grain.

Before describing the different types of grain, it will be as well to
discuss the incorrect uses of this term. These fall under several
heads, *e.g.*, those describing the manner of sawing, those correctly
pertaining to texture, and those referring to width of growth rings.
Of the first, we have quarter, edge, vertical, and comb grain,
referring to timber that is cut parallel to the rays; for such timber
the term '**quarter-** (or **rift-**) **sawn**' is proposed. Timber cut at
right angles to the rays should be described as **flat-** (back-) **sawn**[1]
and not as 'flat grain'. In hardwoods 'coarse' and 'fine grain' are
frequently applied to characteristics that depend on the size of the
elements, and, therefore, are more correctly described as **texture**;
oak, for example, should be described as coarse textured and not
coarse grained. In softwoods, on the other hand, coarse and fine
grained are often used to describe the width of growth rings; the
former to wood with broad rings, and the latter to wood with
narrow rings. Here the feature is neither grain nor texture, and is
better described by the terms **wide-** and **narrow-ringed** or fast-
and slow-grown.

'Even' and 'uneven grain' have been used to distinguish regularity
and irregularity in the width of growth rings. As this character is
neither dependent on the direction of the fibres nor on the size of
cells, but on the rate of growth, it is inaccurate to refer to it as either
grain or texture, and a much clearer idea is given by employing the
phrase '**growth rings regular** (or **irregular**) **in width**'.

Timber that breaks with a short, brittle fracture is frequently
described as *short in the grain*. The description is inapt, as the failure
has nothing to do with the length of the fibres, nor is it connected
with their direction, in relation to the vertical axis of the tree, but
usually with their brittleness, *i.e.*, the readiness with which the
fibre walls fracture at right angles to their length. Brittleness may
be an inherent property of the species, or it may be caused by such

[1] Flat-sawn is often referred to as through-and-through sawing.

factors as fungal decay, 'spongy heart', exceptionally low density (for the species), compression wood, or even maltreatment in seasoning (usually too rapid drying in a kiln at a high temperature and too low humidity). The nearly horizontal fracture, so often described as short in the grain, is not always associated with brittleness of a particular piece of timber or a species. Such horizontal failures may be the result of gross over-loading, often, but not necessarily from an impact load. Tropical timbers tend to be more brittle than temperate-climate woods of similar density because of their higher lignin content. Brittle timbers, however, are stiffer than more flexible ones, and for some purposes this may be an advantage, provided they are not likely to be overloaded, when they would fail suddenly and without warning.

'Grain' is sometimes applied to describe the figure of a timber: *silver grain*, for example, refers to the appearance of timbers with broad rays, cut on the quarter; it is only indirectly connected with the direction of the cells, *i.e.*, the broad plates of ray tissue such as occur in the true oaks and Australian silky oak; to be consistent, figure arising from the presence of broad rays should be described as **silver figure** (Plate 30, fig. 1).

Grain. Using the restricted meaning, six types of grain may be distinguished. **Straight grain** explains itself. In straight-grained timber the fibres and other elements are more or less parallel to the vertical axis of the tree. In addition to being a contributory factor in strength, straight-grained timber makes for ease of milling and reduces waste. On the other hand, it does not give rise to ornamental figure.

Irregular grain. Timber in which the fibres are at varying, and irregular, inclinations to the vertical axis in the log, is said to have irregular grain. It is frequently restricted to limited areas in the region of knots or swollen butts. It is a very common defect and, when excessive, seriously reduces strength, besides accentuating difficulties in milling. Irregular grain, however, often gives rise to an attractive figure. Pronounced irregularities in the direction of the fibres, resulting from knoll-like elevations in the annual rings, produce **blister figure** (Plate 31, fig. 1). The valuable and attractive **bird's-eye figure** (resulting from conical depressions as opposed to the elevations in blister figure) seen on the finished tangential surfaces of selected material of a few species, *e.g.*, maple, is held to be the result of temporary injury to the cambium (Plate 31, fig. 2).

Diagonal grain is a milling defect, and results from otherwise straight-grained timber being cut so that the fibres do not run parallel with the axis of the board or plank; such timber is weaker than that properly sawn.

Spiral grain is produced when the fibres follow a spiral course in the living tree. The twist may be left- or right-handed. The inclination of the fibres may vary at different heights in the trunk, and at any one height the inclination may vary at different distances from the pith. The cause of spiral grain is not definitely known, but there is evidence that it is an hereditary characteristic of individual trees. Although not always readily visible, spiral grain may often be detected from the direction of the surface seasoning checks, often visible for example on telegraph poles. Spiral grain reduces the strength of timber and is, therefore, a serious defect in timber for important structural work.

Interlocked grain, or **interlocked fibre** as it is often called, results from the fibres of successive growth layers being inclined in opposite directions, producing the familiar figure known as **ribbon** or **stripe figure** (Plate 32, fig. 1) on quarter-sawn surfaces. Interlocked grain is relatively uncommon in temperate woods but it is a characteristic feature of a large proportion of tropical timbers. As far as is known, it does not appreciably affect the strength of timber, but it may cause serious twisting during seasoning and, if pronounced, makes the wood difficult to split radially (Plate 33, fig. 2). There is also the added disadvantage that such timber 'picks up', particularly when being planed on the quarter, leaving a very rough finish. In timbers with heavily interlocked grain, *i.e.*, when the pitch (the angle between the fibres and the vertical axis of the tree) exceeds 20° or 30°, and the successive changes in inclination of the fibres occur at intervals of 6 to 12 mm radially, sawing difficulties may be very great: the fibres tend to pull out and wrap themselves round the saw-teeth until the saw becomes buried in the log, bringing all machines driven off one motor or engine to a standstill. With timbers in this class, *e.g.*, keledang, sepul, and terentang, interlocked grain can cause as much trouble in conversion as does the high silica content of other timbers. A reasonably smooth surface can, however, be obtained in sawing and planing with modern machines, employing more set than is ordinarily required, a suitable cutting angle, and a modified rate of feed.

Wavy grain. When the direction of the fibres is constantly changing, so that a line drawn parallel with them appears as a wavy line on a longitudinal surface, the grain is said to be **wavy**. This type of grain gives rise to a series of diagonal, or more or less horizontal, darker or lighter stripes on longitudinal surfaces, because of variations in the reflection of light from the surface of the fibres: this is called **fiddle-back figure** (Plate 32, fig. 3). Wood with wavy grain presents a corrugated surface, as shown in Plate 33, fig. 1, when split. The importance of this type of grain lies in its decorative value, and any reductions in strength are of no consequence. Wavy grain may occur, together with interlocked grain, in one piece of timber, giving rise to a broken 'ripple' on quarter-sawn surfaces, called **roe figure** (Plate 32, fig. 2).

Texture. Just as it was necessary to employ qualifying adjectives to describe the different types of grain, so is it with texture, and we have the terms coarse, fine, even, and uneven texture. The differentiation between coarse and fine texture is made on the dimensions of the vessels, and the width and abundance of the rays. Timbers in which the vessels are large, or the rays broad, are said to be of **coarse texture**, but when the vessels are small, and the rays narrow, the timber is of **fine texture**. Many intermediate grades are met with, and some such classification as the following will be found useful: – *very fine, e.g.,* European box; *fine, e.g.,* sycamore; *medium, e.g.,* birch; *moderately coarse, e.g.,* walnut, mahogany; *coarse, e.g.,* oak. Strictly speaking, all softwoods are fine, or at most only moderately coarse-textured, as their cells are all of relatively small diameter, but a few with particularly thin-walled tracheids, *e.g.,* sequoia, may give a rather rough finish when sawn, and by comparison with the denser pines are distinctly coarse-textured.

The texture of softwoods is influenced by the alternation of the zones of early and late wood. When the contrast between the zones is strongly marked the wood may be said to be of **uneven texture**, *e.g.,* long-leaf pitch pine, Douglas fir, larch; when there is little or no contrast the wood may be said to be of **even texture**, *e.g.,* white pine, true firs, spruce. In this sense the terms may also be applied to hardwoods; ring-porous woods are uneven in texture, but diffuse-porous woods are even in texture unless broad rays or wide layers of wood parenchyma are present, when the texture may be as uneven as that of ring-porous woods.

Figure.   Several different types of figure have been mentioned in the discussion on grain, but many more than these are recognized in trade terminology.   Those recorded are, however, the principal types arising from the type of grain present, and other kinds of figure are mainly modifications of the basic types.   For example, **ram's horn** is a special form of wavy grain in which the waves are comparatively short, so that the resulting horizontal stripes are narrow and close together.   Curls that resemble ostrich feathers are called **feather curl** as in crotch mahogany (Plate 31, fig. 3), and so on.

Figure also arises from the distribution of certain types of tissue in a wood: the broad high rays of the true oaks and 'silky oak' are an example of figure – 'silver figure' – derived from the particular distribution of the ray tissue in these woods.   The alternating layers of dense late wood, and less dense early wood, produce the prominent 'flame' figure of certain softwoods, *e.g.*, Douglas fir, when flat-sawn or rotary peeled.

The distribution of the wood parenchyma in broad conspicuous layers, *e.g.*, species of *Millettia*, gives rise to 'flame' figure, sometimes called **watered-silk** figure, when the timbers are flat-sawn, and similar figure is produced in timbers with alternating layers of different colour, *e.g.*, rengas and the striped ebonies.   The presence of a particular type of grain, or the arrangement of certain types of tissues, is not in itself sufficient to ensure that decorative figure will be apparent, so long as the timber is correctly converted, *i.e.*, quarter-sawn or flat-sawn, depending on the source of the figure. The prominence and decorative effect of figure is dependent on the natural lustre of wood, *vide* page 70.

The Influence of the Direction of the Grain on the Utilization of Wood

When the strength of timber is the primary consideration it is usual to specify that it shall be straight grained.   The importance

PLATE 32

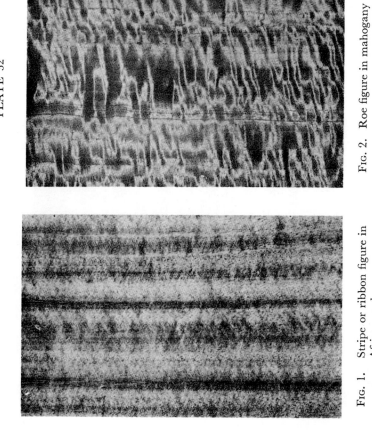

FIG. 1. Stripe or ribbon figure in African mahogany

By courtesy of E. H. B. Boulton, Esq.

FIG. 2. Roe figure in mahogany

Photo by F.P.R.L., Princes Risborough; veneer lent by Messrs. John Wright (Veneers) & Sons

FIG. 3. Fiddle-back figure in walnut

Photo by F.P.R.L., Princes Risborough; veneer lent by Messrs. John Wright (Veneers) & Sons

PLATE 33

Fig. 1. Split block showing wavy grain

Fig. 2. Split block showing interlocked grain

Photos by F.R.I., Kepong, Selangor, Malaysia

of this specification will be seen when it is realized that there is a reduction of about 4 per cent. in bending strength when the slope of the grain is 1 in 25; with a slope of 1 in 20 the reduction is 7 per cent.; with 1 in 15, 11 per cent.; with 1 in 10, 19 per cent.; and with 1 in 5, 45 per cent. The stiffness of a beam is also reduced by sloping grain, but to a less degree; the corresponding reductions in stiffness for the same variations of slope being respectively 3, 4, 6, 11, and 33 per cent. The percentage reductions in bending and stiffness vary somewhat with different species, but the figures quoted give an indication of the general trend. In C.S.I.R. (Australia): *Trade Circular No.* 13, 1933, it is recommended that a slope of grain greater than 1 in 15 should not be permitted in beams; in flooring and in the smaller sizes of joists and rafters, on the other hand, where stiffness is generally of more importance than bending strength, a slope of 1 in 10 is regarded as permissible. Recommendations in the appropriate British Standards are less demanding. In CP 112 (1967) the maximum slope of grain permitted varies with the grade, from 1 in 6 in the 40 grade to 1 in 14 in the 75 grade. In B.S. 1129: 1966: *Timber Ladders, Steps, Trestles and Lightweight Stagings for Industrial Use*, it is recognized that Douglas fir has superior strength properties to other recommended species for ladder components, and, whereas for these other timbers the combined slope of grain shall not be steeper than 1 in 10, in Douglas fir it may be 1 in 8; with treads of step ladders, steps, and platform steps, the slope of grain on edge shall not be steeper than 1 in 20. In ladder accidents, failure in stiles is frequently attributed to excessive slope of grain, if this exceeds the B.S. recommendations, whereas tests on ladder stiles have established that density can be as critical as slope of grain. In timber of average density for the species, a slope of grain in excess of 1 in 10 may be acceptable, say 1:8. Unfortunately, in Douglas fir, which is the strongest softwood used for the manufacture of ladder stiles, slope of grain over short distances may sometimes appreciably exceed the recognized acceptable limits. In use, which often entails mis-use, such ladder stiles are likely to fail at points where the slope of grain is excessive, *i.e.*, as steep as 1 in 5 or 1 in 6, although over the greater part of its length the slope of grain in the stile may be within acceptable limits.

For certain uses of timber, slope of grain is all-important. Timber for tool handles and sports goods is an obvious example, since a slope of grain of only 1 in 25 causes a reduction of 9 per cent. in impact bending (shock-resisting abilities). Timber for such purposes should be as nearly straight grained as possible, and in no circumstances should the slope exceed 1 in 25. Even greater care is necessary in the selection of timber that is to be steam bent, as satisfactory bends cannot be made from other than straight-grained timber. The requirements of wood for barrel staves for tight cooperage illustrate another aspect of the importance of grain direction. Where the grain slopes from the inside to the outside of the cask there is a likelihood of the contents seeping through the staves, hence, straightness of grain is an essential quality in timber for this purpose.

The direction of vessel lines, gum veins, resin ducts, and seasoning checks are useful in indicating slope of grain in a piece of timber, but in the absence of these features visual inspection alone can be misleading (see also Fig. 13). Direction of the grain may be detected by raising a few fibres with the point of a penknife, or by noting the spread of an ink spot. For accurate determination of slope of grain a specially designed scribe, consisting of a cranked rod, with swivel handle, and an inclined needle at the end of the rod, should be used. Determinations should be made on both the face and edge of a piece of timber. The use of the scribe is explained and illustrated in certain British Standards, e.g., BS 248: 1954, and BS 1129: 1966.

COLOUR

From a practical viewpoint colour is of importance because it may enhance or detract from the decorative value of timber. Ebony, sycamore, mahogany, and walnut are notable instances in which the use of timber has to some extent been determined by its colour or lack of it.

Distinctive colours are caused largely by various infiltrates in the cellwall. The infiltrates in some timbers are extracted for use as dyes, e.g., some infiltrates in logwood. Colour changes may occur if timber is exposed to light, air, or heat; many timbers darken with age, others fade. Mahogany fades under strong sunlight, but darkens in moderate light; grey sycamore turns green in daylight, but not in artificial light. Several timbers, e.g., teak, Borneo white seraya, parana pine, exhibit a variety of colours when freshly planed, but after exposure to daylight the colours even out considerably.

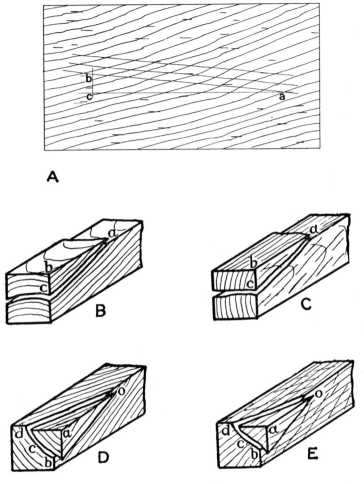

FIG. 13. Measurement of slope of grain: A $= \dfrac{ac}{bc}$; B and C, measurement of slope

of grain when timber is truly quarter- or flat-sawn : slope $= \dfrac{bc}{ab}$; D and E, measure-

ment of slope when the timber is not truly quarter- or flat-sawn: slope $\dfrac{ac}{ao}$, ac being

perpendicular to the line bcd along a growth ring in D, and bcd being perpendicular to a growth ring in E.

A drawn by J. S. Shaw from half-tone plate, Fig. 10 of BS 1860: 1952 *by courtesy of the Director, British Standards Institution.* B, C, D, and E *by courtesy of the Commonwealth of Australia Department of Scientific and Industrial Research*

The moist heat employed during kiln seasoning darkens many woods, so much so that some are steamed purposely to alter the colour, *e.g.*, beech and walnut sapwood. Colour changes are also effected by chemical means, *e.g.*, liming lightens the colour and fuming (with ammonia gas) darkens it, removing the pink or red shades. Bleaching of wood with hydrogen peroxide is also feasible.

In an earlier section it has been mentioned that woods with a high tannin content are often very durable. Such woods, *e.g.*, oak, western red cedar, and merbau, all develop unsightly dark stains if they are allowed to come in contact with iron under moist conditions. Incidentally, they have a bad effect on the iron too, which is important in museum cases containing metal exhibits.

Odour and Taste

Many timbers have a characteristic odour, which is apparent when they are worked in a fairly fresh condition, but which usually disappears as the wood drys out. Perhaps the most outstanding examples are the characteristic resinous odour of the pines, the spicy aroma of sandalwood and Central American cedar, and the camphor-like odour of Formosan camphorwood. Certain Australian timbers of the *Acacia* group possess an odour not unlike violets, coachwood is reminiscent of new-mown hay, West Indian satinwood of coconut oil, and Queensland walnut has an objectionable foetid odour that disappears as the wood dries.

The taste of wood is closely related to odour, and can probably be traced to the same constituents. Both properties influence the utilization of timber: the choice of woods for food containers is, for obvious reasons, restricted to those without pronounced odour or taste, as it is undesirable that any odour or taste should be imparted to the food itself; the flavour of tobacco, on the other hand, is alleged to be improved when stored in Central American cedar boxes, although manufacturers would appear to attach little importance to the merit of the wood since they usually cover it with copious quantities of paper; and camphorwood is used for clothes-chests in the East because it is reputed to repel insect pests.

IRRITANTS

The infiltrates and cell contents of several timbers may give trouble to some wood-workers, and, in extreme cases, they may be the cause of certain illnesses in individuals. The most common complaint is that the 'dust' of several woods irritates the mucous membrane, causing more or less violent sneezing, *e.g.*, sneezewood. Many woods induce dermatitis, which may be so severe as to incapacitate the worker for some days. Exceptionally, serious nose bleeding, or glandular swelling, may be induced, or asthma in those susceptible to this complaint. There is an ever-growing list of timbers known to specialists in the field of Industrial Medicine as responsible for certain occupational diseases, although reaction to many woods is very much an individual idiosyncrasy. It is usually possible to isolate a particular substance, occurring as one of the ingredients among the infiltrates, as the causal agent of the deleterious properties of the wood.

Among timbers that have figured in the medical records are abura, agba, East Indian satinwood, iroko, makoré, mansonia, obeche, opepe, peroba, and teak; the foregoing does not pretend to be in any way an exhaustive list. Timbers likely to head any 'black list' are mansonia and makoré, although those who have had experience with them would no doubt rank dahoma, rengas, and tali high among timbers troublesome to wood-workers.

Improvements in dust-extraction plant in modern wood-working shops are assisting in minimizing inconvenience to operatives. Alternatively, it is often practicable to plan work so that relatively small quantities of the more objectionable timbers are put through the machines in one day. In this way mills have succeeded in cutting considerable quantities of mansonia without any inconvenience to their operatives. Exceptionally, it may be possible to protect exposed parts of the body with barrier creams or oil, as is done by Chinese operatives handling rengas and other timbers of the family *Anacardiaceae*. In rengas 'the irritant' is the black sap that occurs in the radial intercellular canals. Sawdust from the sawing of unbarked logs of a few species, *e.g.*, melawis, is liable to cause intense skin irritations during sawing. Irritation from melawis is caused by the fine, needle-pointed cells in the inner bark.

Lustre

Lustre depends on the ability of the cell walls to reflect light. Some timbers possess this property in a high degree, *e.g.*, East Indian satinwood, lauan, and sapele, but others are comparatively dull, *e.g.*, hornbeam.　As a general rule, quarter-sawn surfaces are more lustrous than flat-sawn, and if stripe, fiddle-back, or roe figure is present the figure is considerably enhanced in timbers possessing a natural lustre.　Although lustre is an asset in a cabinet timber, from a practical viewpoint the capacity for taking a good polish is quite as important, and the two do not necessarily go hand in hand.

CHAPTER 5

The Identification of Timbers

THE PROBLEM

THE identification of timbers may, at first sight, appear to be a comparatively simple matter; when it is realized that there are over 20,000 woody species in the world it will be appreciated that in some cases correct identification may be exceedingly difficult. Actually it is not always possible to arrive at the correct specific name from the examination of a single sample of wood, although it is usually possible to narrow down the identification to a group of related species, and this may be sufficient for most practical purposes. Moreover, although there are so many species that produce woody stems, only a small proportion grow to timber size. Even so, the number of species producing commercial timber runs into some hundreds. The characters available for distinguishing woods are not numerous, and identification should be based on an examination of features that are known to be reliable, rather than on the more obvious characters, *e.g.*, colour and weight, that tend to be far from consistent.

THE PROCEDURE

The average timber user handles relatively few timbers and can usually recognize those with which he is familiar by a cursory glance; he is not, however, in a position to name timbers with which he is not familiar. On the other hand, it is often possible to arrive at the identity of an unfamiliar timber by a process of elimination along certain well-established lines. Each timber or group of very closely related timbers possesses a characteristic end surface; that is, the cells are so arranged as to produce a distinctive pattern. In theory, identification calls for the memorizing of these distinctive patterns.

But just as one would make little headway in learning Chinese characters without an understanding of the roots from which the characters are built up, so with timber identification, the components that give rise to the distinctive cell patterns must be thoroughly understood.　The next stage is the development of sorting devices, which make use of the separate components, a similar problem to that facing fingerprint experts: without a sorting device, identification of other than the few common timbers is often a difficult matter even for the expert.

The apparatus required for identifying timbers may include a high-power microscope and the complete paraphernalia of a laboratory devoted to the study of wood, but with most hardwoods a sharp penknife or a razor blade and a small pocket lens, giving an area magnification × 10 or × 15, are all that is necessary.　The first step is to prepare a small area of end surface by making a clean cut with the knife.　The importance of using a really sharp knife and obtaining an absolutely clean-cut end surface cannot be overstressed.　A blunt knife simply obscures structural details and a notched edge produces scratches, which may be mistaken for rays or lines of parenchyma.　From quite a small, clean-cut area of cross section it is possible with the aid of a lens to see a large amount of detail that is not visible on a rough surface with the naked eye.　The method of making the cut is all important; it must be made in one action and not as a series of small jabs; the knife blade must be as nearly parallel with the end surface of the piece of wood as possible to ensure that the prepared surface is truly transverse and not oblique (a sharp knife tends to dig into the wood, producing a surface intermediate between the transverse and a longitudinal face); and the cut must be made along the rays from the bark towards the centre of the tree.　A piece of wood about the size of a matchbox is convenient to work with.　A preliminary cut may be necessary to establish the direction of the rays on end surface.　When this has been done the piece of wood should be held in the fingers, by the more nearly tangential faces, and a single cut made along the rays. If the cut surface is blurred under a lens, although the knife blade used was sharp, it is probable that the cut was made in the direction from the inside (of the tree) outwards.　To overcome this, the piece of wood should be turned through an angle of 180°, and the cut repeated, when it will be along the rays, but in the correct direction, *i.e.*, from the bark towards the centre of the tree. If no face of the wood is sufficiently radial it may be necessary to trim the block to ensure a transverse cut being made along the rays. After the end surface is prepared, the subsequent procedure depends on several circumstances, and only general principles can be discussed here.

The relative positions of the three surfaces of wood, transverse, radial, and tangential, often confuse the beginner: Plate 34, fig. A, clarifies this point. Figs. B, C, and D of Plate 34 show how different the three surfaces of the same wood can appear.

In every case the first point to decide is whether vessels are present or not; this settles to which of the two main classes of timber (hardwoods or softwoods) the sample belongs. Next, the type of growth rings, or their absence, and then the presence or absence of resin canals are helpful features. The examination of a sample for these three features alone narrows the range appreciably. Other features, such as the type of rays, the distribution of the parenchyma and vessels, weight, hardness, and colour, are used in turn. In addition to resin canals, exclusively solitary vessels, diffuse-in-aggregates parenchyma, and prominently storeyed rays, are readily recognizable features of particular diagnostic value, although storeyed rays are not always a consistent feature in all material of a species, *e.g.*, the genus *Swietenia*. Such features as wood parenchyma and rays are often more apparent if the cut-surface of the wood is moistened.

The procedure outlined above is the basis of all **keys** to the identification of timbers. Keys are artificial devices, leading to correct identification by arbitrary means, *i.e.*, a consideration of unrelated features in the most advantageous sequence. The commonest form of key in general use in botanical and entomological work is the **dichotomous key**, whereby successive pairs of mutually exclusive conditions are so arranged that, by a process of elimination, one is led step by step to the identity of the specimen. Such keys are suitable for a restricted number of timbers, plants, or insects: they become unmanageable if their construction is attempted for too many different individuals, because it is frequently necessary to use features that are subject to considerable variation within a single species, and it becomes increasingly difficult to find pairs of characteristics that are mutually exclusive. In some circumstances it is often necessary to include the same timber, plant, or insect in more than one section of a key to ensure covering variation within

TABLE 1. *Tabular list of hardwood features visible to the naked eye or with the aid of a hand lens*
(The number of features used may, of course, be considerably enlarged)

Name of timber	Vessels							Parenchyma											Other features			Rays				Physical properties		
	Exclusively solitary	Radial multiples	Pore clusters	Scalariform perforations	Distinct to naked eye	Distinct only with lens	Barely visible with lens	Absent or indistinct	Distinct to naked eye	Terminal prominent	Apparently terminal only	Vasicentric	Aliform	Confluent	Banded	Broad conspicuous bands	Fine lines	Reticulate	Normal vertical canals	Ring porous	Storeyed (not rays)	< ¾ width of vessels	Wider than vessels	Storeyed	Aggregate rays	White	Yellow or brown	Red or purple
Ash	–	–	–	–	+	–	–	–	–	+	–	+	±	±	–	–	–	–	–	+	–	–	–	–	–	+	–	–
Dahoma	–	–	–	–	+	–	–	–	+	+	–	+	+	–	–	–	–	–	–	–	–	–	–	–	–	–	+	–
Ebony	–	+	–	–	–	+	–	–	–	–	–	–	–	–	–	–	+	–	–	–	–	–	–	–	–	–	+	+
Ekki	–	–	–	–	+	–	–	–	–	–	–	–	–	–	+	+	–	–	–	–	–	–	–	–	–	–	–	+
Elm	–	–	+	–	+	–	–	+	–	–	–	–	–	–	–	–	–	–	–	+	–	–	–	–	–	–	+	–
Gabon	–	–	–	–	+	–	–	+	–	–	–	×	–	–	–	–	–	–	+	–	–	–	–	–	–	+	–	–
Gurjun	+	–	–	–	+	–	–	±	–	×	–	+	+	+	–	–	–	–	–	–	–	–	–	–	–	–	–	–
Idigbo	–	–	–	–	+	–	–	–	+	–	–	–	–	–	–	–	–	–	–	–	–	–	–	–	–	+	+	+
Iroko	–	–	–	–	+	–	–	–	+	+	–	+	+	+	–	–	–	–	–	–	–	–	–	–	–	–	+	–
Mahogany – Central American	–	–	–	–	–	+	–	–	+	+	–	×	–	–	–	–	–	–	–	–	–	–	–	±	–	–	–	+
Mahogany – African	–	–	–	–	+	–	–	+	–	–	–	×	±	±	–	–	–	–	–	–	–	–	–	–	–	–	–	+
Makoré	–	+	–	–	–	+	–	–	–	–	–	–	–	–	+	–	–	+	–	–	–	–	–	–	–	–	–	+
Meranti	–	–	–	–	+	–	–	–	–	–	–	+	+	+	–	–	–	–	+	–	+	–	–	–	–	–	–	+
Obeche	–	–	–	–	+	–	–	+	–	+	–	+	–	–	–	–	+	–	–	–	–	–	–	–	–	+	+	–
Opepe	–	–	–	–	+	–	–	–	–	–	–	–	–	–	–	–	–	–	–	–	–	–	–	–	–	–	+	–
Oak	+	–	–	–	+	–	–	–	–	–	–	–	–	–	–	–	+	–	–	+	–	+	+	–	–	–	±	±
Sapele	–	–	–	–	–	+	–	–	+	+	–	+	+	+	–	–	–	–	–	–	–	–	–	+	–	–	+	+
Teak	–	–	–	–	+	–	–	–	–	×	–	+	–	–	±	–	–	±	–	+	–	–	–	–	–	–	+	–
Walnut – African	–	–	–	–	–	+	–	±	–	–	–	+	–	–	–	–	–	–	–	–	–	–	–	–	–	–	+	+
Whitewood – American	–	+	–	+	–	–	+	–	–	+	+	–	–	–	–	–	–	–	–	–	–	–	–	–	–	–	+	–

a species, or the lack of really satisfactory, mutually exclusive, characters.

As a preliminary to the construction of a dichotomous key, it is an advantage to list the features present in every timber to be included in the key; to facilitate subsequent work the list should be in tabular form. Because the features used are so different, it is necessary to prepare separate lists, and therefore to construct separate keys, for softwoods and hardwoods.

The procedure in preparing a table of features is to allot one column for each feature, either across or along the side of a piece of paper, and to have a column for the timbers (Table 1). Opposite each timber, and in the appropriate column, a cross is put if the feature occurs and a minus sign if the feature is wanting. If the occurrence of the feature is variable in different specimens of the same timber an alternative sign, e.g., ±, may be used to denote this. In such cases it is necessary to bring out the timber in two places in the key if this feature is used before the timber has been excluded on some other score. A multiplication sign × may be used to indicate that a particular feature is indistinct.

In preparing a dichotomous key the table provides an easy means of seeing at a glance those timbers that can be run down quickly: the columns with few crosses indicate features that will eliminate a few timbers in the early stages of the key. The subsequent sequence is immaterial, but it will be found more satisfactory to use the more clear-cut features before the more ambiguous ones, i.e., features that are easy to recognize and not ones that are open to personal interpretation. The dichotomous key that follows has been constructed from the data in Table 1 – more than one key could, of course, be constructed from the same data by taking successive features in a different order.

1	Wood ring-porous	2
1	Wood not ring-porous	5
2	Pore clusters present (parenchyma absent or indistinct)	Elm
2	Pore clusters absent	3
3	Rays wider than vessels (parenchyma in fine lines)	Oak
3	Rays not wider than vessels	4
4	Wood white (no odour)	Ash
4	Wood brown (distinctive odour)	Teak
5	Vessels exclusively solitary	6

5	Vessels not exclusively solitary	7
6	Normal vertical canals present	Gurjun
6	Normal vertical canals absent	Opepe
7	Scalariform perforation plates distinct	American white-wood
7	Scalariform perforation plates indistinct or absent	8
8	Ripple marks distinct	9
8	Ripple marks indistinct or absent	10
9	Confluent parenchyma present	Sapele
9	Confluent parenchyma absent	Central American mahogany
10	Vessels in radial groups	11
10	Vessels not in radial groups	12
11	Parenchyma banded, wood red-brown	Makoré
11	Parenchyma not banded, wood yellow or black, or streaked brown-black	Ebony
12	Normal vertical canals present	13
12	Normal vertical canals absent	14
13	Canals in short tangential series	Gurjun
13	Canals not in short tangential series	Meranti
14	Tissue other than rays storeyed	Obeche
14	Tissue other than rays not storeyed	15
15	Parenchyma in broad conspicuous bands	Ekki
15	Parenchyma not in broad conspicuous bands	16
16	Parenchyma distinct to naked eye	17
16	Parenchyma not distinct to naked eye	20
17	Wood white or yellow (terminal parenchyma indistinct)	Idigbo
17	Wood not white, terminal parenchyma prominent	18
18	Parenchyma apparently terminal only	Central American mahogany
18	Parenchyma not apparently terminal only	19
19	Confluent parenchyma present	Iroko
19	Confluent parenchyma absent	Dahoma
20	Wood walnut-brown	African walnut
20	Wood not walnut-brown	21
21	Wood pink-brown to red-brown, rays just visible to naked eye	African mahogany
21	Wood light pink, rays distinct only with lens	Gabon

In using this key one takes the first pair of conditions: if the sample for identification is ring-porous, one proceeds to question 2, and if diffuse-porous, to question 5. It will be seen that the same question

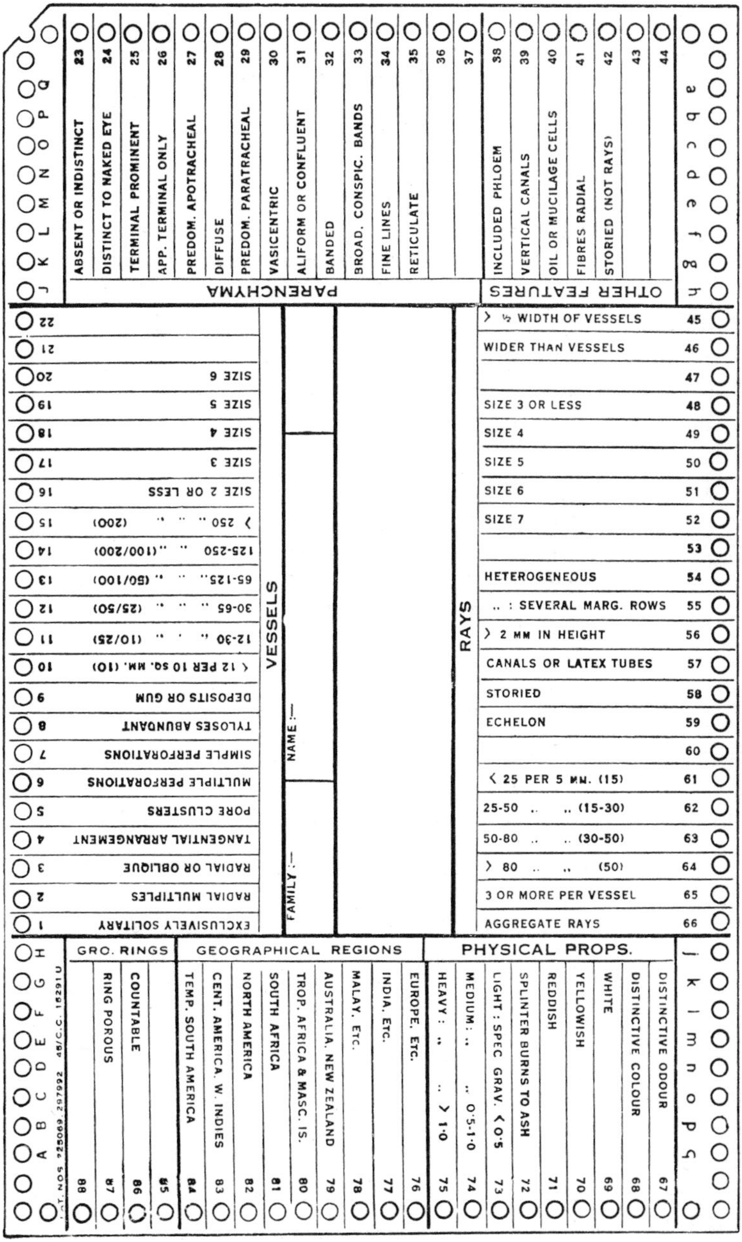

Fig. 14. A Paramount card used in multiple entry card keys. The card illustrated is for a key based on lens features

By courtesy of Messrs Copeland-Chatterson Co., Ltd.

can be used more than once in different sections of the key, *e.g.*, gurjun is brought out under exclusively solitary vessels and normal vertical canals, and meranti under vessels not exclusively solitary, but normal vertical canals present, much later in the key.

Timbers are so numerous, and the differences between many are so small, that it is impossible to construct a workable dichotomous key to embrace all the timbers in the world. Several good keys exist that are restricted to the timbers of particular countries or localities. For example, Chalk and Rendle's key to British hardwoods,[1] Record's key to North American timbers,[2] Dadswell's keys to Australian timbers,[3] and Brown's key to Indian timbers,[4] are excellent for the timbers they embrace. All dichotomous keys have the disadvantage that additional timbers cannot be included in the key without reconstruction of large sections, if not the greater part, of the key.

Mr S. H. Clarke, C.B.E., M.Sc., when a wood anatomist at the Forest Products Research Laboratory, adapted the **Paramount sorting** system to timber identification in his aptly named **multiple entry key.** Special cards, patented by Messrs Copeland-Chatterson Co., Ltd, containing punched holes along their four sides, are employed; each hole is used for one feature (Fig. 14). Every timber to be included in the key requires a separate card and, if certain features are variable in different samples of the same timber, two or more cards may be necessary, exceptionally as many as eight cards. The card is completed for each timber by punching the holes for the features present in the wood with a special punch that cuts a V-shaped slot from the original punched hole. When all the cards are prepared they are sorted so that all are arranged the same way round and the key is ready for use. To facilitate rapid sorting it is convenient to cut one corner of each card on the splay, *e.g.*, the top right-hand corner as in the cards patented by Messrs Copeland-Chatterson.

Mr B. J. Rendle, B.Sc., and his colleague Dr E. W. J. Phillips

---

[1] *British hardwoods*, Forest Products Research Bull. No. 3, H.M. Stationery Office.

[2] *Timbers of North America*, by S. J. Record. John Wiley & Sons, 1934.

[3] Bulls. Nos 67, 78, and 90, Council for Sci. Ind. Research, Commonwealth of Australia, Melbourne.

[4] *An elementary manual of Indian wood technology*, by H. P. Brown. Calcutta, Gov. India. Central Publ. Branch, 1925.

of the Forest Products Research Laboratory have prepared the data for a set of cards for more than 400 commercial timbers; this information has been published as F.P.R. Bulletin No. 25. Similar data for microscopic features, collected by Messrs J. D. Brazier, B.Sc., and G. L. Franklin and their colleagues at the Laboratory, have been published as F.P.R. Bulletin No. 46. Blank cards for these keys are available from H.M. Stationery Office. The definitions of the different anatomical features given in the Bulletins should be studied, as certain features are used in a slightly different sense from the definitions given in Chapter 3. Photomicrographs of the majority of the timbers figuring in these keys have been published as F.P.R. Bulletin No. 26.

To identify a timber, any feature present in the specimen is selected and a steel needle threaded through the punched hole for the selected feature (Plate 35). The cards for those timbers in which the selected feature occurs drop out as the needle is shaken. These cards are again sorted so that they are all the same way round, a second feature is selected, the needle threaded, and the shaking process repeated. The cards that drop out on the second occasion are again sorted and the process continued with different features in turn until only one or a few cards remain: provided a card for the specimen to be identified has been prepared it will be among those finally eliminated. With some timbers the card-key system results in the elimination of all cards but one, although more often a selection has to be made from two or three cards, because it is not possible to include all features on the card, and interpretation of a few features is largely a matter of personal opinion. However, when selection is narrowed to two or three timbers, correct identification can usually be arrived at by matching the unknown timber with authenticated specimens of the few alternative choices. In working with a card-key of this type three precautions must be observed: (1) in shaking the pack, care must be taken to ensure that all the cards free to drop out do, in fact, drop out; (2) care must be taken that the correct pack is used after each sorting, i.e., if a feature is being used positively the cards that drop out are used for the next operation, and vice versa; and (3) the cards must be kept in good condition, i.e., should any of the punched holes become torn a new card must be made out.

The two special advantages of the card-key system are: (1) the simplicity with which new timbers can be added to the key – all that is required is an additional card, and (2) any sequence can be adopted that promises speedy identification.

It is usual to confine keys to features visible only with the aid of a microscope or to features visible to the naked eye or with the aid of a low-power hand lens, but there is no reason why a key should not combine both classes of features, although such keys can, of course, only be used in the laboratory. In a lens key it is desirable to record features as they appear, whether or not the observations are in accord with the facts. For example, the line of tissue bordering a growth ring in beech is not terminal parenchyma, but it is convenient to record it as such because, with a lens, it is rather difficult to establish that it is not true terminal parenchyma. Actually, the line consists of a few rows of radially-flattened fibres, wanting in diffuse parenchyma, and this zone contrasts with the remainder of the fibre-diffuse parenchyma background to give the effect of a line of distinctive tissue.

The successful use of keys necessitates some experience in the examination of small samples of wood, and this can only be obtained by practice. Readers who are anxious to be in a position to identify any but the few common timbers in everyday use would be well advised to make their own keys from a study of a collection of authenticated samples of timbers. As a preliminary to a concerted attack on the problem of identification, there is no better method than the preparation of scale drawings of the end surfaces of different woods. The procedure is to prepare a clean-cut portion of the end surface, and to mark on this a square with one-centimetre sides. Next, a sheet of paper with a square of five-centimetre sides is required. The details visible in the marked square can then be transferred to the paper, more or less to correct scale. It is usually helpful to commence with the rays, and then to draw in the vessels, and subsequently the other details. The method is admittedly laborious, but the preparation of thirty or forty drawings fixes the distinctive patterns of the woods in the mind, and is of great help in mastering the technique of timber identification.

Finally, it may not be inappropriate to comment on the different ways of expressing magnification. It is not always easy to appreciate the significance of ×10, ×15, ×30. An added complication is introduced by makers inscribing the magnification of pocket lenses

PLATE 34

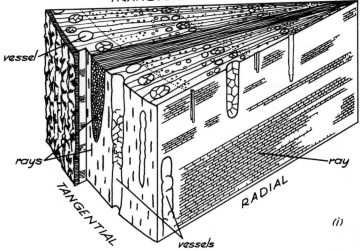

FIG. 1. Wedge of wood

(i) a wedge-shaped piece of oak showing the relative positions of transverse, tangential, and radial surfaces (ii) transverse surface of oak × 30 (iii) tangential surface of oak × 30 (iv) radial surface of oak

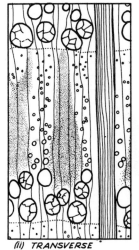

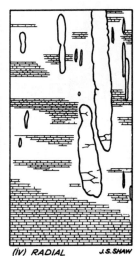

FIG. 2. Transverse
surface (× 5)

FIG. 3. Tangential
surface (× 10)

FIG. 4. Radial
surface (× 5)

PLATE 35

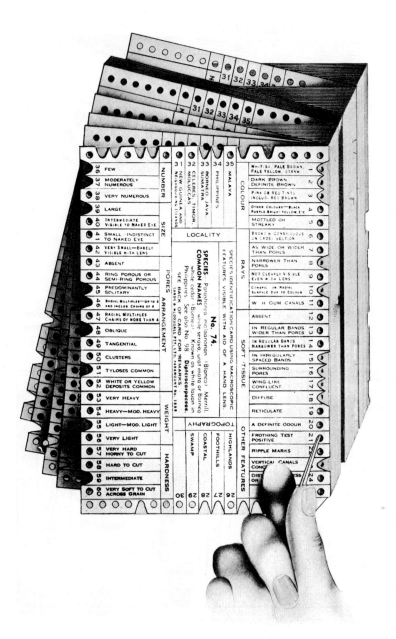

The multiple entry card-key in operation

By courtesy of the Cleaver-Hume Press, Ltd

on a different system from other optical equipment. Whereas it is usual for the magnification of microscopes and textbook illustrations to be in terms of linear dimensions, pocket lenses are usually inscribed in terms of area magnification. Thus a ×10 pocket lens is a linear magnification of $\sqrt{10}$, or just over 3 linear, whereas a text illustration ×10 is equivalent to an area magnification of 10^2 or 100. The frontispiece will, it is hoped, clarify the problem of scale.

The top three squares are designated in terms of area magnification, and the bottom four squares in terms of linear magnification. In both series the squares on the extreme left are 'natural size', or ×1 magnification. The next square to the right in both series is enlarged, *to scale*, at the magnification given. That is, the square in the top series marked ×10 is the size the square marked ' ×1 ' becomes when magnified ×10 area magnification; that marked ×15, the size when the ' ×1 ' square is magnified ×15 area magnification. Similarly, the first two squares to the left of the 'natural size' square in the lower series are drawn magnified ×2 and ×5 linear magnification respectively, and the portion of the ×5 square outlined in white is drawn to scale, magnified ×30. It will be observed that ×5 linear magnification is an appreciably higher magnification than ×10 area magnification – the relation is 5 to, $\sqrt{10}$ or 5 to 3·16. The largest magnification, using a hand lens that it is convenient to work with for any continuous period is an area magnification of ×15; an area magnification of ×20 is the extreme limit of magnification for hand lenses – beyond this the field is too small and the hand is insufficiently steady to permit of keeping the 'object' being examined in focus. Comparison of the amount of detail visible in the section of oak ×5 linear magnification (*i.e.*, ×25 area magnification) with that visible at ×10 linear magnification (Plate 19, fig. 3) will make it apparent that there are very real limitations to the anatomical study of wood with a pocket lens. Nevertheless, with practice, it is possible to see a great deal, often not as clear-cut as could be desired, but sufficient for purposes of identification.

Finally, a word of caution should be sounded. It is just not possible always to determine the specific identity of even a common commercial timber from extremely small fragments, such as sawdust. For example, for certain timbers it is essential to have transverse, radial, and tangential sections *from the same fragment* to be in a position to establish the identity of the fragment, and no degree of expertise or of elaborate laboratory equipment will surmount the problem of too small sections of these different surfaces being available for purposes of identification.

Descriptions of Some of the More Important Commercial Hardwoods

SEVERAL regional studies have been published, providing descriptions of different timbers, and their properties, for particular regions. These works, which are usually not confined to timbers of commercial importance, are ordinarily available only in university and specialist libraries. It seemed appropriate that, following a discussion of the problem and the procedure in timber identification, detailed descriptions of a representative selection of timbers should be included in this book. The difficulty was to decide which timbers should be included in a 'representative' selection.

The important softwoods have been omitted because their identification is based not on lens characters but on microscopic features, visible only in specially prepared sections, which call for laboratory equipment not possessed by the ordinary timber user. This still left a very wide choice as it was impracticable to include all hardwoods now accepted as 'commercial' timbers – there are appreciably more than a hundred timbers in this category. In the circumstances a more or less arbitrary selection was unavoidable, but with certain definite guide lines. The selected timbers embrace examples of the well-defined anatomical features used in identification of timbers with a pocket lens. In many cases there was more than one choice to meet this requirement, and hence additional considerations have influenced selection. For example, the five commercial timbers with abnormally low shrinkage have all been included because of their importance to a large number of consumers. Room has been found for the timbers widely used for furniture that are being used more and more as substitutes for high-class softwood joinery, together with six timbers, little known before the war, that are now well established for a wide range of uses, and seven important U.K. and European hardwoods. Among these thirty-six timbers are the fifteen hardwoods included in the T.R.A.D.A. syllabus for third-year students in Timber Technology.

The illustrations in this chapter are at a magnification of × 15.

LENS FEATURES USED IN THE IDENTIFICATION OF HARDWOODS

Growth rings distinct	European beech
Vessel distribution	
Ring porous	Ash, elm, oak, sweet chestnut
Falsely ring porous	European cherry
Exclusively solitary	Opepe, kapur, keruing
Radial multiples	Makoré, punah, abura (mostly)
Tangential arrangement	Silky oak
Clusters	Elm
Radial arrangement	American red oak
Parenchyma	
Terminal	Central American mahogany, Burma teak, afzelia, European beech (apparently terminal, but actually bands of flattened fibres), European birch
Diffuse	Punah
Diffuse-in-aggregates	Obeche, abura
Fine lines	Jelutong
Banded	Guarea
Broad conspicuous bands	Ekki
Sparsely paratracheal	African mahogany, African walnut
Vasicentric	Ash
Aliform	Afzelia, ramin
Aliform-confluent	Afara, iroko
Confluent	Sapele, utile
Vasicentric to confluent outwards in each growth ring	Muninga
Rays wider than vessels	European cherry, oak
Ripple marks	Afrormosia, Central American mahogany, kapur, muninga, sapele
Other features	
Canals: solitary	Agba, mersawa
multiples of 3 to 7	Keruing
tangential	Kapur, light red meranti, dark red meranti
series	
traumatic	African walnut
Tyloses	European oak, Burma teak
Latex canals	Jelutong

I Timbers with abnormally low shrinkage frequently suitable where teak was formerly exclusively used.

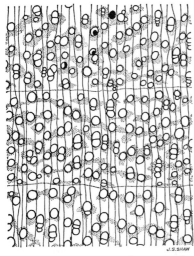

AFRORMOSIA

Pericopsis elata van Meeuwen (syn. *Afrormosia elata* Harms).

Family: *Leguminosae.*

Other Names: Kokrodua (Ghana, Ivory Coast, France); asamela (Ivory Coast); bossanga (Congo).

Weight: 710 kg/m³.

Spot Characters: storeyed rays, confluent parenchyma, and light teak-brown colour are characteristic; growth rings distinct with lens, demarcated by narrow layers of terminal parenchyma; vessels moderately small to medium-sized, radial pairs and multiples of 3 or more predominate in some areas of cross section, gum-like deposits sometimes numerous; parenchyma abundant, in narrow terminal layers visible only with lens, and more or less continuous confluent layers visible to naked eye; rays storeyed, usually producing distinct 'ripple marks'; grain typically interlocked; wood yellow-brown with darker streaks, darkening appreciably on exposure to resemble Burma teak; sapwood distinct, about 25 mm wide.

Durability: very resistant to decay; logs free from pin-worm; liable to stain black if brought into contact with iron when wet (because of tannin content); will not corrode metals under dry, temperate-climate conditions, but liable to accelerate corrosion under high temperature/humidity tropical conditions as does teak.

Strength Properties: no data are available; abrasion tests at F.P.R.L. have shown afrormosia to be comparable in wearing qualities with teak for flooring and decking.

Working Qualities: works well with hand and machine tools, causing little dulling of cutting edges, and only slight tendency for the grain of quarter-sawn material to 'pick up' in planing; takes nails, screws, and glue satisfactorily, and all finishing treatments; can be sliced to yield thin veneers.

Seasoning: air- and kiln-dries excellently, and with little degrade even with exacting kiln schedules; shrinkage figures from 'green' to 12 per cent. M/C: (based on small-scale tests at F.P.R.L.): 1·3 per cent. radially and 2·0 per cent. tangentially. Low hygroscopic properties characterize afrormosia, teak afzelia, iroko, and muninga, and explain the stability of these timbers.

Uses: as a teak substitute, including decking and flooring, where the risk of iron stains do not exist or are immaterial; afrormosia does not bleach as does teak. It is also used for high class furniture.

Supplies: imported as round logs and sawn timber.

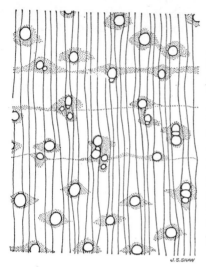

J.S.SHAW

AFZELIA

Afzelia spp., principally *A. africana* Smith, *A. bipindensis* Harms, *A. pachyloba* Harms; and *A. quanzensis* Welw. (East and South Africa).

Family: *Leguminosae.*

Other Names: aligna, apa, ariyan (Nigeria); doussié (Cameroons and France); lingué (Ivory Coast); papo (Ghana); uvala (Angola); chanfuta mussacossa (Mozambique); chanfuti (South Africa); mkora, mbembakofi (Tanzania).

Weight: 830 kg/m³.

Spot Characters: conspicuous aliform parenchyma and narrow terminal layers characterize *Afzelia* and the timbers of several other genera of the *Leguminosae*; timbers individually distinguished by presence or absence of storeyed rays and colour; resemble some *Moraceae, e.g.*, iroko, but parenchyma in latter typically aliform-confluent. Growth rings distinct with lens; vessels medium-sized to moderately large, open or filled with white or yellow deposits; aliform parenchyma distinct to naked eye, also narrow terminal layers visible with lens; wood brown or red-brown, darkening on exposure; sapwood pale straw, distinct, narrow; grain interlocked.

Durability: extremely resistant to decay; stated to be resistant to termite attack and to offer some resistance to marine borers; sapwood susceptible to *Lyctus* attack.

Strength Properties: full-scale test data not available; stronger in bending, stiffness, compression, and shock resisting abilities than jarrah, but less hard and appreciably less resistant to splitting.

Working Qualities: because of its hardness, requires considerable power in conversion but causes only moderate blunting of cutting edges; quarter-sawn material 'picks up' unless the cutting angle is reduced; splits in nailing, necessitating pre-boring; polishes satisfactorily but requires filing; saw dust stated to be irritating.

Seasoning: can be air- and kiln-dried satisfactorily, but slowly, with little distortion and only slight checking; its low hygroscopic properties put it in the same class as teak, iroko, afrormosia, and muninga for small movement and stability once seasoned, *vide* F.P.R.L. leaflet No. 44, and unnumbered leaflet March 1951; shrinkage figures from 'green.' to 12 per cent. M/C: 1·0 per cent. radially and 1·5 per cent. tangentially.

Uses: in countries of origin for indoor and outdoor constructional purposes where durability and good appearance are required, and also for furniture; weight and price likely to restrict the market for Afzelia overseas to work demanding outstanding strength, durability and stability; suitable for garden furniture, draining and switch boards, ship-building (other than decking), chemical vats, and flooring, joinery and fittings in public buildings.

Supplies: *Afzelia* is a widely distributed genus in tropical Africa, especially in the dryzone forests, but the trees are only of moderate height. Imported in log form and sawn timber.

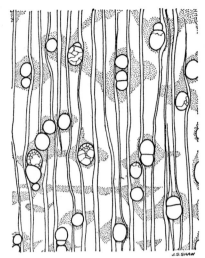

IROKO

Chlorophora excelsa Benth. and Hook. f. and *C. regia* A. Chev.

Family: *Moraceae.*

Other Names: kambala (Belgian Congo and French West Africa); moreira (Angola); mvule (Uganda); odum (Ghana); tule, intule (Mozambique). Formerly called 'African teak'.

Weight: 660 kg/m³.

Spot Characters: light olive-brown colour, darkening rapidly to teak-brown or dark brown, aliform-confluent parenchyma distinct to naked eye, and discontinuous layers of terminal parenchyma characterize iroko; growth rings distinct with lens, demarcated by narrow, interrupted layers of terminal parenchyma; vessels medium-sized to moderately large, tyloses and chalky-white solid deposits abundant; parenchyma in terminal layers, and aliform-confluent (often vasicentric or aliform at beginning of growth rings); grain interlocked; wood light olive-brown when freshly cut, darkening rapidly to teak-brown or dark brown on exposure; sapwood light-coloured, distinct.

Durability: very durable: resistant to decay, termites, and marine-borer attack; logs sometimes attacked by pin-hole borers; sapwood susceptible to *Bostrychid* and *Lyctus* attack.

Strength Properties: inferior in most strength properties to teak and home-grown oak, but harder than teak and more resistant to shear; data available.

Working Qualities: rather variable, sometimes working appreciably easier with hand and machine tools than teak, and causing only moderate dulling of cutting edges, but material containing a high proportion of calcium carbonate harder to work than teak and causing rapid dulling of cutting edges; grain liable to 'pick up' in planing; takes nails, screws, and glues satisfactorily; and all finishing treatments, but requires filling. Large stone-like calcareous deposits cause serious damage to saws if not detected.

Seasoning: air- and kiln-dries very well, with little degrade, and very stable once seasoned because of its low hygroscopic properties; shrinkage figures from 'green' to 12 per cent. M/C: 1·0 per cent. radially and 1·5 per cent. tangentially.

Uses: the most sought-after constructional timber in West Africa; used in this country as a substitute for teak as draining boards, in boat building, laboratory benches, high-class joinery, interior finishings, and flooring.

Supplies: a tree of large girth but only moderate height, occurring in the rain and deciduous forest belts of West Africa, and in East Africa, but as scattered trees, never abundant; imported from West Africa in log form, and from West and East Africa as sawn timber.

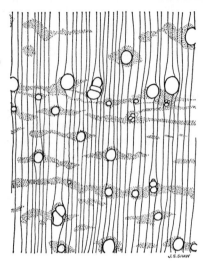

MUNINGA

Pterocarpus angolensis DC.

Family: *Leguminosae.*

Other Names: bloodwood, sealing wax tree (Rhodesia, Zambia, and Malawi); kiaat, kajat or kajatenhout (S. Africa); mninga, muninga (Tanzania); ambila (Mozambique); mlombwa (Malawi); mukwa (Rhodesia).

Weight: 640 kg/m³.

Spot Characters: vessels variable in size, with ring-porous tendency, confluent parenchyma, storeyed rays, and brown or red-brown colour with streaks of red or dark brown characterize muninga; growth rings demarcated by narrow layers of terminal parenchyma, with an irregular row of very large or extremely large vessels tending to be ring-porous; vessels very variable in size from moderately small to extremely large; parenchyma abundant, visible to naked eye, in narrow terminal layers, and confluent, but sometimes aliform-confluent or aliform at beginning of growth ring; rays storeyed; wood either brown (almost golden brown) with darker brown or grey-black streaks or red-brown with bright red streaks; faintly aromatic when freshly sawn; sapwood white to light yellow, sharply defined, 2·5 to 5·0 cm wide.

Durability: resistant to decay, and to termite attack; sapwood susceptible to *Bostrychid* and *Lyctus* attack.

Strength Tests: two logs tested at the Forest Products Research Laboratory, Princes Risborough, gave widely different strength values.

Working Qualities: works well with hand and machine tools, with only moderate dulling of cutting edges; planes to a good finish, turns well; takes nails, screws, and glue satisfactorily, and all finishing treatments.

Seasoning: requires moderate care in air-drying to minimize surface checking, but it dries excellently with very small shrinkage whether air- or kiln-dried, and once seasoned holds its shape well in the most adverse service conditions (*vide* radiator casings at the Commonwealth Forestry Institute, Oxford); shrinkage figures from 'green' to 12 per cent. M/C: 0·8 per cent. radially and 0·9 per cent. tangentially (the lowest recorded figures for any hardwood or softwood so far tested).

Uses: for ship-building in Malawi, also furniture, wagon construction, and domestic purposes; in U.K. for furniture and panelling (Commonwealth Forestry Institute, Oxford), and panelling (Princes Risborough); internal and external joinery, draining boards, flooring.

Supplies: a small tree in South Central Africa, from Angola to Mozambique Province; imported in small quantities as round logs and sawn timber, but shipments irregular.

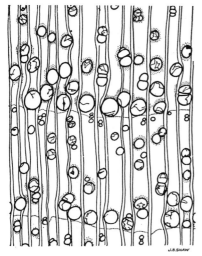

BURMA TEAK

Tectona grandis Linn. f.

Family: *Verbenaceae.*

Other Names: often classified by port of shipment or locality whence obtained; genuine teak is confined to Burma, India, Indo - China, Indonesia and Thailand, and has been planted also on a small scale elsewhere.*

Weight: 660 kg/m³ (Burma teak); Thailand teak is lighter in weight and some plantation-grown teak is heavier.

Spot Characters: distinct growth rings (usually distinctly ring-porous), tyloses and yellow solid deposits in the vessels, terminal and vasicentric parenchyma, distinctive colour, odour, and greasy feel characterize teak; wood ring-porous, pore ring of large oval vessels, latewood vessels medium-sized grading to moderately small, tyloses and light yellow solid deposits sometimes abundant; parenchyma terminal and vasicentric; wood distinctive yellow-brown colour, often with darker streaks, darkening to uniform dark brown colour on exposure; sapwood distinct, 'white' or pale yellow, narrow; greasy feel; characteristic odour of old leather.

Durability: very resistant to decay and most forms of insect attack, but only moderately resistant to termite and marine borer attack.

Strength Properties: comparable when green with oak but 30 to 40 per cent. stronger in bending and in stiffness; when seasoned, a little inferior to oak and only 10 to 20 per cent. stronger in bending and in stiffness.

Working Qualities: works well with hand and machine tools, but dulls cutting edges appreciably; nails and screws well; greasy nature makes gluing difficult; polishes satisfactorily; causes skin irritations among wood-workers.

Seasoning: air seasons well but slowly; little tendency to check, split, or warp; exceptionally stable once seasoned, attributed to abnormal hygroscopic properties; variation in M/C with changing atmospheric conditions about three-quarters of that of most woods; easy to kiln-dry, but drying rates of different pieces in one charge may vary appreciably.

Uses: shipbuilding, especially decking; exterior structural woodwork, joinery, and garden furniture; interior joinery, fittings, and flooring; and many special purposes, *e.g.*, draining boards and sinks, where stability and durability in adverse service conditions are required. Burma teak, being harder, is stronger and superior for all structural purposes to supplies from Thailand.

Supplies: supplies normally adequate but now adversely affected by political difficulties; imported as logs, squares, and sawn square-edged timber.

* Many other timbers have been called teak that have no claim to the name, *e.g.*, 'Philippine teak' (*Dipterocarpus* spp.), but only 'Rhodesian teak' (*Baikiaea plurijuga* Harms) in this category continues generally to masquerade as a 'teak'; suitable alternatives to teak are: afzelia, iroko, afrormosia and muninga (*Pterocarpus angolensis* DC.)

II Timbers widely used for furniture, high-class joinery, exhibition stands, and certain specialist purposes.

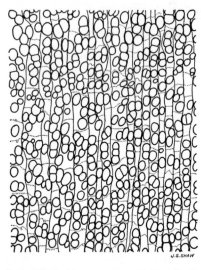

ABURA

Mitragyna ciliata Aubrév. and Pellegr. (West Africa); *Mitragyna* spp. (Uganda).

Family: *Rubiaceae.*

Other Names: *M. ciliata*: bahia (France and Ivory Coast); subaha (Ghana). *Mitragyna* spp.: nazingu (Uganda).

Weight: 580 kg/m³.

Spot Characters: numerous vessels, frequently in radial multiples, and diffuse-in-aggregates parenchyma characterize abura; growth rings sometimes distinct, demarcated by a narrow layer of darker tissue without vessels or wood parenchyma; vessels frequently in radial multiples, moderately numerous to numerous, moderately small to medium-sized; parenchyma often indistinct, diffuse-in-aggregates, sometimes reticulate; wood light-brown with pink tinge; sapwood usually not distinct; characteristic unpleasant odour when worked green. The appearance of some timber is spoiled by black spots sometimes so numerous as to suggest sap-stain fungal infection.

Durability: not resistant to decay; stated to be particularly liable to pin-worm infestation in the log and even after conversion if dried slowly; sapwood liable to sap-stain infestation but not to *Lyctus* attack; regarded as resistant to acids.

Strength Properties: a medium-hard timber, comparing favourably in strength properties with Canadian yellow birch and common elm.

Working Qualities: works well with hand and machine tools, with little dulling of cutting edges; a good timber for mouldings, provided sharp and thin cutting edges are used; poor bending qualities; takes nails, screws, and glue well; can be rotary-peeled satisfactorily after a standard softening treatment.

Seasoning: air seasons rapidly, with little tendency to split or distort; kiln-dries well; shrinkage figures: 'green' to 12 per cent. M/C: 2·5 per cent. radially, 5·2 per cent. tangentially.

Uses: general utility purposes, carving, and battery boxes in West Africa; furniture, mouldings, and fittings in U.K.; a possible alternative to beech when good steam-bending qualities are not important.

Supplies: semi-gregarious in fresh-water forests; imported as round logs and sawn square-edged timber.

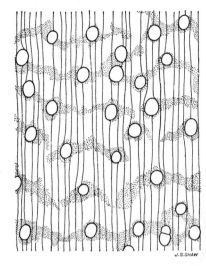

AFARA

Terminalia superba Engl. and Diels.

Family: *Combretaceae.*

Other Names: limba is the recognized alternative standard name of timber of this species originating in Belgian and French territory, also in the Congo and Angola; ofram (Ghana); akom (Cameroons); limbo, chêne - limbo, noyer du Mayombe (France); white afara (Nigeria); fraké (Ivory Coast); korina (a registered name in the U.S.A.).

Weight: 560 kg/m³.

Spot Characters: characteristic odour when freshly sawn, discontinuous wavy layers of apparently terminal parenchyma, and aliform-confluent parenchyma characterize the wood; growth rings not always distinct, undulating, demarcated by discontinuous layers of initial parenchyma; vessels medium-sized to moderately large, widely spaced, tyloses sometimes abundant; parenchyma initial and aliform or aliform-confluent, visible to naked eye; wood pale yellow with light green tinge, centre core of some logs provide timber with ornamental grey-brown or black streaks; sapwood not distinct.

Durability: not resistant to decay; logs subject to pin-hole borer infestation; sapwood liable to sap-stain infestation and to *Bostrychid* and *Lyctus* attack; 'spongy heart' present in some logs.

Strength Properties: data from standard full-scale tests not available; timber from the streaked core said to be brittle.

Working Qualities: works well with hand and machine tools, with only moderate dulling of cutting edges; tendency for the grain to 'pick-up' in planing; tendency to split in nailing; peels well and glues satisfactorily.

Seasoning: can be air- and kiln-dried satisfactorily, but liable to stain and to dote' if not dried rapidly; shrinkage figures not available.

Uses: general utility purposes and interior joinery; extensively used for rotary-peeled plywood; the streaked 'heart' yields ornamental panels.

Supplies: a large, buttressed tree, common on the fringe of the rain forests; imported in log form and as sawn timber.

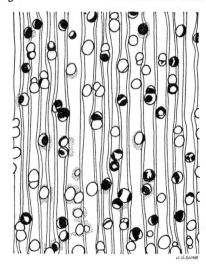

AFRICAN MAHOGANY

Khaya spp. *K. ivorensis* A. Chev. and *K. grandifoliola* C.DC. (West Africa); *K. anthotheca* Welw. C.DC. (West and East Africa); *K. nyasica* Stapf. ex Baker f. (East and Central Africa); *K. sengegalensis* (Dear.) A. Juss. (West and Central Africa).

Family: *Meliaceae.*

Other Names: Ports of shipment and district names are used to qualify the name 'mahogany'; acajou, qualified by port of shipment, used for French consignments; khaya (U.S.A.); ngollon (*K. ivorensis*), (France); mangona (Cameroons); krala (France and Ivory Coast), munyama (Uganda) for *K. anthotheca*.

Weight: 530 kg/m³, *K. grandifoliola* 720 kg/m³ and *K. senegalensis* 800 kg/m³.

Spot Characters: growth rings usually absent (contrast true mahogany, *Swietenia* spp.), no characteristic anatomical features, and mahogany colour characterize African mahogany; vessels medium-sized to moderately large, frequently filled with dark-coloured gum-like deposits; parenchyma absent or indistinct; sparsely paratracheal; traumatic canals not uncommon; wood very variable in colour from pink to quite dark red-brown; sapwood distinct; grain typically interlocked, producing stripe figure when quarter-sawn.

Durability: moderately resistant to decay; logs liable to become infested by pin-hole borers; sapwood susceptible to *Bostrychid* and *Luctus* attack.

Strength Properties: similar to those of Central American mahogany, but more resistant to splitting; has been tested at Madison.

Working Qualities: works moderately easily with hand and machine tools, but rather harder to work than true mahogany; a reduced cutting angle, and thin sharp cutting edges are necessary to secure a smooth finish; nails, screws, and glues well; takes all finishing treatments.

Seasoning: air seasons with little tendency to split and only slight distortion; can be kiln-dried fairly rapidly with little degrade; shrinkage figures from 'green' to 12 per cent. M/C: 1·8 per cent. radially and 2·5 per cent. tangentially.

Uses: an accepted substitute for genuine mahogany; used for furniture, interior joinery, and finishings; also for veneers and plywood and printers' blocks.

Supplies: widely distributed in West and East Africa, but nowhere abundant; exported as round and squared logs, and as sawn timber.

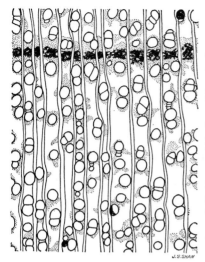

AFRICAN WALNUT

Lovoa trichillioides Harms syn.: *L. klaineana* Pierre ex Sprague.

Family: *Meliaceae.*

Other Names: Port and district names are used to qualify 'walnut'; bibolo (Cameroons); dibétou (Gabon); dibétou, noyer d'Afrique, noyer de Gabon (France); apopo, sida (Nigeria); alona wood, congo-wood, lovoa wood (U.S.A.).

Weight: 560 kg/m³.

Spot Characters: walnut-brown colour with darker streaks, prominent ribbon figure in quarter-sawn material, and lack of distinctive anatomical features characterize the wood; growth rings usually more distinct with naked eye than lens, demarcated by layer of darker tissue and occasionally by terminal parenchyma; vessels medium-sized, sometimes markedly oblique in arrangement, dark coloured deposits often numerous; parenchyma paratracheal, sparse and indistinct, with occasional terminal layers visible to naked eye; traumatic canals frequently present, visible as dark lines on all surfaces.

Durability: reputed to be resistant to decay; logs liable to pin-worm infestation, and sapwood to *Lyctus* attack.

Strength Properties: about 20 per cent. less resistant to impact bending loads than American black walnut, comparable with African and Central American mahogany in strength properties.

Working Qualities: excellent working qualities with hand and machine tools, causing appreciably less dulling of cutting edges than American black walnut; grain of quarter-sawn material liable to 'pick up' in machining; takes nails, screws, and glues well, polishes effectively.

Seasoning: air seasons satisfactorily if rapid drying is avoided; kiln seasons readily without serious degrade; shrinkage figures from 'green' to 12 per cent. M/C: 1·7 per cent. radially, 3·3 per cent. tangentially.

Uses: a cabinet wood, and for furniture, interior fittings, high-class joinery, and veneers; is being used for flooring blocks, for which it is a little on the soft side, although satisfactory for ordinary domestic wear.

Supplies: wide but irregular distribution, preferring regions of high rainfall; imported as round and square logs and as sawn timber.

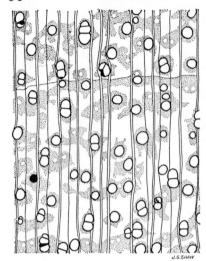

J.S.SHAW

AGBA

Gossweilerodendron balsamiferum Harms.

Family: *Leguminosae.*

Other Names: Nigerian cedar and pink mahogany (G.B.); moboron (Nigeria); tola (Congo); tola branca, white tola (Angola).

Weight: 510 kg/m³.

Spot Characters: intercellular canals smaller than vessels distributed singly throughout the growth ring, parenchyma terminal, and vasicentric tending to become aliform, characterize agba; intercellular canals present, appreciably smaller than vessels, filled with dark-coloured gum or empty, scattered singly throughout the growth ring; growth rings countable, demarcated by narrow layers of terminal parenchyma distinct with lens; vessels medium-sized; parenchyma terminal, and vasicentric, tending to become aliform; wood pale yellow-brown with a pink tinge, canals glisten on planed longitudinal surfaces. Sapwood not readily distinguishable from the heartwood.

Durability: stated to be resistant to decay and to termite attack but statement requires confirmation; logs attacked by pin-hole borers; sapwood susceptible particularly to powder-post beetle attack.

Strength Properties: limited tests have been made, indicating the timber is similar in strength properties to Central American mahogany but inferior in stiffness and compression.

Working Qualities: very dependent on the amount of gum present; if not excessive the timber is easy to work with hand and machine tools, and causes little dulling of cutting edges; takes nails, screws, and glue well, and finishes satisfactorily.

Seasoning: can be air- or kiln-dried with little degrade.

Uses: for furniture, and as a softwood substitute for interior joinery, fittings, and finishings. It is important to take adequate precautions to ensure that the timber does not become attacked by *Lyctus* beetles because the sapwood is usually not distinguishable from the heartwood and appears to be particularly susceptible to attack.

Supplies: a rain-forest tree of very large size, occurring in West Africa and extending to the Belgian Congo; imported as round logs and sawn timber.

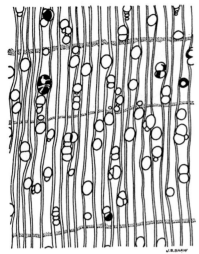

AMERICAN MAHOGANY

Swietenia candollei Pitt. (Venezuela); *S. macrophylla* King (Central America); *S. mahagoni* Jacq. (West Indies); *S. humilis* Zucc. (some Mexican mahogany). The identity of Brazilian and Peruvian is doubtful, but they may be *S. macrophylla*.

Family: *Meliaceae.*

Other Names: baywood is now obsolete. Usually called mahogany, qualified by the country of origin, *e.g.*, Central American mahogany, British Honduras mahogany, but timber of *S. mahagoni* is sometimes called Spanish mahogany. Vernacular names: aguano (Brazil); aguano, caoba (Peru).

Weight: *S. candollei* & *S. macrophylla* 560 kg/m³, and *S. mahagoni* 720 kg/m³.

Spot Characters: mahogany red-brown colour, layers of terminal parenchyma distinct to naked eye (other types of parenchyma not apparent), vessels filled with gum-like deposits, and rays usually storeyed characterize Central American mahogany; growth rings distinct, demarcated by layers of terminal parenchyma visible to naked eye; vessels medium-sized, often filled with dark-coloured gum-like deposits; parenchyma apparently absent (some sparsely paratracheal present) except for terminal layers; rays usually storeyed, producing distinct ripple marks; grain straight, interlocked, or irregular, the last providing many varieties of figure; rather variable in colour, texture, and hardness, depending on source of supply, light red-brown to deep red-brown, darkening on exposure in moderate light, but fading in strong light.

Durability: resistant to decay; logs liable to 'pin-hole' borer damage although a high proportion of clear material is obtained.

Strength Properties: good strength-weight ratio.

Working Qualities: works excellently with hand and machine tools, with little dulling of cutting edges; planes to a smooth finish; takes nails, screws, glue, and all finishing treatments excellently; the most usual standard by which other woods for furniture and cabinet work are judged; extensively sliced to yield decorative veneers.

Seasoning: air- or kiln-dries excellently, with little degrade, and particularly stable once seasoned; shrinkage figures from 'green' to 12 per cent. M/C: 1·9 per cent. radially and 2·6 per cent. tangentially.

Uses: for furniture, cabinet work, and high-class joinery and interior finishings; boatbuilding; and such special purposes as printers' blocks.

Supplies: original sources of supply tending to become worked out, but untapped supplies thought to exist in other parts of Central and South America.

   * Species of *Swietenia* alone provide true mahogany, although timbers of the genus *Khaya* (African mahogany) are sufficiently similar in most properties to be regarded as acceptable alternatives.

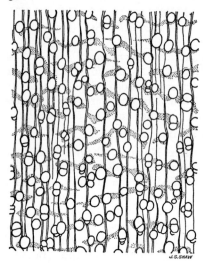

GUAREA

(1) *Guarea cedrata* Pellegr., (2) *G. thompsonii* Sprague and Hutch.

Family: *Meliaceae.*

Other Names: (1) scented guarea (G.B.); obobonufa, white guarea (Nigeria); bossé (France and Ivory Coast); (2) obobonekwi, black guarea (Nigeria). 'Nigerian cedar', 'Nigerian pearwood', 'cedar mahogany', 'African cedar', and 'scented mahogany' have been used for both species but these names are to be discouraged.

Weight: (1) 590 kg/m³. (2) 640 kg/m³.

Spot Characters: banded (apotracheal) parenchyma, medium-weight, and pale pink colour are characteristic; vessels medium-sized, radial pairs and multiples of 3 or more often predominate in parts of cross sections, gum deposits sometimes abundant, causing vessel lines to sparkle; parenchyma abundant, visible to naked eye, in broad conspicuous metatracheal bands (more regular and continuous in *G. thompsonii*, providing a distinguishing feature); grain not characteristically interlocked, frequently wavy or curly (particularly *G. cedrata*), faint 'watered-silk' figure from banded parenchyma when flat-sawn; wood pale pink (*G. thompsonii* usually paler); sapwood distinct.

Durability: stated to be moderately resistant to decay; sometimes heavily attacked by pin-hole borers.

Strength Properties: values for *G. cedrata* lower than for *G. thompsonii* in all properties except shear and resistance to splitting; above those for Central American and African mahogany.

Working Qualities: works well with hand and machine tools, with only slight dulling of cutting edges, comparable with the denser grades of Central American mahogany; tendency for the grain to 'pick up' in planing; takes nails, screws, and glue well, and finishing treatments.

Seasoning: can be air- or kiln-dried easily, with little tendency to distortion, and only slight inclination to check; gum exudations may give some trouble; shrinkage figures from 'green' to 12 per cent. M/C: 1·6 per cent. radially and 3·2 per cent. tangentially.

Uses: as a mahogany substitute; both species peel well and make up into satisfactory plywood.

Supplies: stated to be plentiful; usually imported in log form.

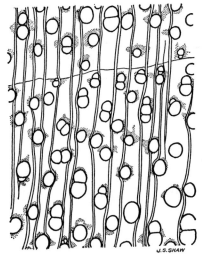

J.S.SHAW

IDIGBO

Terminalia ivorensis A. Chev.

Family: *Combretaceae.*

Other Names: black afara (G.B. and Nigeria); emeri (Ghana); framiré (France and Ivory Coast).

Weight: 560 kg/m³.

Spot Characters: discontinuous, wavy layers of apparently terminal parenchyma, and a superficial resemblance to flat-sawn oak, characterize idigbo; growth rings distinct only with lens; vessels medium-sized to moderately large, radial multiples sometimes predominating; parenchyma less abundant than in afara, in discontinuous initial layers demarcating growth rings, and vasicentric to aliform, rarely aliform-confluent; grain straight or slightly wavy; wood pale yellow when freshly sawn, with slight green tinge, which is less marked than in afara, and changing almost to pale pink-brown on exposure; sapwood usually not distinct.

Durability: resistant to decay; logs liable to pin-hole borer infestation, and 'spongy heart'; sapwood susceptible to sap-strain fungal attack and *Bostrychid* and *Lyctus* beetle attack; regarded as more durable than afara.

Strength Properties: only limited tests have been carried out; similar in most strength properties to iroko, but appreciably less hard and less resistant to splitting.

Working Qualities: works well with hand and machine tools, with little effect on cutting edges; tendency for the grain to 'pick up' in planing; takes nails and screws satisfactorily, and glues well, and can be satisfactorily finished.

Seasoning: can be air- and kiln-dried with little degrade from splitting or distortion; shrinkage figures from 'green' to 12 per cent. M/C: 1·6 per cent. radially and 2·7 per cent. tangentially.

Uses: for interior joinery *in lieu* of joinery grades of softwoods, and furniture; peels well and makes up into good plywood.

Supplies: a moderately large tree, abundant outside the rain-forest zone, through West Africa, including Southern Nigeria and Ghana; imported in log form and more recently as sawn timber.

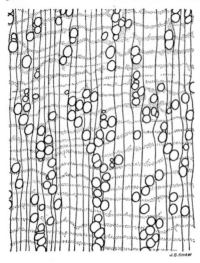

MAKORE

Tieghemella heckelii Pierre ex A. Chev. syn.: *Mimusops heckelii* (A. Chev.) Hutch. et Dalz.

Family: *Sapotaceae.*

Other Names: 'African cherry' (U.S.A.); baku (Ghana); 'cherry mahogany' (G.B.).

Weight: 640 kg/m³.

Spot Characters: vessels in radial multiples, parenchyma in metatracheal bands (banded) visible to naked eye when wetted, and red colour characterize makoré; vessels moderately small to medium-sized, radial multiples usually predominating, tyloses sometimes abundant; parenchyma abundant, in metatracheal layers close together, visible to naked eye on wetted end surfaces, forming reticulate pattern with rays; grain straight, interlocked, and wavy, sometimes wavy and interlocked, producing a mottled, broken stipe figure; flat-sawn faces with 'watered-silk' figure from the banded parenchyma; wood pale pink-brown to deep red, darkening on exposure; sapwood yellow-pink, only moderately sharply differentiated.

Durability: resistant to decay, and stated to be resistant to termite attack; pin-hole borer attack occurs in some logs.

Strength Properties: a few tests have been carried out at the Imperial Institute; appreciably harder and more resistant to splitting than Central American mahogany.

Working Qualities: works satisfactorily with hand and machine tools, but causes appreciable dulling of cutting edges because of the silica in the wood; grain inclined to 'pick up' when planing quarter-sawn material; takes nails, screws, and glue well; slices very well and peels satisfactorily. The sawdust causes irritation to the nose and inflammation of the eyes in some workers, attributed to an irritant saponin in the wood.

Seasoning: can be air- or kiln-dried satisfactorily with little degrade; stable when seasoned; shrinkage figures from 'green' to 12 per cent. M/C: 2·6 per cent. radially and 3·0 per cent. tangentially.

Uses: for furniture, high-class joinery, and interior finishings; figured veneers are suitable for decorative panelling.

Supplies: said to be abundant in the Cameroons and Ghana; imported in log form and small quantities of sawn timber.

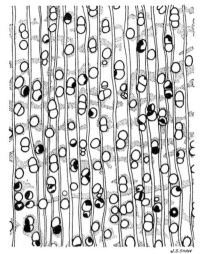

SAPELE

Entandrophragma cylindricum Sprague
The timber of *E. utile* Sprague (utile) and that of other species of *Entandrophragma* are sometimes included with that of *E. cylindricum* as sapele.

Family: *Meliaceae.*

Other Names: aboudikro, acajou aboudikro (France and Ivory Coast); cedar (Ghana); sapele mahogany, scented mahogany (G.B.); sapelli, acajou sapelli (France and Ivory Coast).

Weight: 640 kg/m³.

Spot Characters: prominent confluent parenchyma, storeyed rays, cedar-like odour, — and mahogany-brown colour with prominent stripe figure characterize sapele; growth rings usually distinct, demarcated by moderately broad layers of terminal parenchyma (obscured in samples with numerous broad bands of confluent parenchyma); vessels moderately small to medium-sized, dark-coloured gum-like deposits abundant; parenchyma abundant, in moderately broad terminal layers, and paratracheal, varying from vasicentric to aliform, with occasional confluent layers, to only confluent in more or less continuous layers close together; rays storeyed, usually producing well-defined 'ripple marks'; traumatic canals not infrequently present; grain typically interlocked, producing the characteristic ribbon figure; wood usually deep red-brown, but occasionally paler; sapwood distinct.

Durability: moderately resistant to decay; liable to pin-hole borer attack; sapwood susceptible to *Bostrychid* and *Lyctus* attack.

Strength Properties: superior in all strength properties except bending, but especially in hardness and resistance to splitting, to Central American and African mahogany, and gedu nohor.

Working Qualities: works well with hand and machine tools, with little dulling of cutting edges, but more resistant to cutting than Central American mahogany; liable to 'pick up' when planed on the quarter, unless the cutting angle is reduced to 15°; takes nails, screws, and glue well, and all finishing treatments; makes excellent sliced veneers.

Seasoning: air seasons well but rather slowly; kiln-dries satisfactorily, but liable to distort; shrinkage figures from 'green' to 12 per cent. M/C: 2·1 per cent. radially and 3·3 per cent. tangentially.

Uses: for furniture, panelling, interior fittings, and decorative purposes, and the plain timber *in lieu* of joinery grades of softwoods; also used for flooring; extensively used for sliced veneers. The oval shape of many logs precludes their use for rotary peeling.

Supplies: thought to be adequate to meet any normal demand; imported in log form and as sawn timber.

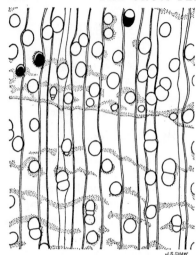

UTILE

Entandrophragma utile Sprague

Family: *Meliaceae.*

Other Names: assié, acajou assié (France and Cameroons); 'cedar' (Ghana); sipo, acajou sipo (France and Ivory Coast).

Weight: 660 kg/m³.

Spot Characters: Resembles sapele, but with less parenchyma and usually lacking storeyed rays, prominent ribbon figure and cedar-like odour; growth rings conspicuous on all surfaces, demarcated by broad bands of terminal parenchyma filled with dark coloured deposits; vessels medium-sized, frequently filled with dark coloured gum-like deposits; parenchyma abundant, in broad terminal layers, and vasicentric, aliform, and confluent, the confluent layers less continuous than in sapele; grain usually straight, and texture milder than in sapele or African mahogany; wood mahogany – red-brown; sapwood distinct.

Durability: no data are available; probably similar to sapele; not immune to 'pin-worm' infestation.

Strength Properties: no data are available; probably intermediate between African mahogany and sapele.

Working Qualities: works excellently with hand and machine tools, with little dulling of cutting edges; machines well with little tendency for the grain to 'pick up' because of the absence of pronounced interlocked grain; takes nails, screws, and glue well, and all finishing treatments.

Seasoning: tendency to develop surface checks and end splits when 'green', but kiln-dries without distortion, and better than sapele.

Uses: for furniture and for all purposes where sapele and African mahogany are used, and preferable to sapele when a plain unfigured wood is required.

Supplies: occurs in both West and East Africa in forests containing species of *Khaya* and *Entandrophragma*; imported in log form from West Africa and as sawn timber from East and West Africa.

III Timbers widely used in their countries of origin and now regularly exported to the U.K. and Europe.

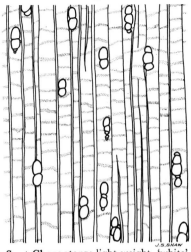

JELUTONG

Dyera costulata Hook. F.

Family: *Apocynaceae*

Other Names: The timber of pulai (*Alstonia* spp.) may occasionally be mixed with jelutong in commercial consignments.

Weight: 470 kg/m³.

Spot Characters: light weight, 'white' colour, vessels in radial multiples, parenchyma in fine lines, and large latex passages, characterize jelutong; growth rings often visible, demarcated by zones without wood parenchyma; vessels medium-sized, radial multiples of 3 predominating; wood parenchyma abundant, in narrow metatracheal layers ('fine lines') close together (more regular and continuous than in pulai); latex passages occur at branch whorls, appearing lens-shaped on flat-sawn faces, about 1·25 cm high, and containing ribbons of shrivelled tissue in dry timber; wood 'white' or pale yellow; sapwood not differentiated.

Durability: not resistant to decay; usually free from 'pin-worm' infestation, but trees damaged by careless topping may be riddled by a longhorn borer, or one of the large ambrosia beetles; liable to powder post beetle attack; very susceptible to sap-stain fungal discoloration.

Strength Properties: no data are available; the timber is likely to be used in circumstances where strength properties are not important.

Working Qualities: works easily with hand and machine tools, and for a soft timber the sawn surface is not woolly; machines to a very good finish; takes nails, screws, and glue well; cuts very cleanly with a sharp pen-knife.

Seasoning: can be air-dried rapidly, with very little degrade other than fine checks extending from the latex passages; unless dried rapidly is very liable to sap-stain discoloration, which usually cannot be prevented in thick material, even when dipped in suitable preservatives; average shrinkage figures from 'green' to 15 per cent. M/C: 2 per cent. radially and 3 per cent. tangentially.

Uses: for all purposes where light weight and good machining properties are more important than strength; also clogs, drawing-boards, and battery separators; imported before the war as a pattern wood; reported to have been exported from Sabah to Japan for pencils; uses likely to be circumscribed by occurrence of latex passages.

Supplies: a large tree, widely distributed throughout Malaysia, Sarawak, and Sabah, where it is valued and protected for its latex (an ingredient of chewing gum); also valued for its timber locally, and supplies for export (as boards, planks, and scantlings) are always likely to be small.

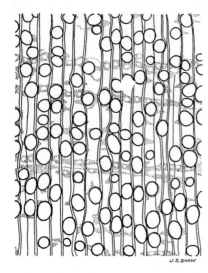

J.S.SHAW

KAPUR

Malaysian kapur: *Dryobalanops aromatica* Gaertn. f., *D. oblongifolia* Dyer.
Sarawak kapur: *Dryobalanops* spp., including *D. aromatica* and *D. oblongifolia*.
Sabah kapur: *Dryobalanops* spp., inc. *D. beccarii* Dyer and *D. lanceolata* Burck.
Indonesian kapur: *Dryobalanops* spp., inc. *D. aromatica, D. oblongifolia, D. beccarii, D. usca* V.Sl., and *D. lanceolata.*

Family: *Dipterocarpaceae.*

Other Names: keladan (*D. oblongifolia* in Malaysia); mahoborn (for timber from Brunei); kapur, kapor, paigie, and formerly Borneo camphorwood (Sabah); kapur and kapoer (Indonesia).
Weight: *D. aromatica* 790 kg/m³, *D. oblongi*-folia 750 kg/m³; and Indonesian kapur 590–830 kg/m³.

Spot Characters: normal vertical canals in more or less continuous tangential series, vessels frequently exclusively solitary, tyloses often abundant, parenchyma paratracheal, rays frequently storeyed, rose-red to deep-red, and camphor-like odour when freshly sawn characterize kapur; vessels medium-sized to moderately large, typically exclusively solitary, tyloses abundant, parenchyma moderately abundant, rather indistinct, forming incomplete borders to vessels, with narrow tangential wings (abaxial rather than aliform), and some diffuse-in-aggregates; rays normally more or less distinctly storeyed in *D. aromatica*, less regular in other species; normal vertical canals in more or less continuous tangential series, much smaller than the vessels, filled with white contents; camphor-like odour when freshly sawn, especially *D. aromatica*; grain interlocked; wood light rose-red to deep-red, becoming rose-red or red-brown on exposure; sapwood light yellow-brown, distinct, about 5 cm wide; liable to stain in contact with iron when wet.

Durability: moderately durable under tropical conditions, resistant to decay under temperate-region conditions; very liable to 'pin-worm' infestation in the living tree, especially *D. aromatica*, and when converted if not properly stacked; not susceptible to *Lyctus* attack; centres of some logs ' spongy '.

Strength Properties: appreciably stiffer as a beam, and superior as a post or strut, compared with teak, but slightly less hard.

Working Qualities: Saws well when green, but dulls cutting edges when seasoned because of silica content; finishes well when seasoned; takes nails, screws, and glue well.

Seasoning: seasons rather slowly, and care necessary to avoid end-splitting and surface checking of thick material, but less recalcitrant than kerung.

Uses: in countries of origin for carcassing and flooring; in U.K. for flooring, carriage stock, waggon planks; suitable, and superior, for all purposes to the gurjun group.
Supplies: semi-gregarious tendencies; locally gregarious (*e.g.* swamp form in Brunei); exportable surplus will always be small if prejudice against 'wormy' material persists.

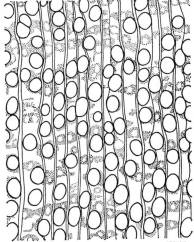

KERUING

Dipterocarpus spp. (31 spp. Malaysia; as many or more in Sabah and Sarawak; species occur from Ceylon to the Philippines).

Family: *Dipterocarpaceae.*

Other Names: apitong: principally *D. grandiflorus*, *D. lasiopodus* Perkins, and *D. vernicifluus* Blanco (the Philippines); dau: *Dipterocarpus* spp. (South Vietnam and Cambodia); eng: *D. tuberculatus* Roxb.; gurjun (Burma and Thailand) principally *D. alatus* Roxb. and *D. turbinatus* Gaertn. f. (Burma); Indian or Andaman gurjun: principally *D. grandiflorus* Blanco, *D. indicus* Bedd., and *D. macrocarpus* Vesque; hora: *D. zeylanicus* Thw. (Ceylon); yang: Thailand, species as for Burma.

Weight: eng 880 kg; hora 740 kg; gurjun 740 kg; yang 740 kg; apitong 740 kg; keruing (all sources) 640–910 kg/m.³

Spot Characters: normal vertical canals in short tangential series of 1 to 7, vessels medium-sized to moderately large, typically exclusively solitary and open, brown or red-brown colour characterize keruing; vessels distinctly oval, medium-sized to moderately large, usually exclusively solitary, tyloses sparse (moderately abundant in some species); parenchyma sparse or absent, diffuse or sparsely paratracheal; normal vertical canals in short tangential series of 1 to 7 throughout the cross section (more or less continuous tangential series may also occur), individual canals much smaller than to as large as or larger than the vessels; grain typically straight; wood light red-brown, with a white 'bloom' in some species, to dark brown or red-brown; sapwood of lighter shades, only moderately sharply differentiated.

Durability: resistance to decay variable; the best are inferior to kapur; moderately durable in temperate regions; sapwood liable to *Bostrychid* beetle and dry-wood termite attack; typically free from 'pin-worm' infestation, unless left lying in the forest.

Strength Properties: stiffer (up to 180 per cent.) and more suitable as a post or strut, than teak, but inferior (except one) as a beam, in shock-resisting ability, and in hardness.

Working Qualities: silica causes rapid dulling of cutting edges, necessitating suitable cutting angles and feed speeds; an initial fibrous finish from the tool necessitates considerable sanding to obtain a good finish taking polish. 'Resinous' content of some logs gums saws, necessitates a jet of paraffin and water on saw during conversion.

Seasoning: high shrinkage calls for care to avoid end-splitting, surface checking, and distortion; kiln dries very slowly; large moisture variations may develop in thick stock; shrinkage from 'green' to air-dry: 3 to 3.5% radially and 6 to 7% tangentially.

Uses: as railway sleepers (treated); in U.K. as waggon planks, telegraph arms (treated), sills, and flooring; recently for work benches and structural beams in place of oak.

Supplies: large quantities available as strips, boards, planks, and squares; also in log form from Sarawak and Sabah; freedom from 'spongy heart' and 'pin-worm' are characteristic features of the group.

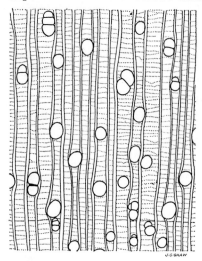

OBECHE

Triplochiton scleroxylon K. Schum.

Family: *Sterculiaceae* (sometimes referred to *Triplochitonaceae*).

Other Names: wawa (Ghana and Ivory Coast), ayous (Cameroons), arere (Nigeria), samba (Ivory Coast), and African whitewood (obsolete U.K. name).

Weight: 390 kg/m³.

Spot Characters: 'white' colour; diffuse-in-aggregates parenchyma (distinct × 10), and tissue other than rays storeyed characterize obeche; growth rings distinct with lens, demarcated by fine layers of apparently terminal parenchyma; vessels medium-sized to moderately large; parenchyma abundant, diffuse-in-aggregates, distinct with lens; tissue other than rays storeyed; grain interlocked, producing distinct ribbon figure in quarter-sawn material; wood creamy white.

Durability: not resistant to decay, very susceptible to blue-stain fungal discoloration; readily attacked by pin-hole borers (said to be subsequent to felling), particularly Ghana wawa (probably because of long over-land haulage), and susceptible to powder-post beetle attack (*Lyctidae* and *Bostrychidae*).

Strength Properties: similar to American whitewood, but inferior in stiffness.

Working Qualities: works very easily with hand and machine tools, requires thin-edged cutting tools to give a smooth finish; nails and screws do not hold too well, but timber glues well.

Seasoning: air seasons moderately rapidly and can be kiln-dried exceptionally rapidly, with little distortion and only slight tendency to split, but very rapid kiln-drying induces tension set conditions, with marked tendency to spring when subsequently worked; shrinkage figures: 'green' to 12 per cent. M/C: 1·4 per cent. radially, 3·2 per cent. tangentially.

Uses: in West Africa for household articles and boxes; in Europe for corestock of plywood, block-boards, furniture, parts of motor bodies, and as an alternative to joinery grades of softwoods for interior joinery.

Supplies: semi-gregarious in some parts of West Africa; trees attain 45 m in height, and 150 cm in diameter above the high buttresses. Previously exported in log form, but increasing supplies of sawn timber are becoming available from some countries.

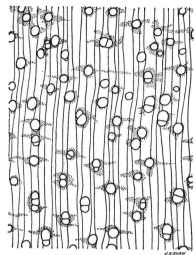

RAMIN

Gonystylus spp., principally *G. bancanus* (Miq.) Kurz.

Family: *Gonystylaceae.*

Other Names: ramin telur (Sarawak); melawis (Malaysia).

Weight: 510 kg/m³.

Spot Characters: pale yellow colour and aliform-confluent parenchyma (narrow wings and layers) characterize ramin; growth rings sometimes distinct, demarcated by narrow layers of lighter-coloured tissue (fibre tracheids); vessels medium-sized to just moderately large; parenchyma moderately abundant, aliform, with narrow wings often extending to link up several vessels, producing aliform-confluent parenchyma; wood pale yellow, often with a centre core up to 150 mm in diameter grey-green to almost black, which is usually riddled with galleries of a large 'pin-hole' borer; sapwood not distinct.

Durability: not resistant to decay or to termite attack; logs left in the forest liable to 'pin-hole' borer infestation, but freshly-felled timber usually clear, except for the heartwood core; usually not susceptible to *Lyctus* attack, but may be attacked during seasoning by a small *Bostrychid* beetle; very susceptible to sap-stain fungal infection, often when the surface is clean and bright.

Strength Properties: small-scale tests have been carried out in Malaysia.

Working Qualities: saws with moderate ease, with only slight dulling of saw teeth (rather more than with red meranti); machines excellently, and mouldings take sharp arrises; nails, screws, and glues satisfactorily, but inclined to split in nailing; takes finishing treatments well. Bark should be removed prior to sawing as it contains sharp-pointed needle-like hairs, which produce temporary intense skin irritation, removed by washing affected parts with water.

Seasoning: air seasons readily with only slight tendency for cupping in flat-sawn boards, but difficult to secure sufficiently rapid drying to eliminate sap-stain unless chemically dipped; liable to end-splitting and deep surface checks unless carefully piled, and with adequate end protection; shrinkage figures from 'green' to about 16 per cent. M/C: 1 per cent. radially and 3·6 per cent. tangentially.

Uses: little used in Malaysia, but has proved very suitable for furniture and mouldings in U.K.; stained timber should be suitable for greenhouses if pressure-treated – the timber takes preservatives readily.

Supplies: plentiful in fresh-water swamp forests in Malaysia and Sarawak; imported as boards and planks, squares, and round logs.

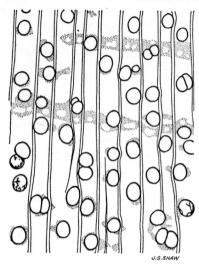

J.S.SHAW

LIGHT RED MERANTI

Shorea spp. (more than 12 spp. in Malaysia and the same and additional species in Sabah, Sarawak, and the Philippines).

Family: *Dipterocarpaceae.*

Other Names: most species have separate vernacular names ; supplies from Malaysia, Sarawak, and Indonesia are marketed as light red meranti in the U.K. and supplies from Sabah as light red seraya. *Shorea almon* Foxw., some *S. squamata* (Turcz.) Dyer, *Parashorea plicata* Brandis, and some species of *Pentacme* provide comparable timber from the Philippines known as light red lauan or white lauan (in part).

Weight: 550 kg/m³.

Spot Characters: normal vertical canals in more or less continuous tangential series, vessels medium-sized to moderately large, diagonal in arrangement (contrast yellow and white meranti), and pink or light red-brown colour characterize light red meranti; vessels medium-sized to moderately large, diagonal in arrangement, tyloses occur (contrast Central American mahogany); parenchyma sparse to moderately abundant, usually rather indistinct, all forms of paratracheal parenchyma may occur; normal vertical canals in more or less continuous tangential series, usually much smaller than the vessels, filled with yellow-white contents; grain interlocked; wood pink, light red, to dark red-brown (exceptionally yellow); sapwood distinct.

Durability: moderately resistant to decay under temperate-climate conditions; frequently attacked by 'shot-hole' borers in the living tree, and by 'pin-hole' borers if logs are left in forest; not susceptible to powder-post beetle attack; sapwood susceptible to staining; 'spongy heart', containing numerous compression failures, prevalent.

Strength Properties: about two-thirds that of teak in most strength properties, but only a little less stiff, and appreciably less hard.

Working Qualities: excellent with hand and machine tools, giving a good finish in most operations, but quarter-sawn material 'picks up' in planing; nails, screws, and glues well, and takes all finishing treatments; an excellent wood for rotary peeling.

Seasoning: dries moderately fast, with little tendency to split, but thin stock requires careful piling to prevent cupping; kiln-dries easily and well; shrinkage figures from 'green' to 12 per cent. M/C: 1·4 to 2·4 per cent. radially and 4·4 to 4·6 per cent. tangentially; once seasoned, it holds its shape well in temperate climates.

Uses: used in the East for all purposes for which softwoods are used in temperate climates; in U.K. for furniture, interior fittings, and, the lower grades, for joinery and flooring; makes excellent plywood.

Supplies: abundant, imported as sawn timber from Malaysia, Sabah, and Sarawak, and as round logs from the last two countries; proportion of 'wormy' timber will always be high.

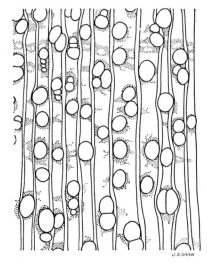

DARK RED MERANTI

Shorea spp., principally *S. pauciflora* King, *S. curtisii* Dyer, and *S. acuminata* Dyer (in part).

Family: *Dipterocarpaceae.*

Other Names: most species have separate vernacular names in their countries of origin, but supplies from Malaysia, Sarawak, and Brunei are now marketed in the U.K. as dark red meranti, and supplies from Sabah as dark red seraya. Comparable timbers, but of other species of *Shorea*, provide red lauan of the Philippines.

Weight: 710 kg/m³.

Spot Characters: normal vertical canals appearing as prominent white lines on all surfaces, vessels moderately large, tyloses and aliform-confluent parenchyma abundant, and purple-red-brown colour characterize dark red meranti; vessels moderately large, tyloses often abundant; parenchyma aliform-confluent to confluent; normal vertical canals in numerous, more or less continuous, tangential series, individual canals larger than in most species of *Shorea*, sometimes not much smaller than the vessels, conspicuous as yellow-white lines on all surfaces; grain interlocked; wood dark red-brown with purple tinge; sapwood distinct.

Durability: moderately resistant to decay under temperate-climate conditions; less frequently attacked by ambrosia beetles than most types of meranti, and then usually by 'shot-hole' borers; not susceptible to *Lyctus* attack.

Strength Properties: stiffer than teak and similar in shear; only 84 per cent. of teak in bending and shock-resisting ability and 75 per cent. as hard.

Working Qualities: works well with hand and machine tools, with only slight dulling of cutting edges; tendency for the grain of quarter-sawn material to 'pick up' in planing; takes nails, screws, and glue well, but the prominent resin canals may be troublesome when a polished finish is required.

Seasoning: dries moderately fast, with little tendency to split, but thin stock requires careful piling to prevent cupping; kiln-dries easily and well; dark red meranti appears to shrink rather less than other forms of red meranti; figures from 'green' to 12 per cent. M/C: 1·9 per cent. radially and 4·0 per cent. tangentially.

Uses: in the countries of origin for all purposes for which softwoods are used in temperate climates; in U.K. for furniture, interior fittings, and joinery; an excellent timber for rotary peeling.

Supplies: the local demand, which takes no notice of 'worm', is likely to restrict the quantities available for export; shipped as boards, planks, and scantlings.

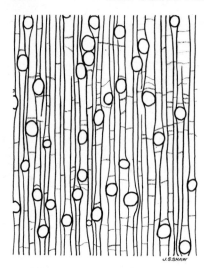

J.S.SHAW

OPEPE

Nauclea diderrichii (De Wild et Th. Dur.) Merrill (Syn. *Sarcocephalus diderrichii* De Wild).
Family: *Rubiaceae.*
Other Names: badi (Ivory Coast); bilinga (Cameroons); kusia, kusiaba (Ghana).
Weight: 750 kg/m³.

Spot Characters: vessels exclusively solitary, medium-sized, open or with gum deposits; parenchyma indistinct, diffuse, tending towards diffuse-in-aggregates; distinctive orange-yellow colour, sapwood pink.

Durability: resistant to decay and reputed to be resistant to some marine borers; pin-hole borer damage may occur.

Strength Properties: appreciably harder than Burma teak; stronger in compression along the grain, and in bending and stiffness, compared with teak; data are available.

Working Qualities: works with moderate ease in most hand and machine operations; quarter-sawn material is liable to 'pick up' unless the cutting angle is reduced to 10°; there is a tendency to split in nailing, but it takes screws and glue moderately well.

Seasoning: refractory, particularly if flat-sawn; air-dries moderately rapidly, but it is liable to split and check—thin stickers are recommended; kiln-drying data are available; shrinkage figures from 'green' to 12 per cent. M/C: 2·3 per cent. radially and 3·8 per cent. tangentially.

Uses: in W. Africa for canoes, planks, and exterior construction; it has been used for framing in railway coaches in the U.K.; suitable for interior fittings and flooring.

Supplies: a semi-gregarious tree, attaining 48 m in height and 180 cm in diameter, with a cylindrical bole; widely distributed in Guinea, Liberia, Ivory Coast, Ghana, Nigeria, and Cameroons. Usually imported in log form and as squared-edged timber, but supplies are not always available.

IV Commercially important home-grown timbers.

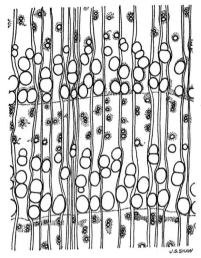

EUROPEAN ASH

Fraxinus excelsior L.

Family: *Oleaceae.*

Other Names: English ash, French ash, etc., are used to indicate the country of origin. American ash from Canada and the U.S.A. is provided by three other species of *Fraxinus*, and Japanese ash by *F. mandshurica* Rupr.

Weight: 710 kg/m³.

Spot Characters: the creamy white colour, with pink tinge when freshly cut, ring-porous structure, with marked contrast between early and late wood, producing a decorative figure when flat-sawn or rotary peeled, and terminal and vasicentric (tending to become aliform) parenchyma, characterize ash; growth rings distinct, demarcated by a layer of terminal parenchyma and the pore ring usually of two to several rows of oval early wood vessels, often with several radial pairs; vessels of the pore ring medium to moderately large in size, late wood vessels only individually distinct with a lens; wood parenchyma distinctly yellow in colour against the background of pink-brown fibres, terminal, vasicentric, and aliform, and tending to become confluent outwards in the growth ring; rays fine; wood typically creamy white with a pink tinge when freshly sawn, but sometimes pink-brown; sapwood and heartwood usually not sharply defined, but logs containing irregular dark brown or black heart are occasionally found.

Durability: the timber is classified as perishable; it is susceptible to pin-hole borer, *Lyctus*, and furniture beetle attack.

Strength Properties: the outstanding property of ash is toughness; in other strength properties it is comparable with oak, but about 20% harder and stiffer than that timber.

Working Qualities: there is a tendency to bind on the saw when converted 'green', but when air-dry it is reasonably easy to work and finishes smoothly. It has excellent steam-bending qualities. It takes all finishes well.

Seasoning: the timber is liable to distort if dried too rapidly but it is not prone to split or check. In kiln-drying shrinkage and distortion may be excessive but the timber responds well to a reconditioning treatment.

Uses: selected good quality ash is one of the best timbers for sports equipment, tool handles, and agricultural implements, and for all purposes where toughness and medium weight are desirable qualities.

Supplies: the demand for good quality ash exceeds the available supplies from the U.K. and Europe.

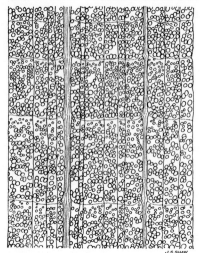

J.S.SHAW

EUROPEAN BEECH

Fagus sylvatica L.

Family: *Fagaceae.*

Other Names: English beech, French beech, Slavonian beech, etc., are used to indicate the country of origin. *Fagus grandifolia* Ehrh. provides American beech from Canada and eastern U.S.A., and *F. crenata* Bl. and allied species Japanese beech.

Weight: 720 kg/m³.

Spot Characters: white or light brown (pink when steamed) colour, with distinct fleck on longitudinal surfaces from the broad rays, diffuse porous structure, distinct growth rings. vessels very small and numerous, and rays wider than the vessels, characterize beech; growth rings distinct, demarcated by a zone of fibres without vessels or diffuse parenchyma, giving the appearance of a band of parenchyma of the same colour as the rays; diffuse porous, vessels very small, individually visible only with a lens, very numerous except in a zone at the end of the growth ring; multiple perforations occur, but because of the small size of the vessels these are only visible in microscopic sections; parenchyma diffuse, only visible in microscopic slides, and wanting in a zone at the end of each growth ring, producing, with the absence of vessels in this zone, apparent bands of terminal parenchyma, but distinguishable from true terminal parenchyma because, with a pocket lens, the inner margin of the 'band' is seen to be indeterminate; rays broad and more than 2 mm. high, wider than the vessels, and producing a distinctive fleck on longitudinal surfaces; wood typically white or light brown, sometimes with a pink tinge, and deep pink when steamed; sapwood not differentiated from the heartwood.

Durability: the timber is classified as perishable. It is occasionally attacked by pinhole borers; it is immune to *Lyctus* infestation but is susceptible to furniture beetle attack.

Strength Properties: the strength properties of unseasoned beech are similar to those of oak, but after seasoning it is about 20 per cent. superior to oak in most properties, and about 40 per cent. more resistant to impact loads.

Working Qualities: beech works to a good finish in most hand and machine operations, but when dry it is somewhat difficult to saw, tending to burn in cross-cutting and in drilling. It turns well and is exceptionally good for steam bending. It glues well and takes all finishes satisfactorily.

Seasoning: a moderately refractory timber in seasoning, with a tendency to check, split and warp; shrinkage is appreciable.

Uses: both home-grown and imported beech are extensively used for furniture and cabinet work, turnery, shoe heels, domestic wood ware, broom handles. It has been used on a small scale for flooring. Impregnated with a solution of copper sulphate, was extensively used in France for railway sleepers.

Supplies: adequate, U.K. and continental sources.

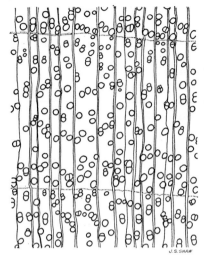

J.S.SHAW

EUROPEAN BIRCH

Betula pendula Ehrh. and *B. pubescens* Ehrh.

Family: *Betulaceae.*

Other Names: English birch, Finnish birch, Swedish birch, etc., are used to indicate the country of origin. Karelian birch, masur birch, and flame or ice birch are used for special figured material. Species of *Betula* occur in Canada and the U.S.A.; *B. alleghaniensis* Britt. and *B. lenta* L. provide yellow birch; *B. papyrifera* Marsh. paper birch; and *B. papyrifera* Marsh. var. *occidentalis* Sarg. western paper birch.

Weight: 670 kg/m³.

Spot Characters: white or light brown in colour, the growth ring boundaries being prominent on flat-sawn faces as darker-coloured lines, diffuse porous, growth rings distinct, parenchyma terminal, and vessels very small to moderately small, characterize birch; growth rings distinct, demarcated by layers of terminal parenchyma as wide as or wider than the rays; diffuse porous, vessels very small to moderately small, solitary and in radial groups of two or more; multiple perforation plates present but visible only in microscopic preparations because of the small size of the vessels; wood parenchyma confined to the layers of terminal parenchyma, distinct to the naked eye on all surfaces; rays fine; wood white to light yellow-brown, sapwood and heartwood usually not distinct.

Durability: the timber is classified as perishable. It is sometimes attacked by pin-hole borers; it is probably immune to *Lyctus* infestation, but is susceptible to furniture beetle attack.

Strength Properties: the strength properties of unseasoned birch are comparable to those of oak and, when seasoned, they are superior.

Working Qualities: birch works relatively easily in most hand and machine operations and it planes and moulds to a good finish. It turns and peels well. It takes all finishes well and glues satisfactorily.

Seasoning: there is a tendency for the timber to warp in both air- and kiln-drying; there is a need to accelerate drying by using thick stickers because the timber is particularly prone to fungal infection.

Uses: turnery, brush backs; imported material from Northern Europe is used in the furniture industry and it is extensively used in Northern and Eastern Europe for rotary cut veneers in plywood manufacture. Canadian material is used in flooring, both as strips and wood blocks, because of its high resistance to wear, but considerable shrinkage is a source of trouble.

Supplies: adequate supplies of birch are available to meet all normal demands.

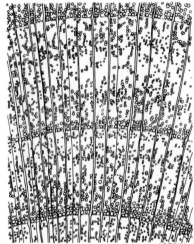

EUROPEAN CHERRY

Prunus avium L.
Family: *Rosaceae*
Other Names: cherry, wild cherry, gean, mazzard.
Weight: 630 kg/m³.

Spot Characters: falsely ring-porous because of the aggregation of vessels in bands at the beginning of each growth ring, rays distinct to the naked eye and wider than the vessels, producing a decorative fleck rather than silver figure on quarter-sawn surfaces, and the warm red-brown colour characterize European cherry; growth rings distinct on cross and tangential surfaces because of the aggregation of the vessels in bands at the beginning of each growth ring; vessels usually small, and only individually visible with a pocket lens, solitary and in radial multiples of 2 to 4 or more, white deposits plentiful; parenchyma sparse, not visible with a lens; rays distinct to the naked eye, wider than the vessels; wood warm red-brown in colour; sapwood light red-brown, not always easily distinguishable from the heartwood.

Durability: it is stated (*Handbook of Hardwoods*) that the timber is probably moderately durable and it is almost immune from powder-post beetle attack, but liable to attack by the common furniture beetle. The writer has encountered one case where shop fittings in several stores belonging to one firm, and widely distributed in many parts of the country, all became heavily infested by *Lyctus*. The fittings were all made from the same parcel of timber, which had been in contact with a parcel of oak, no doubt with infected sapwood. It is presumed that the vessels in this parcel were rather larger than normal and hence the timber would be susceptible to attack.

Strength Properties: described as a tough timber, being as tough as ash in the un-seasoned state, but some 30 per cent. less tough than that timber when seasoned although superior to oak in toughness when dry; 'the bending strength, stiffness, crushing stress along the grain, and hardness of the air-dry timber are about equal to those of oak, but it is about 25 per cent. more resistant to splitting' (*Handbook of Hardwoods*, H.M. Stationery Office).

Working Qualities: works well with most saws, but flat-sawn material may distort in drying; cross-grained material may be difficult to plane, with a tendency to tear, leaving a rough finish; polishes well and glues satisfactorily.

Seasoning: the timber is said to season fairly readily, but with a pronounced tendency to warp; total shrinkage from 'green' to 12 per cent. M/C is rather high, being 3·5 per cent. radially and 6·5 per cent. tangentially.

Uses: European cherry is essentially a cabinet wood, and is used for furniture, domestic ware, and toys. Supplies are small and not regularly available.

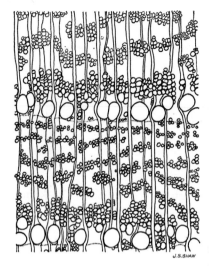

ENGLISH ELM

Ulmus procera Salisb. (syn.: *U. campestris*. Auct. angl.) and *U. hollandica* Mill. var. *hollandica*.
Family: *Ulmaceae*.
Other Names: red elm, nave elm (*U. procera*), Dutch elm (*U. hollandica* var. *hollandica*). Other species of *Ulmus* provide: white elm (*U. americana* L.); wych elm (*U. glabra* Huds. non Mill.), also called mountain elm and Scotch elm; smooth-leaved elm (*U. carpinifolia* Gleditsch); rock elm (*U. thomasii* Sarg.); Japanese elm (*Ulmus* spp.).
Weight: 560 kg/m³; *U. glabra* 690 kg/m³.

Spot Characters: wood dull red-brown, with a purple tinge, ring-porous structure, the marked contrast between early and late wood producing a decorative figure when flat-sawn, growth rings distinct, late wood vessels in clusters, forming wavy tangential lines, characterize elm; growth rings distinct, demarcated by a pore ring of one to two rows of moderately large to large early wood vessels; late wood vessels very small and barely individually visible with a lens, occurring in large clusters and forming more or less continuous tangential layers, sometimes mistaken for broad bands of parenchyma; parenchyma indistinct, predominately paratracheal; rays fine; wood dull red-brown, with a purple tinge; sapwood distinct from the heartwood.

Durability: the timber is classified as non-durable; it is susceptible to furniture beetle and death-watch beetle attack, and the sapwood to *Lyctus* infestation.

Strength Properties: English elm is inferior to oak in strength properties, and Dutch elm is similar except that it is about 40 per cent. tougher than English elm.

Working Qualities: the timber is somewhat refractory in conversion because of a tendency to bind on the saw. It usually picks up in planing, making it difficult to obtain a smooth finish, although Dutch elm is better than English elm in this respect. The timber takes finishing treatments satisfactorily, holds nails well, and has good gluing qualities.

Seasoning: a refractory timber to season because of its marked tendency to distort. It is necessary to place stickers close together in both air- and kiln-drying, and stocks require adequate weighting.

Uses: Formerly extensively used for carcassing purposes in period cottages, and as wide boards for flooring and as weather-boarding. Now extensively used for chair seats, turnery, including fruit bowls and similar articles, and for coffins, and also in boat-building for timbers permanently in the water.

Supplies: adequate supplies are available to meet the restricted uses to which this timber is put.

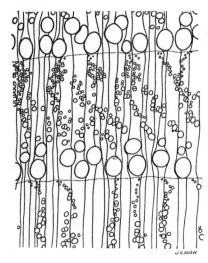

J.S.SHAW

SWEET CHESTNUT

Castanea sativa Mill. (syn. *C. vesca* Gaertn.).

Family: *Fagaceae.*

Other Names: Spanish chestnut, European chestnut (*C. sativa*); American chestnut (*C. dentata* Borkh.).

Weight: Sweet chestnut: 560 kg/m³, American chestnut: 480 kg/m³.

Spot Characters: cream to light brown, ring porous structure, late wood vessels mostly solitary, radial in arrangement, and fine rays, characterize sweet chestnut; growth rings distinct, demarcated by a pore ring of one to two rows of moderately large to very large early wood vessels containing abundant tyloses; late wood vessels very small to moderately small, mostly solitary, and radial in arrangement; parenchyma indistinct, diffuse; rays fine; wood cream to light yellow brown, with narrow, distinct sapwood; it resembles oak but without the silver figure of that timber.

Durability: the timber is classified as durable; logs are occasionally damaged by pin-hole borers; the sapwood is susceptible to *Lyctus* infestation; it is liable to attack by the common furniture beetle and the death watch beetle.

Strength Properties: seasoned timber is only about half as hard and half as tough as oak, and in bending, shear, and stiffness the strength properties are about 80 per cent. of oak in these properties. Unseasoned its resistance to splitting is about 30 per cent. that of oak.

Working Qualities: the timber is easy to work and finishes well with all hand and machine tools. It takes all finishing treatments satisfactorily.

Seasoning: the timber seasons slowly and is liable to collapse and honeycombing in drying. Collapsed timber is stated not to respond well to reconditioning treatments.

Uses: it was formerly used as an alternative to oak for structural work and panelling. It is now used for furniture, coffin boards, and turnery work, and for such external purposes as fence posts, rails, and gates. It is still grown in some parts of the country on a 15-years coppice rotation to provide cleft chestnut fencing, stakes, and hop poles.

Supplies: adequate supplies to meet all normal requirements are likely to be available, and cleft fencing and dimension stock are also imported from France.

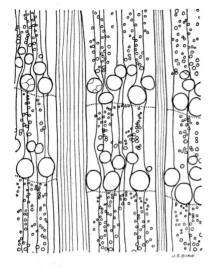

EUROPEAN OAK

Quercus robur L. (syn. *Q. pedunculata* Ehrh.) and *Q. petraea* Liebl. (syn. *Q. sessiliflora* Salisb.).

Family: Fagaceae

Other Names: English oak, French oak, Slavonian oak are used to indicate the country of origin. Other species of *Quercus* provide: American red oak (see page 116); American white oak (principally *Q. alba* L., *Q. prinus* L., *Q. lyrata* Walt., *Q. michauxii* Nutt); Japanese oak (principally *Q. mongolica* Fisch. ex Turcz. var *grosseserrata* Rehd. et Wils.).

Weight: European oak 720 kg/m³, American white oak 770 kg/m³, and Japanese oak 670 kg/m³.

Spot Characters: the light yellow-brown colour, ring porous structure, with marked contrast between early and late wood, broad high rays, producing silver figure on quarter-sawn material, and late wood vessels occurring in triangular-shaped areas between the growth rings, characterize oak; growth rings distinct, demarcated by the pore ring of moderately large to very large early wood vessels, one or two rows wide, and containing abundant tyloses, late wood vessels solitary, very small, only individually visible with a lens, diagonal in arrangement, and occurring in triangular zones between the growth rings; parenchyma predominantly apotracheal, occurring in fine lines (diffuse-in-aggregates) between the rays, variable in distinctness; rays of two distinct sizes, the broad high rays, frequently exceeding 25 mm in height, conspicuous to the naked eye on all surfaces, and the fine rays visible only with a lens on transverse surfaces; wood yellow-brown with distinct white-coloured sapwood.

Note: the anatomical structure of all the true oaks, other than holm oak (*Q. ilex* L.) is essentially similar, but the differences in the American red oak justify describing the spot characters seperately, *vide* page 116.

Durability: the heartwood is durable, but the sapwood is not, and it is particularly susceptible to *Lyctus* attack. The timber may be attacked in the log by pin-hole borers. In suitable service conditions, usually associated with actual fungal decay, both the sapwood and the heartwood may be attacked by the death watch beetle.

Strength Properties: the strength properties are well known, being frequently the standard by which other timbers are judged.

Working Qualities: not difficult to cut when green, but, when seasoned, working qualities vary, the mild timber from the high altitudes of Central Europe working well in all operations, whereas the denser home-grown timber is much more difficult to work.

It holds nails and screws well, but requires pre-boring; it corrodes iron and steel nails and screws, and if brought into contact with iron when green or under damp conditions black stains (from the tannin content) result. It takes all finishing treatment, other than paints, well, and it glues satisfactorily.

Seasoning: the timber is decidedly refractory in seasoning, being liable to checking, splitting, and distortion unless dried slowly and with adequate care.

Uses: the milder timber is extensively used for high-class joinery and cabinet work, panelling, and flooring, and the heavier and harder home-grown timber is used for heavy constructional work, fencing, and wagon construction.

Supplies: adequate supplies of oak from Europe, Japan and America are usually available to meet all normal needs, but high-grade parcels are often in short supply.

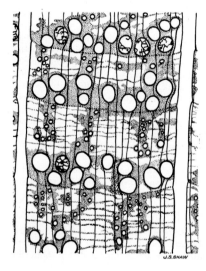

AMERICAN RED OAK

Principally (1) *Quercus rubra* L. emend. De Roi, (2) *Q. falcata* Michx. f. var *falcata*, (3) *Q. falcata*, Michx. f. var. *pagodaefolia* Ell., (4) *Q. shumardii* Buckl.

Family: *Fagaceae*.

Other Names: (1) northern red oak, true red oak; (2) southern red oak, Spanish oak; (3) swamp red oak, swamp Spanish red oak; (4) shumard red oak.

Weight: 790 kg/m³.

Spot Characters: Similar to those of European oak, American white oak, and Japanese oak, except for the distinct pink to light red-brown heartwood; larger late wood vessels, conspicuously solitary and radial in arrangement; tyloses usually wanting; wood parenchyma in wider and more distinct bands than in other forms of oak; and the rays frequently under 25 mm high.

V Timbers selected to show features used in lens identification not illustrated in Section I–IV.

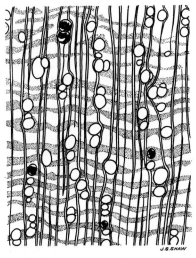

EKKI

Lophira alata Banks ex Gaertn. (syn. *L. alata* Banks var. *procera* Burtt Davey & Hoyle, *L. procera* A. Chev.).
Family: *Ochnaceae.*
Other Names: azobé (Ivory Coast); bongossi (Cameroons); eba (Nigeria); kaku (Ghana); red ironwood (B.W. Africa).
Weight: 1070 kg/m³.

Spot Characters: light-coloured solid deposits in vessels, broad conspicuous bands of metatracheal parenchyma, chocolate-red-brown colour characterize ekki; vessels medium-sized to moderately large, often filled with light yellow-brown solid deposits; parenchyma distinct to naked eye, in broad conspicuous metatracheal bands, about the radial diameter of the vessels apart; grain interlocked, producing prominent ribbon figure on quarter-sawn faces; wood chocolate-red-brown, speckled with yellow-brown (the solid deposits in the vessels); sapwood distinct, about 50 mm wide.

Durability: very resistant to fungal decay, and reputed to be resistant to termites and marine borers; not reputed to be susceptible to pin-hole borer infestation.

Strength Properties: strength properties are comparable to the weight of the timber; it is more than twice as hard as teak, and double the strength in shear of that timber.

Working Qualities: extremely hard to work with hand tools, and requires considerable power for conversion with machine tools; grain very liable to 'pick up' in planing; a smooth finish can be obtained with care, and, when filled, ekki polishes effectively; requires pre-boring before nailing.

Seasoning: an extremely recalcitrant timber, drying very slowly, and liable to distort appreciably in drying; total shrinkage is high, but the differential shrinkage is low; shrinkage figures from 'green' to 12 per cent. M/C: 4·5 per cent. radially and 5·5 per cent. tangentially (F.P.R.L. leaflet No. 44).

Uses: wharf construction, railway sleepers, and bridge decking in West Africa; exported before the war to the Continent for parquet flooring; suitable for harbour work, wagon planks, and heavy constructional purposes.

Supplies: a large tree, semi-gregarious in the wetter parts of the heavy-rain forests of West Africa; exported as logs and as sawn timber.

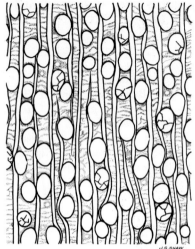

J.S.SHAW

MERSAWA

Anisoptera spp

Family: *Dipterocarpaceae.*

Other Names: kaunghmu (Burma); kayu pengiran (Sabah); krabak (Thailand); mascal wood (for *A. scaphula* (Roxb.) Pierre in India and Burma); mersawa (Malaysia and Indonesia); palasopis (Philippines); ven ven (Vietnam). Localized vernacular names: malai (Kelantan); sanai (Perak); terbak (Kedah).

Weight: mean figures for different species 580–720 kg/m³.

Spot Characters: normal vertical canals scattered singly throughout the cross section, vessels medium-sized, oval, typically exclusively solitary, parenchyma diffuse-in-aggregates or vasicentric to aliform, and the light yellow colour, with or without a pink tinge, characterize mersawa; vessels medium-sized to moderately large, usually markedly oval and exclusively solitary, tyloses sparse in some forms to abundant; parenchyma diffuse-in-aggregates in some, mainly vasicentric to aliform in others; normal vertical canals scattered singly through the fibres, variable in size from much smaller than to as large as the vessels, variable in number in different species, and occasionally in more or less continuous tangential series; rays of two sizes, the large rays distinct to the naked eye; grain interlocked; wood light yellow or yellow brown, sometimes with a rose tinge; sapwood usually not distinct.

Durability: moderately resistant to decay when fully dry, but liable to become doty if close-piled too soon; particularly susceptible to *Bostrychid* beetle attack during seasoning, and susceptible to *Lyctus* attack when dry; freshly felled logs typically free from defects, particularly 'pin-worm', but liable to attack if left in the forest; sapwood susceptible to staining.

Strength Properties: comparable with light red meranti in most strength properties, although similar in weight to heavy red meranti; low in shock-resisting ability, but good in stiffness and shear, *vide* Malayan Forest Record No. 14, p. 124.

Working Qualities: there is appreciable variation between the timbers of the different species, dependent on the silica content; difficult to hand-saw, and to power-saw wide dimensions in narrow thicknesses; planes to a smooth finish, but causes rapid blunting of cutting edges.

Seasoning: extremely slow drying—in 25 mm thickness mersawa takes up to five times as long to air-dry as red meranti under Malaysian conditions, and up to 6 weeks to kiln-dry 25 mm material; shrinkage figures for one species from 'green' to 15 per cent. M/C: 1·3 per cent. radially and 4·5 per cent. tangentially.

Uses: for furniture, particularly when a plain timber capable of taking stains well is required, but uses limited by reason of slow drying and rapid dulling of tools.

Supplies: a large tree, widely distributed; supplies should be adequate for all normal demands; imported as sawn timber. It has been substituted for chengal (*Balanocarpus heimii* King) by unscrupulous contractors.

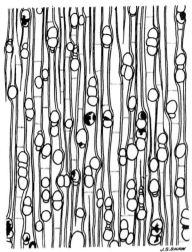

J.S.SHAW

PUNAH

Tetramerista glabra Miq.

Family: *Marcgraviaceae* (also referred to *Ternstroemiaceae* or *Ochnaceae* by some botanists).

Other Names: none.

Weight: 720 kg/m³.

Spot Characters: vessels oval, moderately large, typically in radial multiples, chalk-like deposits sometimes common, parenchyma diffuse, individual strands of large cross-section, giving the end surface a speckled appearance, and the light yellow-brown colour, with an orange tinge, characterize punah; vessels oval or truncated, mostly in radial multiples of 2 to 3, gum-like or chalky-white solid deposits often abundant; parenchyma abundant, diffuse and tending to become diffuse-in-aggregates, individual strands of large cross-section, distinct × 10; grain usually straight; wood light yellow or yellow-brown, with a pink, rose, or orange tinge; sapwood only distinct in seasoned timber.

Durability: not resistant to decay in exposed conditions, but durable under cover; not susceptible to powder-post beetle attack; normally free from pin-worm infestation; sapwood susceptible to sap-stain fungal discoloration.

Strength Properties: not so strong in bending or compression, and notably less resistant to impact-bending loads, than heavy red meranti and medium-heavy keruing, but more resistant in shear; harder than heavy red meranti, *vide* Malayan Forest Records No. 15, p. 319.

Working Qualities: works very well with hand and machine tools, with very little dulling of cutting edges; rather more difficult to plane to a smooth finish; liable in thin dimensions to split in nailing, but thicker timber holds nails well.

Seasoning: dries fairly rapidly, but care is necessary to avoid end splits and cupping; shrinkage figures from 'green' to 15 per cent. M/C: 2·7 per cent. radially and 8 per cent. tangentially.

Uses: in Malaysia as a constructional timber; also popular in India; unlikely to make any special appeal to the U.K. market in normal times because the local demand raises prices to too high a level for a timber of this type.

Supplies: abundant in the fresh-water swamps; a large tree, usually remarkably free from defects, but occasional large heart shakes filled with an opaque, curd-like white substance occur; imported in small quantities in plank and scantling sizes.

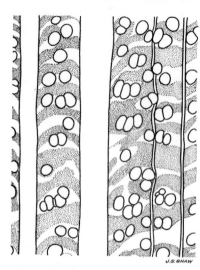

J.S.SHAW

'SILKY OAK'

Grevillea robusta A. Cunn.

Family: Proteaceae.

Other Names: 'silky oak' of Queensland is now provided by *Cardwellia sublimis* F. Muell., and the commercial supplies of *G. robusta* are from plantations in East Africa, where it has been extensively grown as a shade tree.

Weight: 580 kg/m³.

Spot Characters: pink-brown colour, very broad, high rays, vessels in tangential multiples, parenchyma in broad metatracheal bands characterize 'silky oak'; vessels medium-sized, in tangential multiples, appearing suspended from the bands of parenchyma; parenchyma abundant, distinct to naked eye, as broad conspicuous metatracheal bands, in scallops between the broad rays, and some paratracheal parenchyma; rays of two distinct sizes; broad, high rays conspicuous on all surfaces, and producing prominent 'silver figure' in quarter-sawn material, and fine rays, visible only with a lens on end surface; wood pink-brown, with occasional darker streaks; sapwood distinct; traumatic canals, appearing as dark-coloured lines on all surfaces, are not infrequent.

Durability: experimental data are not available, but the sapwood is known to be particularly susceptible to powder-post (*Lyctus*) beetle attack, and to sap-stain fungal infection; where used outdoors for gates, etc., not in contact with the ground, it should be at least as durable as softwoods in similar circumstances.

Strength Properties: no data are available.

Working Qualities: works well with hand and machine tools, without excessive dulling of cutting edges; clean-cut mortices are obtained, and the wood planes to a smooth finish; it takes nails, screws and glue satisfactorily.

Seasoning: no experimental data are available; it appears to air-dry rather slowly, retaining a damp feel for a long time; more prone to cupping than checking in thin dimensions; an air-dried 25 mm flooring strip and a short length of scantling size, stored for 9 months behind a radiator, kept its shape well.

Uses: formerly being the commercial source of 'silky oak' from Queensland, the wood was essentially a furniture and cabinet wood; supplies are now only available from East African plantation-grown trees, which are not of great size, and yield too small a quantity of wide, quarter-sawn material with silver figure for it to be suitable as a decorative wood; used as a softwood substitute for joists, t. & g. flooring, gates and outdoor, all-timber buildings.

Supplies: supplies are limited, but shipments in the early post-war years were more regular than supplies of most East African timbers; imported in board, plank and scantling sizes.

PART III

The Properties of Wood

CHAPTER 7

Moisture in Wood

DETERMINATION OF MOISTURE CONTENT

THE timber of living trees and freshly felled logs contains a large amount of water, which often constitutes a greater proportion by weight than the solid material itself. The water has a profound influence on the properties of wood, affecting its weight, strength, shrinkage, and liability to attack by some insects and by fungi that cause stain or even decay.

Since the properties of timber depend so much on the amount of moisture it contains it is frequently necessary to know the exact **moisture content** of a particular sample, *i.e.*, how much water is present in the sample. The amount of moisture present in converted wood varies appreciably in different circumstances, but the dry weight of wood substance in a given sample is constant. Hence, it is usual to express the variable – moisture content – as a percentage of the constant – dry weight of the sample. The ratio is simply:

$$\frac{\text{Weight (or volume) of water present}}{\text{Dry weight of wood substance}} \times 100$$

There are several ways of determining the moisture content of wood, but by far the most satisfactory for most purposes is the **oven-dry method** described below.

Oven-dry method. In this method, the moisture content of the moment is obtained as follows:

$$\frac{\text{Initial wt of sample – dry wt of sample}}{\text{Dry weight of sample}} \times 100$$

The initial weight of a sample is the actual weight at the time of test, and the dry weight is the weight of the sample after the moisture has been expelled.

Apparatus. The apparatus required is a simple balance, and some form of drying oven that can be maintained at a more or less constant temperature. The type of balance suitable in commercial practice is one of the self-registering type, weighing to an accuracy of 0·5 gram and up to a maximum of about 500 grams. The metric system is recommended, as it simplifies the calculations; otherwise Imperial units would do equally well. Various types of drying ovens are on the market, but all are essentially similar in principle, in spite of makers' claims to special advantages. The essential points to look for are: (1) the capacity of maintaining an even temperature between 95° and 105° C. (205°-220° F.); (2) good ventilation (a sample heated in a closed box would, of course, not lose moisture once the surrounding air space had become saturated); and (3) source of heat. For heating, electricity has the advantage of simplicity, but gas, oil, or steam may equally well be used; it is entirely a matter of convenience. Designs of simple ovens, one electrically and the other steam heated, are given, and their method of construction described, in leaflets issued by the Forest Products Research Laboratory, Princes Risborough. The one illustrated in Plate 36, manufactured by Messrs W. and J. George & Becker Ltd, is to the Princes Risborough Laboratory's design.

Sampling. Provided attention is paid to the essential points enumerated above, the method of selecting test blocks is of greater importance than any particular features of the apparatus used for weighing and drying the samples. It is essential that the sample shall be representative, not only of the board or plank from which it is cut, but also of the parcel as a whole. For example, the moisture content of the sapwood of some species varies appreciably from that of the heartwood, so that the proportion of sapwood in the test blocks should be similar to that in the whole parcel. In practice, the moisture content of a large quantity of timber should not be based on a single sample, but on two or three selected at random. When taking sample boards from a stack of timber, outside ones should be avoided, as these often differ appreciably in moisture content from those in the interior. Having selected the samples, off-cuts of the full cross section of each sample should be taken not less than 22·5–30 cm (preferably 60 cm) from either end, and 0·94 to 1·56 cm along the grain. Larger pieces take much longer to dry, and are not necessary. Each sample should be reasonably free from knots:

gram (handwritten)

although their actual influence on moisture-content determinations is not known, they are not typical of the wood as a whole; in some softwoods, for instance, they may have a high resin content, and the resin may run out during oven drying.

Procedure. Once the test block has been cut out, rapidity of weighing is essential to minimize the chance of the sample picking up, or losing, moisture between the time of cutting and that of weighing, since small moisture losses or gains during this interval will introduce appreciable errors in the calculated moisture content, and such losses or gains are much more rapid in the small-sized samples used for moisture-content determinations than they are in the boards or planks from which the samples are cut. After the initial weighing the samples should be transferred to the drying oven. This should be run at a temperature of 60° C. for the first few hours to prevent the moisture in the centre of the samples from being sealed in, as a result of case-hardening.[1] The temperature can be raised to 102° afterwards, and the samples left in the oven overnight. They should be re-weighed first thing on the following morning, and again some hours later. Rapidity of weighing is of particular importance when the samples are oven dry, as, in this state, they will absorb moisture in a very short space of time. If there is no appreciable difference between the last two weighings, the lower may be taken as the oven-dry weight. If, however, the second weighing shows an appreciable drop, drying must be continued for a further period. Drying in an oven does not expel all the moisture, but the small discrepancy – the last one per cent. or so – is not of practical importance.

Example. A test block with an initial weight of 88·7 grams weighed 76·7 grams twenty-four hours later, and 76·6 grams four hours later still. Accepting the second of these re-weighings as the dry weight, the moisture content of the sample was:

$$\frac{88 \cdot 7 - 76 \cdot 6}{76 \cdot 6} \times 100, \text{ or } \frac{12 \cdot 1}{76 \cdot 6} \times 100 = 15 \cdot 8 \text{ per cent.}$$

Single calculations of moisture content may be sufficient for determining whether a stack of timber is suitable for a general purpose, *e.g.*, indoor or outside use, but occasions will often arise when it is desirable to determine moisture contents of the same

[1] See page 251 for a definition of case-hardening.

material from time to time over a period. This is the case, for example, when studying the progress of seasoning in a stack. Such work can be greatly simplified if **test boards** are so arranged in the stack that they can be withdrawn without disturbing the stack on each occasion. This can be achieved by notching the **stickers** (the slats of wood used for separating the layers of timber in a stack) above the test boards, so that the latter are free to move. Having estimated the moisture content of the test boards from samples cut in the normal way at the time of the first test, the moisture contents can be calculated on subsequent occasions from re-weighing of the boards alone.

Example. The initial weight of a test board is 23 kg, and its moisture content, as determined from samples, is 27·5 per cent. Let its dry weight equal 'x', then $x + \dfrac{27·5x}{100} = 23$, and $x = 18$ kg. After a lapse of a week the board is re-weighed and found to weigh 21·9 kg. The dry weight is known from the previous calculation, so that the moisture content of the moment can be determined by substitution in the formula

$$\frac{\text{Present weight} - \text{dry weight}}{\text{Dry weight}} \times 100$$

in this case

$$\frac{21·9 - 18·0}{18·0} \times 100 = 21·7 \text{ per cent}$$

The process can be repeated as often as is required during the seasoning period.

Distillation method. The presence of oils or resins introduces an error in the calculated moisture content by the oven dry method because these substances, being volatile, are lost in the process of drying, and are counted as moisture, so that the calculated figures are too high. This should be borne in mind when dealing with such timbers as gurjun, apitong, keruing, and resinous samples of long-leaf pitch pine or timber impregnated with an oil preservative such as creosote. Moisture-content determinations in such circumstances can be accurately determined by the distillation method.

In this method the sample takes the form of about 50 grams of chips, borings, or sawdust. The sample is placed in a flask containing a water-insoluble oil of low density; xylol is the one most often

used. The apparatus employed is illustrated in Fig. 15. It consists
of a flask, with suitable heating arrangements, a reflux condenser
discharging into a graduated trap which collects the condensed water
from the wood and returns the solvent oil to the flask. Distillation
is continued until no more water collects in the trap. The volume

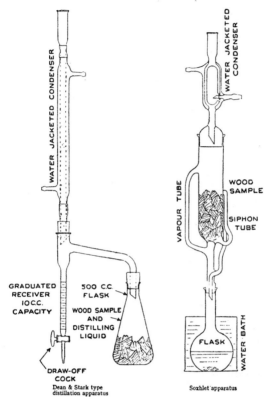

FIG. 15. Distillation method of moisture-content determination:
alternative types of apparatus

By courtesy of the Director. F.P.R.L., Princes Risborough

of water collected is read direct in cubic centimetres. As one cubic
centimetre of water weighs one gram, the weight of water in the
sample is obtained automatically. With samples containing only
natural oils or resins, the moisture content is arrived at simply as follows:

$$\frac{\text{Wt in grams of water collected}}{\text{Initial wt of the wood sample} - \text{Wt in grams of water collected}} \times 100$$

= per cent. moisture content of sample

Samples of impregnated wood contain, in addition to wood substance and water, an unknown weight of preservative. This weight must next be determined; this is done by extraction of the preservative, with a suitable solvent, from the liquid remaining in the flask at the end of the initial distillation process. The moisture content of the treated wood can then be calculated:

$$\frac{\text{Wt in grams of water collected}}{x - (\text{Wt of preservative} + \text{Wt in grams of water collected})} \times 100$$

= per cent. moisture content of sample

where x is the initial weight of wood sample.

The distillation method probably gives more accurate results than the oven-drying method for any wood, but the delicate apparatus required, and the risk from fire that heating xylol entails, render it suitable for use only in properly equipped laboratories. Such laboratories are not normally to be found, nor are they required, in ordinary commercial layouts. Even with timbers such as gurjun, the oven-drying method gives moisture-content values within about 1·5 per cent. of the true values, which is accurate enough for most purposes for which such timber is used. Moreover, it is easy for errors to creep in with the distillation method: chips of wood, borings, and sawdust, unless carefully and rapidly weighed after collection, may quickly lose moisture, so that the calculated moisture content is below that of the piece from which the sample was taken. It is, however, necessary to use chips or their equivalent to ensure that all the contained moisture is completely and rapidly extracted. The small size of the sample makes complete extraction of water essential; even minute quantities retained in the wood would appreciably magnify the errors in the calculated moisture content. A sample of 50 grams has been found convenient to work with; smaller ones might magnify errors unduly, and larger ones involve inconveniently cumbersome apparatus. Besides accuracy, the distillation method has the advantage of rapidity: distillation should take only three or four hours.

Moisture meters. The moisture content of wood may be determined indirectly by measuring some other property that varies proportionately with changes in moisture content. For example, the electrical resistance of wood is an index of its moisture content.

Fortunately, the range over which a close relationship exists is that from about 6, up to 24, per cent. moisture content, and this fact makes the measurement of electrical resistance or conduction by means of **electrical moisture meters** practicable.

Electrical moisture meters have been developed beyond the laboratory stage: they have been in commercial use for many years. Two general types of instruments are made: one evaluates moisture content by measuring electrical resistance, and the other determines the electrical capacitance of wood.

The resistance type of machine has electrodes in the form of spikes, which are driven into the wood. The electrical conductivity of the wood is used as an index of moisture content, which, in some instruments, is directly read from a scale. In the capacity type of instrument surface plates are clamped on opposite faces of the timber and the capacitance between the plates is balanced against a variable condenser.

Instruments can be obtained that will indicate by the frequency of flashing of a neon tube whether the wood is drier or wetter than a predetermined standard; or, alternatively, the moisture content may be indicated directly on a scale.

Several electrical moisture meters of the resistance type are available on the market; an instrument of British manufacture is illustrated in Plate 37, fig. 1. Any equipment capable of measuring the very high resistances involved, when fitted with a suitable electrode assembly for making contact with the wood, and calibrated, can be used within the limitations of this method of moisture-content determination. For example, an ordinary insulation testing set, capable of reading up to 1000 megohms, can be adapted to read moisture contents from about 11 per cent. to fibre saturation point.[1]

Resistance varies with the temperature, increasing with a falling temperature, and decreasing with a rising temperature. Moreover, the resistance for any given moisture content and temperature is not constant for different species. Hence, corrections have to be made for temperature and species. This is fortunately not difficult to do with sufficient accuracy for practical purposes; instruments are usually supplied by the manufacturers with the necessary correction data.

Besides the question of cost, and the fact that a different scale, or

[1] For definition of fibre saturation point see page 131.

correction factor, has to be used for each species, resistance moisture meters have other limitations. In the first place, the figure read off the scale is the moisture content of the piece to the depth of penetration of the electrodes, or, with contact instruments, the surface of the piece. In the early stages of drying, the surface moisture content of a board or plank is likely to be very different from that of the interior, and with thick timbers this is likely to be so always, so that the meter reading gives too low a figure for the piece as a whole. On the other hand, a surface film of water left by a shower of rain will cause the reading to be too high. In the second place, the instruments are delicate, and require careful handling and considerable technical knowledge to maintain them in an efficient condition. Moisture meters, however, have distinct commercial possibilities: they give results almost instantaneously, and when appropriately used will give readings within one to two per cent. of the true values. Moreover, they provide the only practicable means of determining the moisture content of finished woodwork *in situ* without damaging the wood. Greater accuracy can be obtained by using longer than normal electrodes, *e.g.*, 12 to 25 mm long, but, so equipped, the moisture meter no longer provides instantaneous readings, and the relatively large diameter of the longer electrodes may be an objection—the diameter of a 25 mm long electrode leaves a far from inconspicuous hole. Moisture meters are particularly suitable where comparative rather than absolute figures are required.

Capacitance-type moisture meters are not restricted to the moisture-content range of resistance instruments, because the electrical capacitance of wood varies directly with the amount of moisture it contains, from the green to the oven-dry state. Further, the effect of temperature is so small as to be of no importance for ordinary purposes, and different species do not introduce variations in the measured moisture contents. The apparent advantages of the capacitance-type meters are, however, more than offset by this important disadvantage: the instruments measure the weight of water in wood, which can only be converted to a moisture content if the specific gravity of the wood is known. In practice, capacitance-type meters are calibrated on the average specific gravity of a species, which may well not be the actual specific gravity of the piece being tested for moisture content. Contact with the wood is made by various types of suitably insulated condenser plates; in one form two plates are placed on opposite sides of the piece of wood under test, and in another four quadrant-shaped plates, assembled in the form of a flat circular disk several centimetres in diameter, are pressed against one surface only of the wood; the latter arrangement is used with capacitance-type meters in commercial production in America.

The Occurrence of Water in Timber

It has been mentioned that the cells of wood are hollow, and that in the living tree many of the cell cavities are filled with water. Moreover, the solid material of which the cell walls are composed is itself saturated with water, much in the same way as seaweed that has just been uncovered by the tide. As might be expected, the 'free' water in the cell cavities has very little influence on the properties of wood other than its weight; it may be compared with water in a bottle. If the 'free' water were removed from the cavities of the wood the properties of the timber would not be greatly changed, any more than are those of a bottle emptied of its contents. It is actually impossible to remove all the water in the cell cavities without removing some from the cell walls, bu., as a starting-point of a discussion, it is convenient to imagine the theoretical state where the cavities are empty and the cell walls are saturated; this state is known as the **fibre saturation point.**

In describing the cell-wall structure of woody tissue it was explained that the walls consist of several concentric layers, and that the layers are composed of fibrils, which were pictured as minute, needle-like units. The water in the cell walls appears to be in films between these units, more or less like mortar in a brick wall, but also inside the fibrils themselves, in some form of physico-chemical composition with the molecular structure of cell wall substance. There is a limit to the thickness of the films of the water, and consequently to the amount of water held in the cell walls. In most timbers the walls can hold about 25 to 30 per cent. of their dry weight; when this amount is present the wood is at fibre saturation point.

Reverting to the analogy between seaweed and wood, it may be recalled that the former is sometimes used as a weather guide because it is **hygroscopic**; that is, it is able to absorb moisture from a humid or damp atmosphere, and allows water to evaporate when the atmosphere is dry. Wood behaves in a similar way, in that there is a constant interchange of water between wood and air depending on which is the wetter. When 'green' wood is exposed to dry air it loses water to that air. Free water in the cell cavities is first given up, and, when this has been removed, water in the cell walls is gradually absorbed by the surrounding air. The cell-wall water is conveniently termed **hygroscopic moisture**. With loss

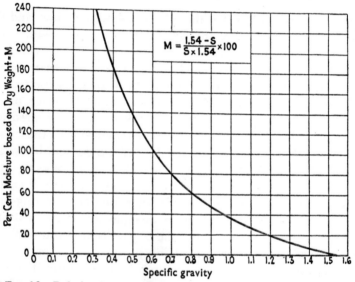

$$M = \frac{1.54 - S}{S \times 1.54} \times 100$$

Fig. 16.   Relation between specific gravity and theoretical maximum moisture content. For explanation see text

By courtesy of A. Koehler, Esq.

of water from the cell walls, wood shrinks, and, like seaweed, becomes stiffer and harder. If dry wood is placed in a humid atmosphere it absorbs water, and swells, and as it does so it becomes less rigid.

The amount of free water that a piece of wood can retain is governed by the volume of the cell cavities and intercellular spaces. The volume of these depends on the amount of cell-wall substance in any given volume of wood: the greater the proportion of wall substance, the greater the amount of water that can be absorbed by it, but the smaller the proportion of free water, and also of the total amount of water in a given piece of saturated wood. This relation is illustrated theoretically by the curve in Fig. 16. The graph is only approximate, because infiltrates of different specific gravity from wood substance influence the position. The presence of infiltrates of higher specific gravity than wood substance causes the percentage of water read from the graph to be too low and the presence of those of lower specific gravity causes the percentage to be too high.

'MOVEMENT' IN WOOD

The tendency of wood to shrink or swell with changes in the moisture content of the atmosphere is a factor inseparable from the material. This characteristic behaviour of timber is popularly called **'working'** or **'movement'**; it cannot be eliminated by any particular method of seasoning or storage, although the deleterious results can be minimized by taking certain precautions. Certain chemical treatments modify the hygroscopic properties of wood, thereby permanently reducing movement in service of such treated timber, but the treatments are decidedly costly and hardly applicable to the everyday use of timber. These measures are discussed later in this chapter.

It is a matter of general observation that timber shrinks hardly at all along the grain: in drying from the green to the oven-dry state shrinkage in this direction is only a few tenths of 1 per cent. In the radial and tangential directions, however, movement is appreciable: in drying from the green to the oven-dry condition shrinkage in the radial direction may amount to 7 per cent., and in the tangential to as much as 14 per cent.; average figures for several timbers are as high as 4 and 8 per cent. respectively. The movement of 'seasoned' timber occurring under normal atmospheric variations in temperatures and humidities is, of course, much less than occurs over the wide range from the green to the oven-dry state.

Tangential shrinkage is usually about twice as great as the radial, although in some species it may be as much as eight times as great. It will be appreciated that this difference is a matter of importance in the utilization of timber, and the **differential shrinkage**, as the ratio of tangential to radial shrinkage is called, may determine the suitability of a timber for a particular purpose. Advantage is commonly taken of the fact that quarter-sawn material of most timbers shrinks appreciably less than flat-sawn, *e.g.*, for flooring. The exceptionally low differential shrinkage of Central American mahogany, coupled with the low total shrinkage of this wood, makes it particularly suitable for certain special purposes, *e.g.*, cabinet work and backing for engravers' plates. Differential shrinkage is important because it is related to distortion in drying, but low total shrinkage, and, particularly, a low absolute difference between radial and tangential shrinkage, largely determine the amount of movement in seasoned timber, and may in some circumstances be the more important factor, *e.g.*, in joinery and roofing shingles. Exceptionally low shrinkage characterizes certain softwoods, which have in consequence established reputations as joinery timbers, *e.g.*, yellow pine and western red cedar. Total shrinkage in hardwoods is typically greater than in softwoods, but an outstanding exception is teak, the total shrinkage of which is little higher than in the best softwoods. This fact, although the least stressed when the merits of teak are discussed, is one of the most pertinent reasons for the pre-eminent position of teak for so many exacting purposes.

Even more remarkable than teak, or Central American mahogany, is the East African muninga (*Pterocarpus angolensis* D.C.): the shrinkage of this wood is the lowest of any timber so far tested, being only 0·8 per cent. radially, and 0·9 per cent. tangentially, in drying from the 'green' state to 12 per cent. moisture content. Other timbers with a small overall shrinkage, and a low ratio of radial to tangential shrinkage, and therefore comparable in stability with teak, are afzelia (*Afzelia* spp.), iroko, and afrormosia (*Pericopsis elata* van Meeuwen-syn. *Afrormosia elata* Harms).

The fibrillar structure of the cell wall helps to explain why longitudinal shrinkage is negligible, but transverse shrinkage is appreciable. Cell-wall moisture is held between the fibrils, and between the micellae that compose them, and removal of hygroscopic moisture results in these units packing closer together, causing appreciable contraction transversely, but little change in their length. It has already been mentioned that the fibrils are arranged more or less parallel with the longitudinal axis of the cell; they are not completely longitudinal, and any deviation, however slight, gives rise to shrinkage in the longitudinal direction as may be seen from Fig. 17.

The foregoing is a simplification of the facts, imperfectly though they are understood to date. The picture as presented does, however, assist in explaining certain readily observable phenomena: a truly rectangular scantling of green wood may become distorted in seasoning, a flat-sawn board is liable to cup, but a quarter-sawn board will usually remain flat. Appreciable stresses develop in the course of drying, intensifying distortion.

PERMANENT SET

But for differential shrinkage there would be no distortion of converted timber in drying: shrinkage would be the same in all directions, and each piece would merely become proportionately smaller. On the other hand, the true geometrical shapes that

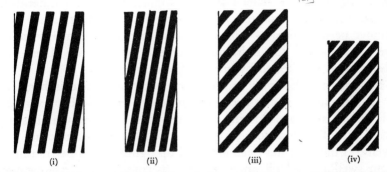

| (i) | (ii) | (iii) | (iv) |

FIG. 17. Fibrillar structure and its relation to shrinkage. (i) Diagrammatic representation of fibrillar structure in green timber: the black lines represent the fibrils and the white the films of water; (ii) the same cell when seasoned: the black lines are slightly narrower as a result of small water-losses from the films surrounding the fusiform bodies, but the white lines are appreciably narrower because of the much larger water-losses from the films surrounding the fibrils (note that the reduction in length of the cell is much less than the reduction in width); (iii) and (iv) diagrammatic representation of fibrillar structure in green and air-dry timbers, in which the pitch of the fibrils is less than in normal wood (note that the reduction in length of the cell as the wall dries out is much greater than in normal wood)

differential shrinkage calls for are usually not attained in practice, because of the influence of drying stresses, and the peculiar property of wood substance of becoming **set**.

When wood dries under a compressive stress it assumes a **compression set**; that is, its final dimensions are less than they would be were it dried under stress-free conditions. The final dimensions are permanent for the particular combination of humidity and temperature conditions of the atmosphere of the moment. Hence the term **permanent set**.   Subsequent changes in the prevailing atmospheric conditions will be accompanied by shrinkage or swelling of permanently set wood, as in stress-free wood, but the starting-point of the dimensional changes is that of the set condition.   Because converted timber rarely dries uniformly throughout, stress-set timber is almost the rule rather than the exception, and this upsets the predictable final dimensions that differential shrinkage alone requires.   Wood can also dry under a tensile stress, when the final dimensions are greater than they would be under stress-free conditions, *i.e.*, **tension set** has been induced.

Ordinary re-wetting, or further drying, of compression or tension set wood does not relieve the set condition; high temperature treatments in a saturated atmosphere, *i.e.*, steaming, alone will remove permanent set.   This is because the stresses that induce set are appreciable: the individual elements or cells are themselves distorted, as is revealed by microscopic examination.

Set does not arise only in the course of drying green timber, that is, in seasoning; it can occur in timber in service.   For example, adequately seasoned flooring, laid in a new building that has not dried out fully, tends to pick up moisture from its surroundings and swell.   If the floor has been clamped up tightly it will be restrained from swelling the full amount, and, in consequence, compression set results.   As the building dries so will the flooring, but, starting from the set state, the flooring ultimately reaches equilibrium (its initially intended state) with smaller dimensions than its initial dimensions.   For similar reasons, wood adequately seasoned for the service conditions to which it will be exposed, and without provision for 'movement', is liable to shrink to a noticeable extent over a period of years.   This explains why provision for 'movement' should always be made in floors, for example, whatever precautions are taken to season wood adequately in the first place.   Compression set is also induced in ladder rungs or axe handles if they are allowed to become wetted: on drying, previously tight-fitting rungs or handles are found to be loose (see also Plate 37, fig. 2).

Variation in the amount of shrinkage (or swelling) in timbers of different species is not explained by the foregoing discussion.   The small shrinkage in the radial direction has been attributed to the restraining influence of the rays, and recent critical studies have added further explanation by reference to the microstructure of the

PLATE 36

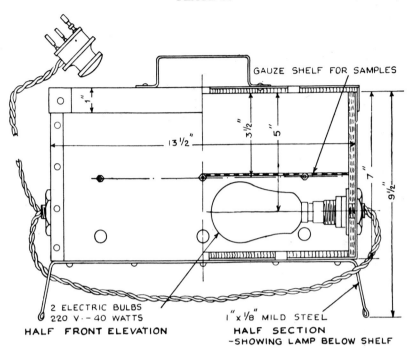

GAUZE SHELF FOR SAMPLES

13 1/2"

3 1/2"   5"   7"   9 1/2"

1"

2 ELECTRIC BULBS
220 V. - 40 WATTS

HALF FRONT ELEVATION

1" x 1/8" MILD STEEL

HALF SECTION
-SHOWING LAMP BELOW SHELF

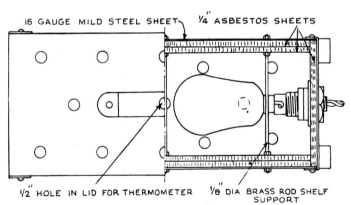

16 GAUGE MILD STEEL SHEET   1/4" ASBESTOS SHEETS

1/2" HOLE IN LID FOR THERMOMETER   1/8" DIA BRASS ROD SHELF
SUPPORT

ALL VENTILATING HOLES 1/2" DIA.

HALF PLAN
(WITH LID)

HALF PLAN
(LID & SHELF REMOVED)

A simple constant-temperature oven for drying samples used
in moisture-content determinations

By courtesy of the Director, F.P.R.L., Princes Risborough

PLATE 37

FIG. 1. Moisture-in-timber meter: resistance type

By courtesy of Messrs Marconi Instruments Ltd.

FIG. 2. Compression set induced experimentally. The left-hand 25 mm cube of mahogany was subjected to 15 wetting and drying cycles. Each time, prior to soaking, the cube was lightly clamped so that it could not expand when picking up moisture. After soaking the cube was dried each time to 12 per cent. moisture content, re-clamped, and re-wetted. At the end of 15 cycles the cube was reduced to the size illustrated on the right.

By courtesy of The Airscrew Company and Jicwood Ltd, Weybridge, Surrey

cell wall. Ritter and Mitchell[1] have sought an explanation in the fibrillar structure of the wall. Pits are more numerous in the radial walls of fibres than in the tangential walls, and pits must cause a displacement of the fibrils, much as the elements of the main stem are displaced by knots. In consequence, the influence of longitudinal alignment of the molecular structure making up the fibrils comes into play, and because this component is introduced more frequently in the radial than in the tangential plane, by reason of the more numerous pits, total shrinkage in drying is less in the radial than in the tangential direction.

ESTIMATION OF MOISTURE CONTENT FROM MEASUREMENT OF SHRINKAGE

The methods of determining moisture contents that have already been described do not demand especially elaborate apparatus, nor appreciable technical skill, but they are suitable only where moisture-content determinations are a routine practice, as they should be in timber yards, joinery works, and other wood-using factories generally. A simple test is available for checking the moisture content of timber without the use of special apparatus; it is based on measurement of shrinkage, and is useful in determining the suitability, as far as seasoning is concerned, of timber on a job.

The procedure is to select a representative scantling, and one that has not been lying on top of the pile, and to cut off a cross section, about 12 mm along the grain, some 600 mm from one end. Measure the width and thickness of the sample accurately to the nearest 0·4 mm. To ensure re-measurement in precisely the same line, it is advisable to put small dots at the points of measurement. Expose the sample in a well-ventilated, inhabited room for a few days, i.e., 4 to 7 days, and then re-measure. In this period the sample should reach equilibrium with its surroundings, and the decrease in width and thickness will represent the amount of shrinkage that will occur when the timber is put into service. If this test is carried out in summer, greater shrinkage than that actually measured may be expected to occur in winter, because domestic heating dries the air more than do summer temperatures. It is possible to obtain an idea

[1] Ritter, G. J., and Mitchell, R. L., *Paper Trade Journ.*, 108: No. 6, 33 (1939).

of the amount of shrinkage to be expected without the necessity for measurement, by matching the test sample from time to time with the identical piece of timber from which it was cut.   Measurement is more satisfactory, however, and more convenient when the timber to be tested is some distance from the office where the experiment is carried out.

This test gives information as to the state of the timber at the time of inspection.   If there is a considerable delay between the time of test, and the time when the timber is put into use, a loss, or increase, in moisture may occur in the stack in the interval, *i.e.*, further seasoning may occur, or, if the timber was kiln-seasoned stock, it may become wetter.   Little change is likely to take place in a period of 3 or 4 weeks if the timber is close piled and covered with a tarpaulin, but if the timber is properly stacked and roofed over, some drying (down to 15 or 16 per cent.) of partially seasoned stock may occur during such a period in the summer months.   Little or no drying, even under the most favourable conditions, is likely to occur out-of-doors in winter.   Timber of low moisture content will absorb moisture while awaiting fixing: up to about 15 to 16 per cent. in summer, and up to 18 to 20 per cent. in winter, if protected from the weather, but if exposed to rain the figures quoted may be exceeded in a short space of time.

Variation in Moisture Content of Green Timber

The amount of moisture in timber of living trees and newly felled logs is primarily a question of species; in some it is only about 40 per cent., and in others it may exceed 200 per cent. of the dry weight. In Douglas fir, for example, the average moisture content of newly felled trees is about 40 per cent., but in the American chestnut it exceeds 120 per cent.   In most species there is usually a marked difference in the moisture content of sapwood and heartwood; particularly is this the case with softwoods.   In long-leaf pitch pine, for example, the moisture content of the sapwood exceeds 100 per cent., and the heartwood ranges from 30 to 40 per cent.   Moisture content may also vary with height in the tree: butt logs of sequoia and western red cedar often sink in water, although the upper logs float.   In species with a marked difference between the moisture content of sapwood and heartwood the position may, however, be reversed, because the upper logs contain a higher percentage of sapwood of high moisture content.

There is no evidence in support of the widely held opinion that timber felled in winter, when the sap is said to be 'down', is drier than that felled in summer, when the sap is said to be 'up'. In fact, such evidence as is available suggests that there is either no difference, or the moisture content is, if anything, somewhat higher in the winter months. In Germany figures have been collected for birch, poplar, and oak, showing that the moisture content of these timbers was lower between June and February than between February and June, the minimum occurring in June or July and the maximum in March, April, or May. Similar figures have been collected for several species in America, and in no case was the moisture content in winter found to be lower than that in summer. In the light of these figures the prejudice against summer felling cannot be sustained on the grounds usually advanced. There is some justification on scientific grounds, however, for preferring winter felling. Chief of these is the reduced activity of insects and fungi during the cool months, and the lower degrade in seasoning resulting from a slowing down of drying from the ends of logs. Moreover, the risk of damage to the bole is probably greater when felling heavy-crowned trees in full leaf. The prejudice no doubt arose, partly as a result of practical experience, which showed that winter felling gave better results, and partly because of economic necessity: agricultural activities absorbed the population in the summer months and left them free for work in the woods during winter.

VARIATION IN MOISTURE CONTENT OF SEASONED TIMBER

Considerable variation in moisture content occurs in different pieces of so-called 'air-dry' or 'seasoned' timber. The moisture content at any given moment will depend on the atmospheric conditions to which the timber is exposed, the stage reached in the drying process, the dimensions of the piece, and the species.

However prolonged the period, or favourable the conditions, drying does not continue indefinitely: a stage is reached when there is no further interchange of moisture between the wood and air; this may be referred to as a state of equilibrium. Any subsequent change in temperature, or increase or decrease in moisture in the air, however, upsets this balance, and there is a further exchange of moisture until a new state of equilibrium is reached. This inter-change occurs because the amount of available water is distributed between air and wood in certain definite proportions when equi-librium conditions are established; *i.e.,* both air and wood have affinities for water, and mutual adjustment is made when both are subjected to stable conditions of temperature and available water supplies for a sufficiently long period.

Much of the so-called 'seasoned' timber on the market today is little more than surface dry, and must be expected to shrink appre-ciably until it eventually comes into equilibrium with ordinary, atmospheric conditions. Timber in old houses and furniture, on the other hand, has already reached the equilibrium state and is, therefore, relatively stable. If, however, the atmospheric conditions are changed, as, for example, by the installation of central heating in a previously unit-heated house, the moisture equilibrium will be disturbed and noticeable movement will occur in the timber. A case is known of genuine Stuart pine panelling that had been in position for 300 years shrinking when re-erected in a centrally-heated room. More recently the installation of air-conditioning plants in buildings has introduced new seasoning problems, particularly in the tropics: if the moisture-equilibrium conditions are changed, and, in practice, they always are, shrinkage or swelling will occur in furniture, fittings, and joinery, previously in equilibrium with ordinary atmospheric conditions. In extreme circumstances, such as those prevailing in the tropics, the difference in the average equilibrium moisture contents of timber in ordinary and air-condi-tioned buildings may be as high as 5 per cent., *e.g.,* from around 15 per cent. to as low as 10 per cent., and furniture, etc., that was previously stable is bound to shrink appreciably, and probably develop serious splits, under the new conditions. If kiln-seasoning facilities are available it is a simple matter to dry timber to any required moisture content, when, provided the wood is not exposed to higher moisture-equilibrium conditions for too long a period in the course of manufacture, it will remain stable under the low moisture-equilibrium conditions to which it was kiln-dried. If timber below atmospheric moisture-equilibrium conditions is re-quired, and kiln-drying facilities are not available, the conditioned chamber in which the timber is to be installed should be used as a kiln. Before the wood required for furniture, fittings, and interior finish for the conditioned room is made up, it should be thoroughly air-dried and then re-stacked in the conditioned chamber, with the plant running, until the required low equilibrium moisture-content conditions are obtained, when shrinkage and splitting after manufacture will be avoided.

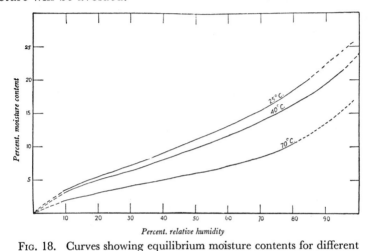

FIG. 18. Curves showing equilibrium moisture contents for different relative humidities at 25°, 40°, and 70° Centigrade

By courtesy of the Director, F.P.R.L., Princes Risborough

The moisture content of air-seasoned timber, at any given temperature and in equilibrium with its surroundings, bears a direct relation to the **relative humidity** of the atmosphere. Atmospheric air normally contains some water vapour: for every temperature there is a definite maximum amount of water vapour that air can hold. When the maximum amount is reached the air is said to be saturated. The relative humidity of the air is the actual amount of water vapour present at any time, expressed as a percentage of the maximum possible for that temperature. The curves in Fig. 18 show the relation between the moisture content of wood and the relative humidity of air during drying, at three different temperatures. From these curves the moisture content of timber in equilibrium with its surroundings can be read for any particular relative humidity, at temperatures of 25°, 40°, and 70° C. Separate curves for any other temperatures could be constructed; they would follow similar trends to those illustrated. The curves illustrated are average figures for several timbers. The curves for one timber, at these and other temperatures, although following the same general trend, would not be absolutely identical, because the equilibrium moisture contents of different timbers for any given set of atmospheric conditions are not the same. For example, the equilibrium moisture content of teak exposed to air at 40° C. and 70 per cent. relative humidity is just under 11 per cent., whereas oak exposed to the

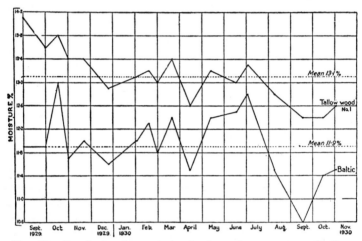

FIG. 19.   Graph showing variation in the moisture content of tallow wood and Baltic deal, *i.e.*, European redwood

By courtesy of M. B. Welch, Esq., and reproduced from
the Journ. Royal Soc. of New South Wales

same temperature/humidity conditions would reach equilibrium at a little over 14 per cent. moisture content. These two timbers provide examples of the extreme range in equilibrium moisture contents, probably with many more timbers approaching the figures fcr oak, rather than those for teak. In effect, teak, and other similar timbers, are less hygroscopic than most woods, which explains their greater stability in use.

The drying curves discussed in the previous paragraph do not represent the position when the process is reversed, and dry wood is gaining moisture from the atmosphere. In such circumstances the actual moisture content, when a new state of equilibrium is reached, is very slightly lower than would be the case if the timber had not previously been dried below this level.

A study of the variation in moisture content of seasoned timber with changes in the relative humidity over a period of fifteen months was made in Australia several years ago. Twenty-six samples, of eleven species, were selected for this purpose. The variations were found to agree closely in the different samples, although the actual equilibrium moisture contents for each species were different. Fig. 19 illustrates the amount of variation observed in two timbers over the period, and it is well to remember that variations of this order occur in seasoned timber in service in every locality.

Similar studies were made by the officers of the Seasoning Section of the Forest Products Research Laboratory, Princes Risborough, before the last war, and repeated subsequently to ascertain whether the restricted use of fuel then was influencing equilibrium moisture contents of timber in service. Test pieces of oak and Scots pine were used for the investigation, and were exposed in different rooms of four houses at Princes Risborough, and in a centrally-heated office at the laboratory. As was to be expected, maximum moisture contents occurred in the winter months in samples exposed in houses not centrally heated, and minimum moisture contents in the summer months, whereas in the centrally-heated office the periods of maximum and minimum moisture contents were reversed; Plate 39 illustrates the seasonal moisture content readings obtained from these investigations. It will be seen that the differences in moisture content at different periods were appreciable, and, further, that the changes in moisture content were not haphazard, but followed a remarkably similar pattern in all the experiments, irrespective of species of timber used or the precise locality where the samples were exposed, showing that prevailing atmospheric conditions were the over-riding factor in determining moisture movements. The figures for the oak samples were consistently about $\frac{1}{2}$ per cent. higher than those for Scots pine in identical circumstances.

Shrinkage and swelling result, of course, from changes in moisture content, but subsequent movement is minimized by selecting in the first place material of a moisture content midway in the range to be expected in service. A series of standards for different purposes, which it is recommended should be adopted in this country, has been worked out at the Princes Risborough Laboratory, Princes Risborough (Tables 2 and 3. This is also shown diagrammatically in Fig. 22.)

Similar figures to suit American conditions have been issued by the U.S. Department of Agriculture. For interior-finish woodwork in dwellings in most parts of the United States a moisture content of 5 to 10 per cent. is recommended, but in the damp southern coastal

TABLE 2. *Moisture-content specifications for constructional timbers in Great Britain*[1]

Timber for	Moisture content not to exceed
	Per cent.
General carpenter's work	25
High-class carpenter's work	20
General joinery work	15
Best joinery, block and strip flooring, panelling, and decorative work	9 to 12*
	10 to 14†

* For centrally-heated rooms and buildings.
† For rooms and buildings not centrally heated.

regions a range of 8 to 13 per cent. is suggested, and in the dry southern regions 4 to 9 per cent. For exterior sheeting, framing, siding, and exterior trim, the corresponding figures are 9 to 12 per cent. in the first two regions and 7 to 12 per cent. in the third.

TABLE 3. *Moisture-content specifications for furniture timbers in Great Britain*[2]

Timber	Environment			Average conditions
	Bedroom	Living-room	Office	
	Per cent.	Per cent.	Per cent.	Per cent.
Oak (American white)	13·6	12·8	12·5	13
Mahogany (Cent. American)	12·8	12·0	11·8	12·3
Scots pine	12·2	11·3	11·2	11·5

In the United Kingdom, and most regions of the United States, thoroughly air-dry timber, seasoned under the most favourable conditions, contains 15 to 18 per cent. of moisture; in the more humid tropics, *e.g.*, Malaysia, 14 to 18 per cent.; and in hot arid regions 8 to 12 per cent. or even less.

The rapidity with which adjustments to changes in the relative humidity of the atmosphere occur depends on the dimensions of the timber and on the difference between the initial, and equilibrium,

[1] Figures from Report of the F.P.R. Board for the year 1930, by courtesy of the Director, F.P.R.L., Princes Risborough.

[2] Figures from *The moisture content of wood with special reference to furniture manufacture*, Bull. No. 5, 1929, by courtesy of the Director, F.P.R.L., Princes Risborough.

moisture contents of wood. The adjustment is more rapid when the difference is large, and becomes extremely slow as equilibrium conditions are approached. But for this, movement in timber would be greater and more frequent than it is.

The equilibrium moisture contents cited in the tables are mean figures for timber in inhabited buildings, and a deviation of ±2 per cent. from these figures in individual pieces would be unlikely to have serious practical consequences. Special precautions, however, are necessary when installing wood-work in new buildings; it is not sufficient merely to select timber of the correct moisture content for subsequent conditions. Immediately on completion, the relative humidity of the air in new buildings is likely to be abnormally high, and timber in equilibrium with such atmospheric conditions would reach equilibrium at a moisture content of between 16 and 20 per cent., according to the season. If joinery and finishings of 10 to 12 per cent. moisture content are installed in these circumstances some swelling, which might give rise to buckling, is to be expected. On the other hand, it would not do to use timber of 16 to 20 per cent. moisture content, because appreciable shrinkage would occur later. Two alternative courses are to be recommended: either temporary heating should be installed, and the building dried out before the joinery and finishings are fixed, or fixing should be delayed for three to six months to give the building a reasonable opportunity of drying out of its own accord. The risk of introducing compression set in timber dried to the correct final moisture content before fixing must not be overlooked (see pages 135 and 136).

The practice of baking buildings in the early days of occupation is thoroughly unsound, and may result in considerable damage to the timber, because even the most carefully seasoned and fitted joinery will shrink and distort if suddenly exposed to much drier conditions than those for which it was prepared. The proper course is to employ no more heat than is necessary for occupational use, and this applies both before and after the joinery is fixed. Pre-baking is likely to give trouble with the carcassing timber and lath-plaster work. An alternative to using direct heat is to use de-humidifying equipment. Some of these units are available on hire, *e.g.*, Antimoist Building Dryer & Moisture Extractor, available from Conveyor and Equipment Co. Ltd, C & E Works, Leeds Street, N.18. This drying machine employs a chemical, calcium chloride, which attracts and absorbs moisture, the latter dripping into a bucket. The machines are of two types: non-heater and heater, the latter being a 2·6 kW unit of 110 or 240 volt capacity. In the early days of drying out a 'green' building, the results are quite dramatic, up to 1 gallon of water per hour being extracted from the structure. The intelligent use of such equipment greatly reduces shrinkage and swelling problems with second fixings and flooring, besides reducing the risk of mould growth problems if buildings have to be occupied very soon after completion.

Size, density, species, and initial moisture content of the timber, and rate of air circulation and its temperature, are, then, the factors that influence the rate of adjustment of the moisture in wood to the relative humidity of air. It may also be mentioned that flat-sawn material will lose moisture more rapidly than quarter-sawn, and sapwood more rapidly than heartwood.

Factors Affecting the Hygroscopicity of Wood

It has been explained that the hygroscopic nature of wood is responsible for causing variations in moisture content of timber following changes in the relative humidity of the atmosphere. Further, it has

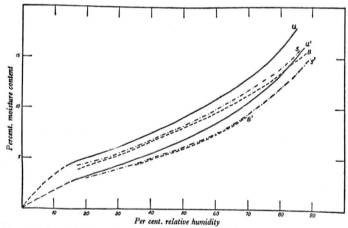

Fig. 20.   Curves showing effect of boiling and steaming on the hygroscopicity of wood. U = untreated, U¹ = the same, but absorbing moisture from the dry conditions; S = steamed, S¹ = steamed wood absorbing moisture from the dry state; B = boiled, B¹ = boiled wood absorbing moisture from the dry state

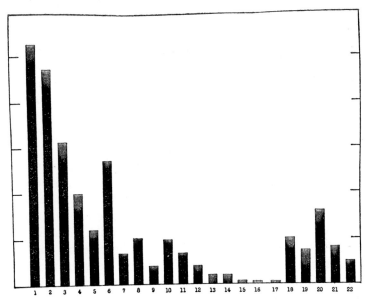

Fig. 21. Relative effectiveness of protective coatings against moisture absorption by dry wood exposed to a saturated atmosphere for 17 days*

1. Untreated wood.
2. Five coats linseed oil, 2 coats of wax.
3. Impregnation with paraffin and gasoline (vacuum and pressure).
4. Cellulose varnish (vacuum and pressure).
5. Three coats of cellulose varnish.
6. Filler and 3 coats of spar varnish. (Poorest of 43 tested.)
7. Filler and 3 coats of spar varnish. (Best of 43 tested.)
8. Filler and 2 coats of enamel (red-lead pigment) plus varnish.
9. Filler and 2 coats of enamel (white-lead pigment) plus varnish.
10. Filler and 3 coats of commercial enamel (average of 11 brands).
11. Filler and 3 coats of commercial enamel (best brand).
12. Filler and 3 coats of shellac and aluminium powder.
13. Five coats of bakelite plus 5 coats of varnish.
14. Metal leaf coatings: filler and shellac or varnish under-coat and varnish size and aluminium leaf and 2 coats of varnish shellac or enamel (average of all types).
15. Ditto (best type).
16. Sprayed with copper or aluminium and 3 coats of varnish.
17. Electroplated with copper.
18. Filler and 3 coats, brushed. (Average of 7 varnishes.)
19. Filler and 3 coats, dipped. (Average of the same 7 varnishes.)
20. Filler and 2 brush coats.
21. Filler and 6 brush coats.
22. Filler and 12 brush coats.

* From A. Koehler, *The structure, properties and uses of wood*, by courtesy of the author.

also been stressed that hygroscopicity cannot be eliminated: it can, however, be permanently reduced by certain treatments. A permanent reduction in the hygroscopic properties of wood is effected by high-temperature treatments, steaming, and boiling, and by certain chemical means. The higher the temperature, and the longer it is maintained, the more is hygroscopicity reduced, but too severe treatments may damage the timber. Boiling and steaming act similarly, as may be seen from the curves in Fig. 20. The solid lines are drying curves for untreated wood, and the dotted ones are those for timber that has been either steamed or boiled. Boiling will be seen to be more effective than steaming.

It is not known how the hygroscopic properties are permanently altered by boiling or steaming, and, in practice, it does not matter whether there is an actual reduction so long as subsequent movement is reduced. Paints, varnishes, and linseed oil, for example, are effective because they reduce the rate at which moisture can be absorbed by timber: as a result of their semi-permeable nature they offer resistance to the exchange of moisture between the air and wood. Fig. 21 illustrates the relative effectiveness of several different protective coatings. It will be seen that the selection of a suitable paint or varnish is an important matter, and can be very effective in reducing movement in timber. Of several different protective coatings tried, aluminium leaf, between a filler and shellac under-coat and two coats of varnish or enamel, was the most efficient. Varnishes, enamels, and paints containing aluminium powder, were less effective, but considerably more efficient than ordinary paints; and linseed oil and wax had little effect at all. Protective coatings in no way change the hygroscopic properties of wood.

Two other methods of controlling movement in wood have been attempted, namely the introduction of hygroscopic substances to change the equilibrium moisture content, and the introduction of materials to occupy the space that would normally be occupied by the sorbed moisture. It will be realized that the first method naturally involves the second in some degree. Creosote and hot paraffin have been found to reduce hygroscopicity, and claims have been advanced that ozone and ammonia reduce shrinkage, but the effectiveness of the former has been questioned. Some substances (*e.g.*, caustic soda) increase hygroscopicity.

The substances most commonly used in moisture control in the past have been common salt and sugar (Powell process). These are supposed to act partly by replacing the sorbed moisture, and partly by their influence on hygroscopicity. If it were possible to introduce sufficient quantities of such materials into wood, shrinkage (or working) would be eliminated; such heavy impregnations would, however, make the treated timber damp and unpleasant, and they would be difficult and expensive to achieve.

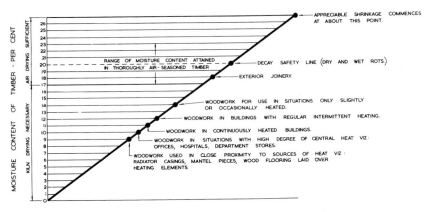

Fig. 22. Moisture content of timber in various environments.
The figures for different species vary, and the chart shows only average values
Crown Copyright Reserved

None of the methods for permanently reducing the hygroscopic properties of wood – as opposed to interposing a semi-impervious barrier between wood and the surrounding atmosphere – discussed in the foregoing paragraphs can be regarded as effective commercial methods for reducing shrinkage, and subsequent movement of wood, to negligible proportions. More recently, the effectiveness of *impregnating* wood with waxes or with synthetic resins has been investigated. Waxes are effective in reducing shrinking and swelling over short periods, and they impart some other desirable properties to the wood, but they do not confer immunity from movement in wood exposed to moisture over long periods. Impregnation with resins appears to reduce hygroscopicity permanently, the resin entering the fine structure of the cell wall, and being fixed there by heat-induced **polymerization**. Polymerization is a chemical process by which molecules of certain organic compounds are induced to unite to form giant molecules. Modern plastics are examples of polymers.

Impregnation with resin to the extent of 20 per cent. of the dry weight of the wood has been found to reduce hygroscopicity, as measured by subsequent shrinking and swelling, by 50 to 75 per cent. Certain other properties are also enhanced, *e.g.*, side hardness, measured by resistance to indentation, has been increased by 50 per cent., and moisture transfusion has been reduced to 10 per cent. of that of untreated wood. Moisture transfusion is a convenient term for describing the property of absorbing moisture and transmitting moisture. A low value for moisture transfusion is indicative of suitable materials for use as moisture barriers: untreated wood is not suitable, but wood impregnated with synthetic resins is. The use of synthetic resins in this connection may be said to be still in the experimental stage, but some commercial processes have been patented. The scope of such treatments is very extensive, and their present relatively high cost should not discourage their full investigation. A development from mild treatments is the use of synthetic-resin products, with very high pressures, to give the so-called compressed wood that found special uses under war-time conditions. With such treatments the properties of the resultant material are very different from normal wood, or even lightly treated timber. Hardness and strength properties, for example, are enormously increased; in fact, an entirely new product results that cannot be compared on any basis with natural wood, but the amount of chemicals used and the cost of the treatment are naturally very high.

The officers of the Wood Preservation Section of the Forest Products Research Laboratory, Princes Risborough, have experimented with the impregnation of wood with various natural oils and resins, in an endeavour to find cheaper products than the synthetic resins for modifying the hygroscopic properties of wood.[1] A treatment that has been developed from this work consists of impregnating wood with a solution of resin in paraffin wax. Such impregnated wood was found to machine remarkably well, and the rate of moisture absorption was reduced, thereby tending to stabilize the dimensions of the treated wood. The method lends itself to ordinary commercial pressure-impregnation processes, whereby heavy impregnations can be obtained with such permeable woods as beech,

---

[1] Forest Products Research, 1939–47: Report of the Director for 1946, pp. 40-44, H.M. Stationery Office, London, 1950.

PLATE 38

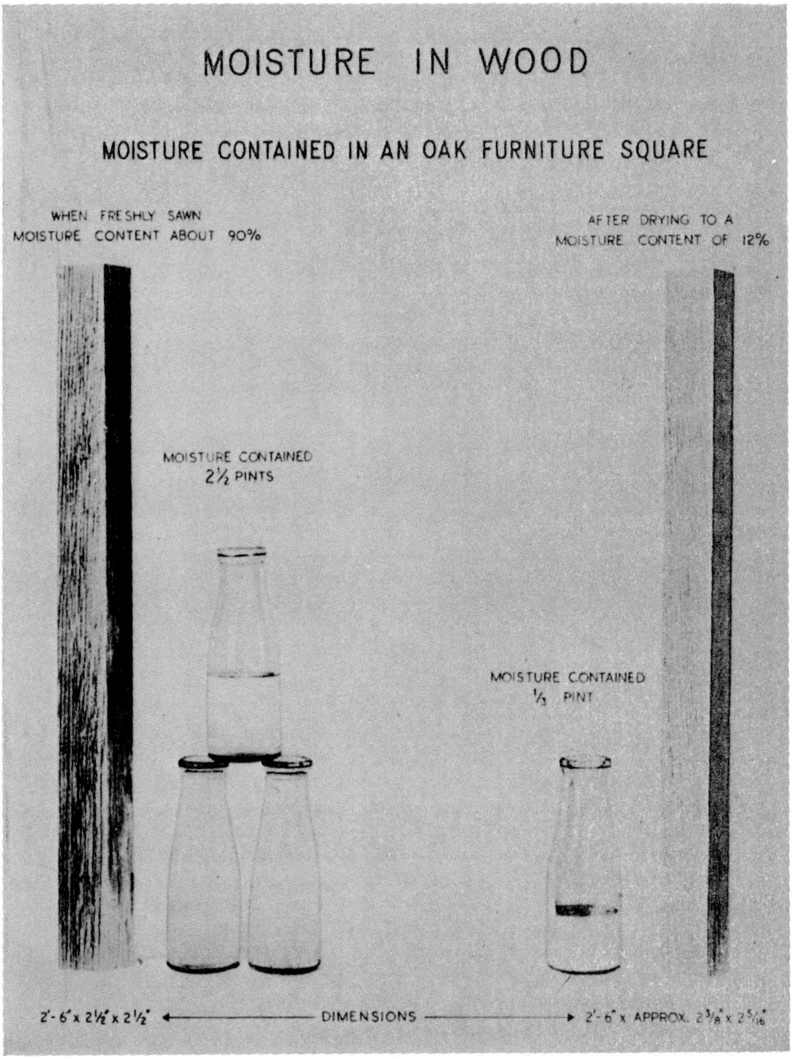

MOISTURE IN WOOD

MOISTURE CONTAINED IN AN OAK FURNITURE SQUARE

WHEN FRESHLY SAWN
MOISTURE CONTENT ABOUT 90%

AFTER DRYING TO A
MOISTURE CONTENT OF 12%

MOISTURE CONTAINED
2½ PINTS

MOISTURE CONTAINED
⅓ PINT

2'- 6" x 2½" x 2½" ←——————— DIMENSIONS ———————→ 2'- 6" x APPROX. 2⅜" x 2⁵⁄₁₆"

By courtesy of the Director, F.P.R.L., Princes Risborough

PLATE 39

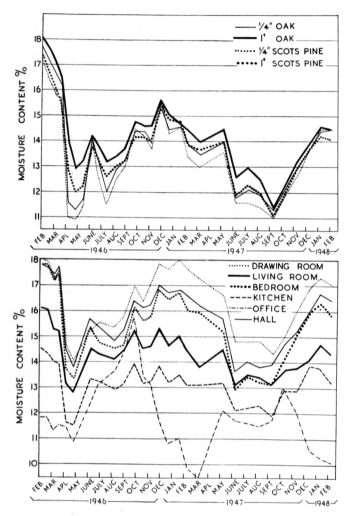

Seasonal variation in moisture content of oak and Scots pine
samples kept in various environments in Princes Risborough
February 1946 to February 1948

hornbeam, lime, sycamore, and alder. Even resin in paraffin wax is not particularly cheap when heavy impregnations are used, since, at 6p per kg, a timber such as beech can take up as much as £24·24 per cubic metre of these ingredients, but, for certain special uses of wood, the process is thought to have commercial possibilities.

The moisture exchange of wood with its surroundings, and consequent dimensional changes, can present almost insoluble problems. For example, when buildings with panelled rooms are undergoing major structural alterations, protection of the panelling often proves difficult. If the building remains unheated throughout a winter there is a considerable risk that the solid framing will pick up moisture and swell, thus pinching the thinner panels, so that they are no longer free to move. Should a hot spring or early summer follow, the thin panels will give up moisture and be ready to shrink before the thicker framing begins to dry out, and if they are being pinched by the swollen framing, they are likely to split from top to bottom. Chemicals that retard moisture exchanges have been suggested, but none has given encouraging results. Ideally, though rarely practicable, the panelling should be dismantled and stored under constant temperature/relative humidity conditions. With painted panelling it helps if, before the advent of warm weather, or the turning on of central heating, paint is stripped from the framing, and a palate knife run round the edges of the panels to break the paint seal. Central heating should be introduced at quite low temperatures: with outside temperatures around −1° C., temperatures indoors should at first be no more than about 7° C. Only when ultimate equilibrium moisture content conditions are reached should the panelling be redecorated.

CHAPTER 8

The Density of Wood

IN Part II certain characters of wood visible to the naked eye were discussed. These were features that in some measure determine the usefulness of wood, or are a guide to identification. In this chapter another character of wood, namely its weight, is considered; this is the best single criterion of the strength of a piece of wood. It is usual to speak of the weight of wood in terms of a standard volume, and this figure is called the **density.** In the United Kingdom the density of wood was formerly expressed in pounds per cubic foot, which units were well understood by practical men. With the adoption of the metric system, density is now expressed in kilogrammes per cubic metre.

DETERMINATION OF DENSITY

Density is defined as the mass of unit volume, and is therefore obtained by dividing the weight by the volume. The weight is determined on a balance or pair of scales, to an accuracy depending on the purpose for which the determination is required. For most practical uses an accuracy of 2 per cent, *i.e.*, 20 g. per kilogram, is adequate. There are several ways of determining the volume. The simplest is a calculation based on the direct measurement of length, width, and thickness, of a squared sample. It is recommended that the block should be not less than $7.5 \times 5 \times 2.5$ cm. For smaller blocks, and those of irregular shape, the following procedure is more suitable. A beaker of water is placed on the pan or balance and counterbalanced by sand or weights. Then the test block, suspended by a needle clamped in a stand, is lowered into the beaker and completely immersed in the water; arrangements are made so that this can be done without any of the water running over, and so that when the block is immersed it is not in contact

with the sides or bottom of the beaker. Weights are then added
to the opposite pan until equilibrium is restored: the weights in
grams added to restore balance are equal to the volume of the test
block in cubic centimetres. It is, of course, necessary to pay regard
to the units in which measurements are made. Provided the metric
system is used, it is a simple matter to convert grams to kilograms and
cubic centimetres to cubic metres by multiplying by 1000, *i.e.*, by
moving the decimal point three places to the right. Fig. 23 from the
earlier editions of this book has been retained as there may be occasions

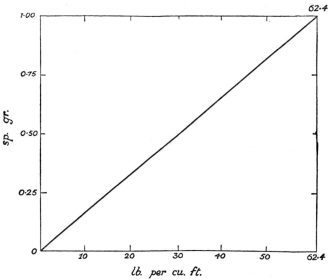

Fig. 23. Specific gravity and pounds per cubic foot

when it would be useful to convert specific gravities to the old units
of pounds per square foot. As wood is a porous substance it is necessary to coat the test block with an impervious material, such as
paraffin wax, if its volume is to be determined by the immersion
method. The block is dipped in a bath of melted wax and quickly
removed, and when the coating has set the surplus wax is scraped off.

The density is found by substitution in the formula:

$$\frac{\text{Weight of block in grams}}{\text{Weight in grams added to restore balance}}$$

To obviate the need for large samples it will be more convenient to work with samples that can conveniently be weighed in grams, their volume being determined in cubic centimetres. In effect, the **specific gravity** is determined, and the density is calculated from it. The specific gravity of a substance is merely the relative density of that substance in comparison with a standard density: usually that of pure water in grammes per cubic centimetre. Water is a particularly useful standard because the weight of one cubic centimetre is one gramme. In consequence, provided the weight of any given volume of water is known, the weight (or density) of the same volume of all other substances can be calculated from their specific gravities. The weight per cubic foot can be read from the graph (Fig. 23), or it can be calculated from the specific gravity by multiplying that figure by 62·4.

Variation in Density of Wood

A piece of perfectly dry wood is composed of the solid material of the cell walls, and the cell cavities, which contain air and small quantities of gum and other substances. The specific gravity, or relative density, of the solid material of the walls has been found to be similar in all timbers, and is in the neighbourhood of 1·5; that is to say, the cell walls are about one and a half times as heavy as water, and a cubic metre of solid wood, without cell cavities and inter-cellular spaces, would weigh roughly 1500 kilograms. Different timbers, however, vary in weight from about 160 to 1250 kilograms per cubic metre. This variation is caused by differences in the ratio of cell wall to air space in different timbers, and to the amount of water in the test piece at the time the density is determined; if it were possible to compress absolutely dry wood into a solid mass the maximum variation in density of different samples of all woods would probably not exceed 4 per cent.

Besides the range in density occurring in timbers of different species, there is a considerable variation in density between different samples of the same species, and containing the same amount of water, expressed as a percentage of the dry weight of wood in the sample (Figs. 24 and 25). This variation occurs between the timber of different trees, and in timber from different parts of one tree. In the former, variation follows no particular pattern, but is influenced by such factors as rate of growth, site conditions, and probably other growth factors as yet not fully investigated. Variation in density within a tree is by no means haphazard.

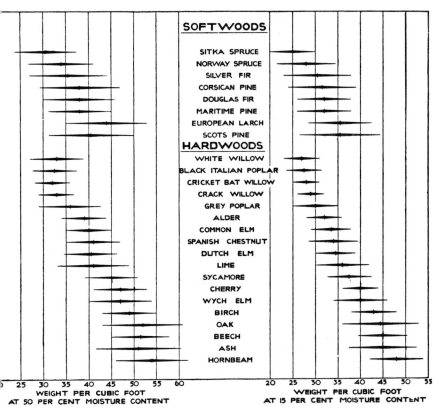

FIG. 24. Mean weights and range in weights per cubic foot of home grown timbers

By courtesy of the Director, F.P.R.L., Princes Risborough

As a general rule the heaviest wood is found at the base of the tree, and there is a gradual decrease in density in samples from successively higher levels in the trunk. At any given height in the trunk there is usually a decrease in density from the pith to the outside of the tree in ring-porous hardwoods, but in softwoods the position is reversed, and the heaviest wood is usually found near the outside. In diffuse-porous woods, however, in passing from the pith to the

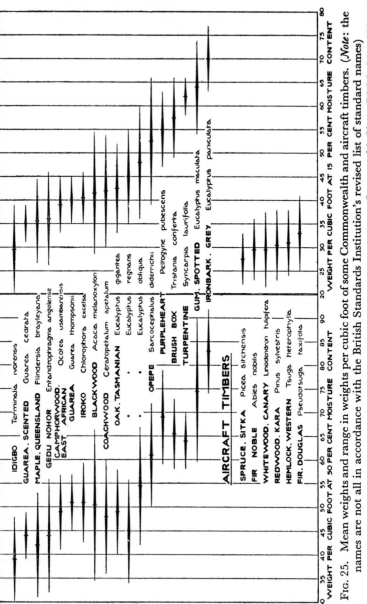

FIG. 25. Mean weights and range in weights per cubic foot of some Commonwealth and aircraft timbers. (*Note*: the names are not all in accordance with the British Standards Institution's revised list of standard names)

By courtesy of the Director, F.P.R.L., Princes Risborough

outside of the tree there is at first a slight increase and then a gradual decrease in density.

THE PRACTICAL SIGNIFICANCE OF DENSITY

The density of wood is of practical interest because it is the best single criterion of strength. This generalization, however, requires qualification. Density is of limited value in determining the strength properties of individual pieces of wood, because of the influence of other factors discussed in more detail in Chapter 9. It is useful for indicating the lower limit for a species, below which a specimen will invariably be weak, compared with average material of that species. A relationship exists between specific gravity and strength because these properties depend to a greater or lesser extent on the thickness of the walls of individual cells, and on the proportions of the different kinds of tissue in each piece of wood. If the cell walls are thicker in one sample than they are in another the ratio of wood substance to cell cavity will be greater and, in consequence, the specific gravity will be higher. The proportions of the different kinds of tissue are important because fibres, for example, have thicker walls than parenchyma cells, and if the proportion of fibres is greater in one sample than it is in another, the specific gravity will be higher.

Two other factors modify the importance of specific gravity as a criterion of strength; namely the arrangement of the individual cells, and the physico-chemical composition of the cell walls. If, for example, the parenchyma is distributed in broad layers these may constitute planes of weakness along which the timber will shear, despite a relatively high density for the sample as a whole. It has now been established that the physico-chemical composition of the cell wall is the major influence in determining the strength properties of individual pieces of wood; in particular, the degree of lignification of the cell walls has a direct bearing on most strength properties. For example, tests have indicated that, for timbers of equal density and moisture content, tropical species are less resistant to shock, but are stronger in compression parallel to the grain, than timbers from temperate regions. This can be explained by differences in the chemical composition of the cell walls, in particular, a higher lignin content in the cell walls of tropical species.

CHAPTER 9

The Strength Properties of Wood

DEFINITIONS

IT requires no special knowledge to appreciate that the strength of a timber has an important bearing on suitability for a particular purpose. A timber for beams, posts, or struts in buildings should possess different qualities from one required for spokes, hubs, or axles of carts; timbers for sports goods and tool handles would not necessarily make good chopping blocks or bearings for machinery, and so on.

The term strength applied to a material such as wood refers to the ability of the material to resist external **forces** or **loads** tending to change its size and alter its shape. The effect of applying external loads to a body is to induce internal forces within the body that resist changes in size and alterations in shape. These forces are called **stresses**; they were expressed in pounds per square foot, but, with the adoption in this country of the International System of metric units (SI units), the data are now expressed in newtons per square millimetre. With the change to SI units the sizes of test specimens have also been changed, but the data available from the earlier tests are still of value because they can be converted into SI units. Conversion factors are given in Table A1 of the Appendix in Bulletin No. 50 (Second Edition, *Metric Units*) published by H.M. Stationery Office, 1969. The changes in size or shape are known as **deformations** or **strains.** If the load is small the deformation is small, and when the load is removed there is a complete or partial recovery to the original size and shape, depending on the **elasticity** of the material. Up to a point the deformation or strain is proportional to the load; this point is called the **limit of proportionality**. Beyond this limit the deformation increases more rapidly than the load. The point beyond which it is impossible to increase the load without establishing a permanent change in shape, or **permanent set**, is called the **elastic limit**. As wood is not a perfectly elastic material it is more usual to determine the load at the limit of proportionality rather than at the elastic limit. If the load applied exceeds the forces of cohesion between the tissues a rupture or **failure** occurs. The load required to cause such failure is called a **maximum load**, *e.g.*, *fibre stress at maximum load* is the greatest resisting stress the fibres are able to exert before failure.

It is important to appreciate that the word strength has little meaning unless qualified in some way; wood has several types of strength, and a timber strong in one respect may be comparatively weak in another. Different strength properties are called into play, for example, in resisting a **compressive stress** tending to crush a timber, a **tensile stress** tending to elongate it, or a **shearing stress** tending to cause one portion to slide over the remainder. In practice, timber is frequently subjected to a combination of these stresses acting together, although one usually predominates. The ability to bend freely and regain normal shape is known as **flexibility**, and the ability to resist bending is called **stiffness**. The **modulus of elasticity** is a measure of the relation between stress and strain within the limit of proportionality, providing a convenient figure for expressing the stiffness or flexibility of a timber: the greater the modulus of elasticity the stiffer the timber, and, conversely, the lower the modulus of elasticity the more flexible it is. For each type of stress there is a separate modulus of elasticity. The term **brittle** is used to describe the property of suffering little deformation without breaking, whether the load necessary to cause deformation is large or small. It may be observed that brittleness does not necessarily imply weakness. For example, both cast iron and chalk are brittle substances, although the loads required to cause them to fail are very different.

Toughness is a property that is often used vaguely, sometimes referring to the difficulty or otherwise of splitting, *i.e.*, **fissibility**; sometimes to high resistance to sudden loads, *i.e.*, **shock-resisting ability**; and sometimes to the type of fracture occurring on failure, *i.e.*, a stringy fracture as in flexible timbers, as opposed to a clean break characteristic of brittle woods. In timber-testing laboratories three separate criteria in combination have been used to give a measure of toughness in wood. These are: **shock-resisting ability**, measured by the height of drop of a hammer; **work**

done to maximum load, which is a measure of the capacity of a substance to store a considerable amount of energy before failure; and **total work in bending**, which provides an estimate of the ability of a substance to sustain a considerable load after the maximum load has been reached. Throughout this chapter toughness is used in the restricted sense, and refers to the ability of a wood to endure suddenly applied loads, exceeding the limit of proportionality. Authorities are not agreed as to what test data are the best indication of toughness. Koehler[1] favours figures for work to maximum load in the static bending test, but the Timber Mechanics Section of the Princes Risborough Laboratory has devised a separate test (see page 169) which gives values fairly closely related to those from the Izod test, and those obtained from the height of drop of a hammer formerly of 50 lb. weight, but now 1·5 kg, using a smaller test specimen.

Hardness, like toughness, may have more than one meaning. It may be used to describe resistance to cutting, which is influenced by such factors as deposits of silica in the storage tissue and interlocking of the fibres; or it may be used with reference to resistance to abrasion or resistance to indentation. The last two properties are inter-related, but only resistance to indentation is easy to measure by standard, readily duplicated, methods. The data obtained from indentation tests have a purely comparative significance, but so long as this is understood the data serve a practical purpose in certain circumstances.

Assessment of Strength Properties

Much empirical knowledge exists regarding the strength properties of a few timbers. For example, the outstanding qualities of oak as a structural timber, the toughness of ash, and the hardness of holly are well known, and similar information is available regarding some of the more common timbers in other countries. An accurate comparison of timbers of different countries, however, can only be made by evaluating their strength properties under standard conditions. The evaluations are based on the measurement of

---

[1] Koehler, A.: *Causes of brashness in wood.* U.S. Dept. Agric. Tech. Bull. No. 342, 1933.

stresses and strains. A stress is expressed in terms of load and sectional area, *e.g.*, in newtons per square millimetre, or simply

$$\frac{\text{load}}{\text{sectional area}}$$

and a strain in linear units in relation to the length of the object undergoing strain, or

$$\frac{\text{deformation}}{\text{original length}}$$

Before the foundation of modern timber research laboratories, available data for the strength properties of wood were more or less confined to figures for the few timbers with established reputations in their own countries. The methods used for evaluating such data were by no means standardized, and the small amount of timber tested was inadequate to cover the normal variation met with in different pieces of the same wood. Increasing knowledge of alternative structural materials to wood, and the arrival of several new timbers on the market, created a need for comparative and accurate figures for the strength properties of woods. This led to the establishment of timber-testing laboratories at several centres, and the special demands of the 1914–18 war years saw these laboratories stabilized into permanent Government institutions. Laboratories are situated at Princes Risborough, England; Ottawa, Montreal, and Vancouver, Canada; Madison, U.S.A.; Melbourne, Australia; Dehra Dun, India; and Kepong, Malaysia.

METHODS OF DETERMINING THE STRENGTH PROPERTIES OF WOOD

Two alternative methods of determining the strength properties of wood are available: service tests and laboratory experiments. Service tests have the advantage that they are carried out under the conditions to which timber is exposed in use, and such conditions, however nearly imitated, cannot be exactly reproduced in the laboratory. On the other hand, the data take much longer to collect, external factors likely to influence strength properties are more difficult to control, and the decentralization of the experiments increases their cost. In the circumstances, laboratory tests provide a practical solution. In the laboratory two classes of tests are made:

tests on small, clear specimens, and tests on timber in structural sizes. The former are of value for comparative purposes, and they provide an indication of the different strength properties of individual timbers. Since the tests are designed to avoid the influence of knots and other defects the results do not indicate the actual loads that structural members can carry, and a reduction factor must be applied to obtain safe working stresses. Tests on timber of structural size more nearly reproduce service conditions, and they are of particular value because they allow for defects such as knots and splits. They have the disadvantage of being costly, because of the large amount of timber required, and the length of time needed to load larger-sized test 'pieces' to the point of failure. Moreover, the personal factor is of more importance in the selection of test material in large sizes of uniform quality, compared with small-sized pieces.

The procedure for tests on small, clear specimens has been standardized. As the strength properties of wood are greatly influenced by its moisture content (see page 178), tests are made separately on green material, i.e., freshly-felled timber, and on material dried to a standard moisture content, usually 12 per cent. moisture content, the timber being brought to this moisture content in special conditioning chambers, or the tests are made on air-dry material of known moisture content and the strength figures obtained are corrected to the standard moisture content. Precautions are taken to eliminate certain other factors liable to cause variation in strength properties. For example, the sizes of test pieces were standardized and these standard sizes have had to be altered with the change to SI units. As it has been shown that the ultimate strength properties of a piece of wood are affected by the rate of strain, all test specimens in each test are loaded at a fixed and constant rate. For a full-scale test it has become usual to select material from five healthy trees of merchantable size, and characteristic of the average in the locality, to allow for variation in different pieces of wood of the same species. If the timber is of sufficient importance, full-scale tests are made on consignments from more than one locality, involving the testing of material from ten, fifteen, or more trees.

Tests on small, clear specimens of many different timbers have now been carried out, and, in spite of the limitations of the data obtained, the tests are of considerable practical importance: the published figures for different timbers are strictly comparable because the method of testing has been standardized, and the figures themselves can be used in calculations of safe working stresses because appropriate correction factors have also been determined. The usual tests, and their practical application, are described below.

DESCRIPTION OF THE TESTS

Compression strength. Plate 40, fig. 1, illustrates the apparatus used for testing compression strength perpendicular to the grain. The test pieces were 6 in. in length along the grain, and 2 in. square in section, with two radial surfaces. The load was applied to one of the radial surfaces through a plate 2 in. wide, placed centrally with the length of the test piece, compressing the latter over an area of 2 in. square. Readings of deflection and load were taken simultaneously. The stress computed was the fibre stress at the limit of proportionality; it was calculated by dividing the load in pounds, over the bearing surface in square inches. The data from tests of compression strength perpendicular to the grain were of the same order as those obtained from indentation tests used for assessing hardness (see page 170). High values for resistance to crushing in compression perpendicular to the grain are indicative of woods suitable for use as sleepers, rollers, wedges, bearing blocks, bolted timbers, and other similar purposes. The tests for determining compression strength perpendicular to the grain are no longer normally carried out in Britain.

Plate 40, fig. 2, illustrates the apparatus used for making tests in compression parallel with the grain. The test pieces were formerly 8 in. in length along the grain, and 2 in. square in section. The test piece was placed on end on the flat surface of a hemispherical bearing, and the load was applied through a plate acting on the full sectional area of the test piece, and parallel to the grain of the wood. The original sized test piece was strained in compression at a uniform speed of 0·6096 mm per second until complete failure occurred; deformation, at regular increments of load, was measured in about 20 per cent. of the specimens, and the maximum load at the point of final failure in all specimens. The calculations made, in pounds per square inch, were (a) the maximum crushing strength, (b) fibre stress at the limit of proportionality, and (c) modulus of elasticity;[1] and, in inch-pounds per cubic inch, (d) the elastic resilience, or work to elastic limit.[2] With the adoption of SI units, the test samples are now 60 × 20 × 20 mm.

High strength in compression parallel with the grain is required of timber used as columns, props, posts, and spokes. As a rule such members have a relatively great length in comparison with their sectional area, and, in consequence, they are likely to fail in bending before the full crushing force is applied. To avoid this difficulty in mechanical tests the samples are of relatively large cross-sectional area for their length. Plate 42, fig. 1, illustrates the types of failure that occur in standard-size test specimens subjected to a compression stress parallel with the grain. Figures from this test are probably the best single criterion of the strength properties of a timber.

Shear strength. Plate 41 illustrates the test piece and apparatus used for determining shear strength parallel with the grain. The test is arranged so that the shearing stress acts radially in half the test pieces and tangentially in the remainder. The load is applied to the top surface of the test piece and shears the specimen in two. The property measured is the load at complete failure, *i.e.*, the total load required to shear the specimen in two. Strength in shear perpendicular to the grain is not measured because timber would fail from other causes before the maximum load could be applied. The present size of the test specimen is a 20 mm cube.

Tensile strength. Plate 42, fig. 2, illustrates the test sample and apparatus formerly used in making tests of tensile strength perpendicular to the grain. The stress measured was that required to pull the specimen in two, and was calculated as follows:– maximum load in pounds to produce failure, divided by the minimum sectional area over which the force is acting, measured in square inches. Strength in tension parallel with the grain was usually not determined, because, although wood is strongest in this property, there

[1] $\dfrac{\text{Load at limit of proportionality} \times \text{original length}}{\text{Area of cross section in square inches} \times \text{total shortening at the limit of proportionality}}$

[2] $\dfrac{\text{Load at limit of proportionality} \times \text{total shortening at the limit of proportionality}}{\text{Twice the volume of the test sample in cubic inches}}$

are difficulties in the way of making the tests, and, in practice, timber would fail from other causes first.

A practical application of tensile stresses is in tie beams loaded from above: failure occurs through bending, which tends to elongate the fibres on the under side of the beam while compressing those in the upper layers. High tensile strength is also of special value in timber subjected to steam bending. The type of failure that occurs depends on the nature of the wood: thin-walled fibres break in two, thick-walled ones pull apart in the neighbourhood of the primary walls.

The test described above has been abandoned in most laboratories because it was discovered that the stress set up in the test piece was not one of pure tension. More accurate tension tests have since been devised, but the preparation of the test pieces is difficult and takes time, so that they are not always included in a standard series of tests.

Static bending (*cross breaking strength* or *strength as a beam*). The three strength properties so far discussed have been considered separately, but static bending tests measure the effect of these stresses operating together. Plate 43, fig. 1, illustrates the apparatus used in such tests. The test pieces were 30 in. in length along the grain, and 2 in. square in section. The test piece was supported at both ends, with the ends free to move, and the heart face was always uppermost; the supports were 28 in. apart. The load was applied at the middle of the span, and at such a rate as to deflect the test piece 0·015 in. per minute; readings of deflections and load were taken simultaneously. The calculations made were: (a) fibre stress at the limit of proportionality, (b) fibre stress at maximum load, (c) modulus of elasticity, and (d) work in bending. The formulae employed in these calculations were as follows:

(a) *Fibre stress at the limit of proportionality:*

$$r = \frac{1 \cdot 5\, P_1 L}{bh^2}$$

where r = fibre stress at limit of proportionality in pounds per square inch,

P_1 = load at limit of proportionality in pounds,

L = span, *i.e.*, distance between the points of support in inches,

b = breadth or width of the test piece in inches,

h = thickness or depth of test piece in inches.

(b) *Fibre stress at maximum load:*

$$R = \frac{1 \cdot 5 \, PL}{bh^2}$$

where R = the fibre stress at maximum load (or modulus of
          rupture) in pounds per square inch,
      P = the load in pounds,
and L, *b*, and *h*, are as in the previous formula.

(c) *Modulus of elasticity:*

$$E = \frac{P_1 L^3}{4 \, Dbh^3}$$

where E = modulus of elasticity,
       D = deflection in inches,
and P_1, L, *b*, and *h*, are as before.

(d) *Work in bending.* Whereas formulae (a) and (b) above give
values for the loads sustained at different stages, work in bending is
the cumulative energy consumed in reaching these stages, and was
expressed in inch-pounds per cubic inch.

With the change to SI units, the test specimens are now 300 ×
20 × 20 mm, supported over a span of 280 mm. 'The load is applied
to the centre of the beam, the loading head descending at a constant
speed of 0·11 mm/s . . . The orientation of annual rings is parallel
to the direction of loading. Load-deflection diagrams are recorded
automatically for all tests, the diagrams being continued until the
specimen either fails to support one-tenth of the maximum load
recorded or deflects by more than 60 mm, whichever occurs first.'

'The strength properties determined from this test are the *modulus
of rupture, modulus of elasticity, work to maximum load,* and *total work.*
Modulus of rupture is the equivalent stress in the extreme fibres of the
specimen at the point of failure, calculated on the assumption that
the simple theory of bending applies' (Bulletin 50 (Second edition,
Metric Units), H.M. Stationery Office, 1969, p. 2).

The test, as described, introduces shear stresses that are unimportant when comparison of bending qualities of different timbers
is made from the data obtained from the test. The influence of
shear stresses should, however, be eliminated when accurate figures
for bending strength are required for purposes of design. This is
done by applying the load at two points, equidistant from the points
of support. In tests on small, clear specimens a beam, 40 in. in

PLATE 40

FIG. 2.

FIG. 1

Illustrating the testing machines, the test pieces, and method of applying the load in compression perpendicular and parallel to the grain

PLATE 41

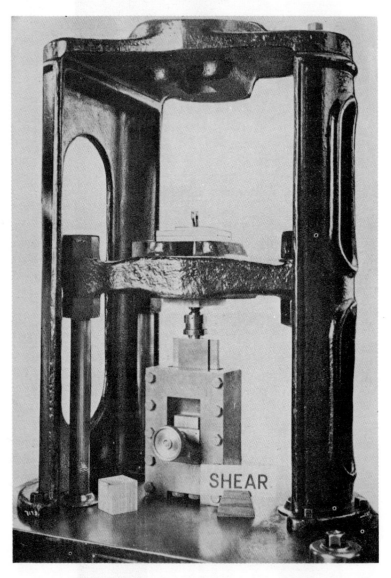

The load is applied over a portion of the cube illustrated on the base of the testing machine. Formerly the specimen was cut away, leaving a pro-jecting lip and the load was applied to the stepped portion

Photo by F.P.R.L., Princes Risborough

PLATE 42

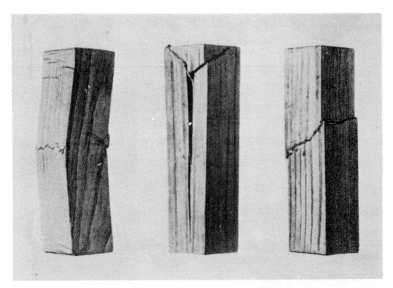

Fig. 1. Types of failure in compression parallel to the grain

Fig. 2. Illustrating the testing machine, the test piece, and method of applying the load in tension perpendicular to the grain

Photo by F.P.R.L., Princes Risborough

PLATE 43

FIG. 1. Illustrating the testing machine, the test piece, and method of applying the load in static-bending tests

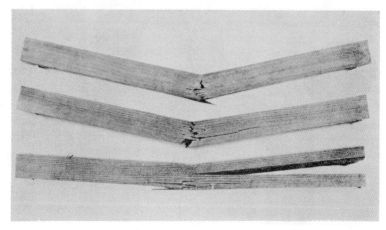

FIG. 2. Types of failure in static bending and impact bending tests

Photos by F.P.R.L., Princes Risborough

length and 2 in. square in section, was substituted for the smaller beam; it was supported similarly; but it was loaded at two points as described. The figures obtained from this modified static bending test were for pure bending, without shear.

Static bending is a measure of the strength of a material as a beam. In the resting position the upper half of a beam is in compression and the lower half in tension. Midway between the upper and lower surfaces is the neutral axis where both compression and tensile stresses are theoretically nil. A shearing stress operates along the neutral axis. The result of applying a load in the middle of the span is to deflect the beam out of the horizontal. This causes a shortening of the fibres on the upper, concave surface, and an elongation of those on the lower, convex surface. As the load increases compression failures develop on the upper surface, and the neutral axis moves towards the lower surface. The subsequent sequence depends on the kind of wood and its physical condition. For example, in unseasoned wood the initial failure is a compression failure immediately below the point of loading, followed by either a tensile failure on the lower surface, or horizontal shear along the neutral axis. Examples of the types of fracture that occur in static bending tests are illustrated in Plate 43, fig. 2. The clean breaks seen in the upper two test pieces (Plate 43, fig. 2), usually described as 'short in the grain', are characteristic of brittle timbers; the bottom test piece (Plate 43, fig. 2) shows a typical shear failure.

The load any member can sustain is dependent on the span, *i.e.*, the distance between the points of support, and on the sectional area of the member. The mathematical relationships between these three dimensions, however, are not directly proportional. For example, the effect of doubling the span is to halve the load that a beam of the same sectional area can carry. The effect of doubling the width of a beam, other factors remaining constant, is to double the load that can be sustained, but to double the depth of a beam is to increase the maximum supportable load fourfold. Because of this, beams are made rectangular, with the greater dimension in depth. There is, however, a practical limit to the magnitude of the ratio of depth to breadth in beams; a ratio greater than 4 to 1 introduces a tendency for the member to twist when loaded.

Loads applied to joists and beams involve bending stresses, but in selecting suitable sections for such members it is often necessary to allow for more than the minimum strength in bending to avoid sagging of floors, and, in particular, the cracking of plaster ceilings beneath.   In other words, adequate strength in bending does not necessarily ensure adequate stiffness when the permissible deflection is very small.   The reason for this is that timber, like most other materials, is subject to fatigue or, more correctly with timber, to creep.   Research has shown that the mechanical properties of wood are appreciably affected by the duration of loading, and stiffness is the property most affected: deflection of a green timber loaded to its ordinary working stress may be two to four times as great under long-term loading, compared with short-term load-ing.   Hence, for many purposes a timber may be sufficiently strong in bending, but not sufficiently stiff, when required to support loads of long duration; unless allowance is made for this, deflection will, in time, become excessive.   To offset this possibility, and there-by to eliminate sagging in horizontal beams, it is recommended that for design purposes provision should be made for accommodating a load equivalent to the live load plus three times the dead load. This ensures that, provided there is no over-loading subsequently, the 'permissible deflection' initially selected will not be exceeded throughout the service life of the beam.

The limitations imposed by the effect of long-term loading on deflection can often be countered, without adding to constructional costs and sometimes even lowering them, by using lower grade timber but of greater depth.   The lower grade timber will have a lower value for fibre stress at maximum load than the superior grade, but, by increasing the depth sufficiently, the resulting modulus of rupture is adequate for the loading conditions to be accom-modated.   Stiffness varies as the cube of the depth, and is less affected by grade of timber, compared with strength in bending. Thus, by allowing greater dimensions to provide for adequate bending strength in a lower grade of timber an even greater increase in stiffness is obtained.

Impact bending.   Plate 44, fig. 1, illustrates the apparatus used in impact bending tests.   The test pieces were of the same size as those used for static bending tests (30 in. in length and 2 in. square in section), and they were supported similarly, except that the ends were not entirely free to move.   The test consisted in dropping a weight

THE STRENGTH PROPERTIES OF WOOD 169

of 50 to 100 lb., depending on the strength of the timber, from successively increasing heights on to the centre of the test piece, the test being continued to the point of complete failure or a deflection of 6 in. In some laboratories a record of height of drop, deflection of the test piece, and permanent set, was obtained automatically until a drop of 12 in. was attained, by means of a recording drum and stylus pen incorporated in the test apparatus. The data collected from these impact bending tests were:– (1) fibre stress at the limit of proportionality, (2) modulus of elasticity, (3) work in bending, and (4) height of drop to cause failure or a deflection of 6 in.

With the change to SI units the test specimens are $300 \times 20 \times 20$ mm: 'supported over a span of 240 mm on chair supports radiused to 15 mm. This span-depth ratio of 12 to 1 conforms to Continental practice for the pendulum type of impact machine used for 20 mm specimen . . . The height from which the hammer drops is increased from an initial drop of $50 \cdot 8$ mm by increments of $25 \cdot 4$ mm until a height of 254 mm is reached and thereafter by increments of $50 \cdot 8$ mm until complete failure occurs or a deflection of 60 mm is reached. The weight of the hammer, also radiused to 15 mm, is $1 \cdot 5$ kg and was determined experimentally to give the same values for *maximum drop* as obtained in the standard 2-in. impact bending test.' (Bulletin 50, page 2.)

Other tests are sometimes used to measure shock-resisting ability. The Timber Mechanics Section of the Princes Risborough Laboratory used specimens $\frac{5}{8}$ in. square in section ($15 \cdot 9$ mm square) and 10 in. long (254 mm); the specimen was freely supported near the ends and broken transversely by a single blow from a falling pendulum. The data from these tests were for total work to complete failure; that is, the energy absorbed in fracturing the test sample. In the Izod test the specimen was $\frac{7}{8}$ in. square in section ($22 \cdot 2$ mm square), and a notch of standard shape was cut across the grain the full width of the piece, and at right angles to its length. The specimen was firmly clamped just below the notch, and held in a vertical position A heavy pendulum was then allowed to strike the specimen and break it, and from the reduction in amplitude of the swing of the pendulum it was possible to calculate the energy absorbed in fracturing the specimen.

The shock-resisting properties of wood are considerably in excess of the maxima for sustained loads. Shock resistance is an essential quality of timber for hammer handles, athletic goods, and similar purposes. Timbers of this class are popularly said to be '**tough**', and in this case the word is used in the same sense as throughout this chapter.

The combination of resistance to impact bending and the character of failing gradually is sometimes of special value. In places where it is impossible to forecast the exact load a wooden member may have to carry, as in mine timbers, it is a distinct advantage to employ a timber that will fail gradually, so that warning of impending collapse is given. On the other hand, it is a mistake to confuse the character of failing gradually with total strength properties. A timber may have high strength properties without being particularly resistant to sudden loads in excess of its maximum fibre stress in static bending: it is stiff rather than flexible. Such a timber will support a much greater total load than one with, for example, lower strength properties but better resistance to sudden loads in excess of its maximum fibre stress. Only in circumstances where it is impossible to pre-determine probable loading, and the warning of impending collapse is invaluable, would the selection of the latter timber, at the expense of the former, be justified.

Hardness.   Plate 44, fig. 2, illustrates the apparatus used for making hardness tests. The test consisted in measuring the force required to embed the hemispherical end of a steel rod 0·444 in. in diameter into a test piece to a depth of 0·222 in. Penetrations were made on the radial, tangential, and end surfaces. This test measured only resistance to indentation. With the change to SI units 'a special hardened steel tool (known as the Janka hardness tool) rounded to a diameter of 11·3 mm (projected area 100 mm²) is embedded to one-half of its diameter into the [20 mm] test piece. This tool has a device embodying an electrical contact which rings a signal bell when the correct depth of penetration has been reached . . . the hardness test is performed . . . on the static bending specimen, after the latter has been tested. Side-splitting is avoided by clamping the test piece between two distance pieces of the same species in a jig to form a block approximately 50·8 × 50·8 mm in cross-section. The rate of penetration of the ball is 0·11 mm/s'. (Bulletin 50, page 4.) It is pointed out in Bulletin 50 that investigations have established 'that there is a very good correlation between the side hardness of timber, as determined by the Janka hardness tool, and its compression

M

strength perpendicular to the grain'. In consequence, by applying the appropriate correlation coefficient between these two properties, 'the compression strength perpendicular to the grain (N/mm²) is generally calculated from the hardness (N) test results, thus avoiding a special test'. (Bulletin 50, page 4.) The popular conception of 'hardness' embraces ease of cutting. This latter property is dependent on the nature of the grain, and the presence of silica and other substances in the cell cavities, quite as much as on the resistance to indentation. Several instances exist, particularly among tropical timbers, of woods that are at the most only moderately hard, measured by resistance to indentation, but which are extremely hard on cutting tools, e.g., white meranti and Queensland walnut.

Hardness is of value in timber for paving blocks, flooring, bearings, and other similar purposes, although for paving blocks and flooring uniform wearing qualities are of greater importance than absolute hardness. Wearing qualities are influenced by the method of conversion: flat-sawn material will not wear so uniformly as quarter-sawn. Moreover, even good wearing qualities do not complete the requirements in timber for paving blocks and flooring. For the former, low absorbent qualities are sometimes at least as important, and for both purposes low shrinkage, with corresponding small 'movement'[1] in service, is most desirable. Further, for flooring in particular, very hard timbers, which are naturally very dense and tend to be of very fine texture, may be too slippery when planed to provide a safe surface, and they are always more noisy to walk upon than the less dense timbers that may have poorer wearing qualities.

Cleavage. Plate 45 illustrates the test piece and apparatus used in cleavage tests. Half the pieces are cut radially and half tangentially, and the cleavage in the two directions is calculated separately. The stress measured was the load in pounds necessary to split the specimen in two, divided by the width in inches of the section at the point of application of the load. With the adoption of SI, this test specimen is 45 mm in overall length and 20 × 20 mm in cross-section. 'The splitting force is applied through grips . . . and the cross-head movement is 0·042 mm/s' (Bulletin No. 50, page 4).

[1] For a definition of 'movement' see page 133.

The results obtained may be considerably influenced by irregularities in the grain of particular samples. Interlocked grain, for example, provides a resistance to splitting traceable to the arrangement in the longitudinal plane of the elements in that sample. In consequence, data from cleavage tests should be regarded solely from the comparative standpoint and not from that of absolute values. The factor that determines resistance to splitting is the arrangement of the different tissues in relation to one another: for instance, broad rays provide planes of weakness in the radial direction, and tangential series of resin canals (as in timbers of the *Dipterocarpaceae*) planes of weakness in the tangential direction. As a general rule straight-grained timbers split more readily radially than tangentially, and more readily dry than 'green', but timbers with markedly interlocked grain split more readily tangentially and are often extremely difficult to split radially.

The readiness or otherwise of a timber to split, which cleavage denotes, has a practical application in certain circumstances. In firewood, and material for the manufacture of tight barrels, charcoal, and hand-split shingles, high cleavage is a very desirable asset; for nail- or screw-holding purposes, as in packing-case manufacture, high resistance to cleavage is an essential quality.

INFLUENCE OF MICRO-STRUCTURE OF CELL WALLS

It has already been indicated that, in general, the strength properties of wood are roughly proportional to specific gravity, that in softwoods and ring-porous hardwoods strength is dependent on the proportion of late wood in the growth ring, and that very slow-grown specimens of timber are below the average for the species in both specific gravity and strength. These conclusions sum the experience of generations of timber users, and have been confirmed by the examination and mechanical testing of many thousands of specimens, but it cannot be too strongly emphasized that they are generalizations, and questions still remain as to how far specific gravity may be regarded as a guide to the properties of the individual piece of timber, and what other properties play a part in determining strength. It has been established, for example, that strength properties vary with position of a sample in the tree, and that some localities produce timber of more than average strength for the specific gravity. Bulletin No. 50, however, underlines that the greatest variation in strength properties is between material from different trees, rather than position in the tree, or material from trees from different localities: 'Thus differences in the properties of material from different trees are generally of greater significance than the differences in material from different parts of the same tree. Differences between sites and even between geographical regions are also often less significant than any differences between trees.' (Bulletin No. 50, page 1.)

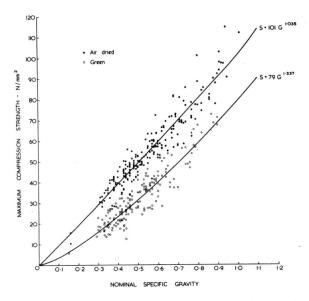

Fig. 26. The relation of maximum compression strength to nominal specific gravity for all species tested, green and air dried.

Differences in the chemical composition of the cell wall were suspected of having a major influence on the strength properties of individual pieces of wood. Preliminary investigations disclosed that such abnormal tissue as compression and tension wood have different strength properties from normal wood of the same species and density. On the other hand, the presence of extractives appeared to have little influence on the strength properties of wood, and it was established that sound sapwood is not inferior in strength properties to sound heartwood of the same species.

S. H. Clarke, when at the Princes Risborough Laboratory, investigated the influence of the microstructure of the cell wall in some detail. Clarke directed his attention in the first place to two strength properties, namely, resistance to longitudinal compression and toughness. In the course of a typical investigation some 1000 specimens of beech, from comparable parts of the tree, and representing 36 trees from 6 localities, were tested to destruction under longitudinal compression: the specific gravity of each specimen was determined and a microscopic examination was made of each. When specific gravity was plotted against strength the points for the various specimens were found to be scattered over an elliptical area. To avoid confusion only a few individual points are shown in Fig. 27; the line AB is the computed line of best fit to all the points. Individual specimens of the same specific gravity were found to differ by as much as 2700 lb. per sq. in., indicating that maximum crushing strength was influenced to an even greater degree by some other factor than density. A point of interest brought out by Fig. 27 is that the points of specimens from the same tree fell fairly close together; actually they were in small ellipses within the large ellipse. A similar relationship was found to hold in ash, oak, willow, and sweet chestnut.

In the ring-porous woods it was shown that the composition of the growth ring explained some of the variation in strength, additional to that attributed to variations in specific gravity (which is itself related to the growth rate), but the major variations were still unexplained.

A comparison of specimens matched in respect of specific gravity and growth characteristics revealed that the wide variations in strength were largely dependent on the composition of the cell wall, and it was shown by means of micro-chemical reagents that a high compression strength was accompanied by a relatively heavy degree of lignification, particularly in the region of the secondary wall of the fibres, while weaker specimens were below the average in this respect. The term lignification is used here in the botanical sense, and refers to a condition of the cell wall revealed by staining reactions and micro-chemical tests. Research has not yet gone far enough to correlate completely the degree of lignification as revealed by micro-chemical tests with the results of chemical analysis, but it is certain that conditions may be recognized that follow the variations in strength.

Variations in toughness were more difficult to explain: while specimens of low density were invariably weaker, it did not follow that those of high specific gravity were tough, and the influence of lignification was not readily apparent. A study of many fractures in specimens that had been submitted to the Izod test indicated that the initial failure usually occurred in the region of the middle lamella, and that failure was usually initiated in a zone of special weakness, for example, in the parenchyma, and that the strength of the secondary walls of the fibres rarely came into play.

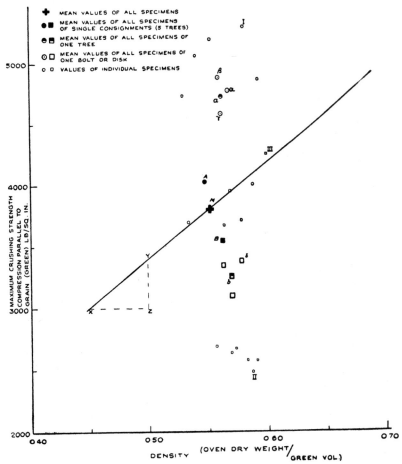

FIG. 27. Relation of strength and specific gravity
By courtesy of the Director, F.P.R.L., Princes Risborough

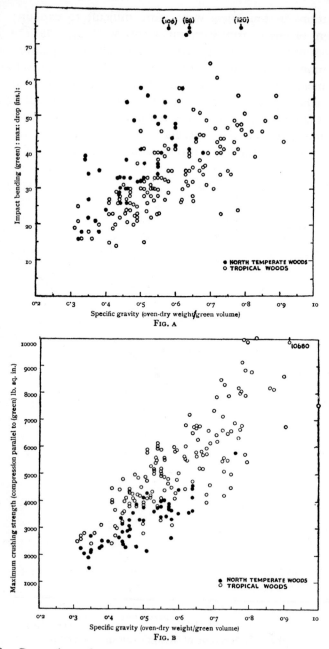

FIG. A

FIG. B

FIG. 28. Comparison of strength properties of temperate region and tropical timbers

By courtesy of the Director, F.P.R.L., Princes Risborough

There were indications, however, of a negative relation between compression strength and toughness, in that there was a tendency for material that was especially strong in compression to be weak in toughness. An interesting comparison may be observed in Fig. 28, A and B, which show the relations between compression strength and specific gravity, and toughness and specific gravity, for some 300 temperate regions and tropical timbers. Each point shows the average value for one species of timber. It is at once apparent that, on the average, tropical timbers are stronger in compression than timbers of temperate regions of the same specific gravity, but that the reverse is true for toughness. Micro-chemical tests revealed the expected difference in the condition of lignification between tropical and temperate region timbers. This was not only true as a general condition, but the difference was also revealed between individual specimens of the same species grown under temperate or tropical conditions.

A further comparison of these strength properties has been made in respect of tension wood and compression wood. The former is, on the average, tougher than normal wood but weaker in compression, whereas compression wood is stronger in compression than normal wood of the same specific gravity but less tough. It is well known that tension wood is less strongly lignified, and compression wood more strongly lignified, than normal wood.

Another point of interest is that as wood of most species dries from a green condition it undergoes a reduction in toughness but an increase in compression strength.

Taken together, these facts point to the importance of the physico-chemical condition of the cell wall as a factor in determining strength properties. It will readily be understood that the composition is determined by growth conditions. Much remains to be discovered of the effect of temperature, water relations, traces of certain elements, etc., in the growing tree on the degree of lignification.

In conclusion, it may be reiterated that for most species there appear to be optimum conditions under which trees make the best growth, and that timber produced under these optimum conditions is superior in strength properties to that produced under less favourable conditions. Further, the chemical composition of the cell wall, probably the lignin content in particular, is influenced by

locality; and the fibrillar arrangement of the cell wall is also influenced by growth conditions, in particular the slope of the stem
when growing.

INFLUENCE OF MOISTURE IN WOOD ON STRENGTH PROPERTIES

An important factor affecting the strength properties of wood is
the moisture content[1] of individual samples.   In general, below the
fibre saturation point,[2] the mechanical properties of timber increase
with decreasing moisture content, although the rate of increase is

TABLE 4.   *Average increase (or decrease) in value of various strength properties effected by
decreasing (or increasing) moisture content 1 per cent. when at about 12 per cent.*[3]

Property	Per cent.
Static bending:	
Fibre stress at elastic limit	6
Modulus of rupture	4
Modulus of elasticity	2
Work to elastic limit	8
Work to maximum load	– 1
Impact bending:	
Fibre stress at elastic limit	4
Work to elastic limit	5
Height of drop of hammer causing complete failure	– 3
Compression parallel to grain:	
Fibre stress at elastic limit	5
Crushing strength	4
Compression perpendicular with grain:	
Fibre stress at elastic limit	6
Hardness – end	3
Hardness – side	1
Shearing strength parallel with grain	4
Tension perpendicular to grain	1

[1] The term 'moisture content' is defined on page 123.

[2] See page 131 for definition of fibre saturation point.

[3] Garratt, G. A., *The mechanical properties of wood*, p. 133, 1931.

not identical for the different strength properties. Toughness is an exception to the general rule, since both in work to maximum load in static bending and height of drop of hammer necessary to produce complete failure in impact bending, there is usually an actual decrease as moisture content decreases; certain softwoods, however, *e.g.*, sitka spruce, show an increase in toughness with decrease in moisture content. In other words, dry wood will support a far greater load than green timber, but it will not bend so far before failure occurs; it is more brittle. It has been found that small, clear specimens of thoroughly air-dry wood (12 per cent. moisture content) have practically twice the strength in bending and endwise compression of the same material when unseasoned. When kiln-dried to approximately 5 per cent. moisture content the increase in strength may be threefold. Table 4 shows the average change in value of various strength properties, effected by an alteration in moisture content of 1 per cent., for timber previously dried to about 12 per cent.

For most purposes the margin of safety in general use is such that the increase in brittleness on drying is not of consequence. Mine timbers, however, are an exception, since it is not always possible to determine what load such timbers will have to carry, and brittleness may become a serious fault.

It must not be overlooked that the development of seasoning defects may offset any increase in strength properties as timber dries. Moreover, the figures given in Table 4 are *average* figures for many species, and in using the figures the warning already given about averages should be borne in mind. They are valuable as a general trend, but cannot be applied directly in determining the precise strength of individual specimens. A further point of interest is that timber once dried below a given moisture content, allowed to absorb moisture again, and then re-dried, has slightly lower strength properties, and is more brittle, than material that has never been dried below the given moisture content.

INFLUENCE OF DEFECTS ON STRENGTH

Differences in the mechanical properties of different timbers are obviously dependent on many factors: in particular, the relative abundance of the different kinds of tissue and the arrangement of the individual elements in relation to one another. The micro-structure of the cell wall, and moisture content, have also been shown to be of paramount importance. Various other factors, however, influence the properties of individual samples, *e.g.*, ir-regularities of grain; splits and checks that develop during season-ing; the presence of rot; and abnormalities in anatomical structure. These different factors are usually classified as defects (*vide* Chapter 12). Many defects reduce the strength properties of wood, but their weakening effect varies with their position in relation to the piece of timber as a whole, and the use to which the timber is put in service, *i.e.*, whether it is used in such a way that it is exposed to bending, compression, or shear stresses, or sudden heavy loads (impact bending) rather than normal, continuous loads. In other words, defects affect different strength properties differently.

The influence of the direction of the grain on the strength properties of wood has already been discussed, *vide* pages 65–6. Slop-ing grain is the most important in this connection, but in ordinary practice the margin of safety is such that an appreciable degree of tolerance in regard to sloping grain is permissible for most purposes, other than in timbers for athletic goods. In some cases, however, the margin of safety is comparatively small, *e.g.*, in the legs of chairs, when freedom from sloping grain is most desirable.

'Spongy heart'[1] reduces the shock-resisting ability of a timber appreciably, but if it is not extensive it may be of little significance in timber used as short columns, and it is only slightly more serious in timber subjected to ordinary static bending stresses. If, how-ever, visible compression failures[1] are also present, strength properties are appreciably reduced: any timber likely to be subjected to even half its allowable working stress should contain no compression failures.

All forms of warping[1] reduce the strength properties of wood because loads applied in service will no longer act directly parallel with or perpendicular to the grain. The most serious of these defects is likely to be bend in a long column loaded in compression parallel with the grain: a bend or curvature of 1 in 1000 might reduce the strength by 20 per cent.; further reductions in strength with in-creases in curvature are not proportional, but appreciably less than a direct ratio.

[1] These terms are defined in Chapter 12.

Splits and checks, that is, actual ruptures of the tissues, naturally affect the strength properties of wood, particularly by reducing resistance to shear. The effect on strength properties is largely dependent on the plane of the splits or checks: they seriously lower the strength of a beam or plank if they occur in a horizontal plane, but they are of little importance in the vertical plane. Hence, checks or splits do not greatly affect the strength properties of a short, straight-grained column.

Knots[1] influence the strength properties of a piece of wood to a varying degree, depending on their size, position, and type. Strength properties are adversely affected not because the wood of which the knot is composed is ordinarily inferior to normal wood, but because of the irregular grain that occurs in the vicinity of knots. Knots do not lower the stiffness of a timber appreciably, but they reduce its tensile strength. Hence, they are of greater significance in joists, beams, and similar timbers, than they are in columns. In beams, the weakening influence of knots is greatest when they occur in the vicinity of the maximum bending stress and on the bottom of the beam; they are of less importance when they occur near the top of the beam, and of still less significance if they occur near the centre of the depth of the beam. Any one knot will be progressively more serious as its position is farther away from the points of support of a beam. Reduction in strength properties, resulting from knots, increases at a faster rate than the proportional increase in area of the knot.

Application of Tests on Small Clear Specimens

The limitations of tests on small, clear samples have already been indicated, but the tests have a very real practical application. For example, in the absence of tests on timber in structural sizes, they are the only sound basis for comparing the relative strength properties of different timbers, and, because external factors are more easily controlled, they are superior to tests on timbers in structural size (see Chapter 12). In the absence of the latter tests, and provided correction factors, *i.e.*, factors of safety, are applied to the figures obtained from tests on small, clear specimens, the data may be used in constructional design. There is no doubt that the tendency in the past has been to use timber of unnecessarily large cross section in building work, whereas, if the information now available were applied intelligently, appreciable economies could be effected. American workers have devised reduction factors that attempt to take into account variation occurring within a species and the presence of what may be accepted as a normal amount of degrade resulting from irregularities of grain, splits and checks that develop during seasoning, and similar causes. These reduction factors, which should be used with the figures from tests on small, clear specimens in the green condition, are as follows:

(a) *In bending:*
      1/6 in dry places under cover,
      1/7 outside, but not in contact with the soil, and
      1/8 in wet places,
of the figures for the fibre stress at the maximum load for green timber.

(b) *In compression parallel to the grain:*
      1/4 in dry places under cover,
      1/4·5 outside not in contact with the soil, and
      1/5·5 in wet places,
of the figure for the maximum crushing strength for green timber.

(c) *In shear:*
      1/10 of the average of the figures for the radial and tangential shear stresses.

(d) *In side compression:*
      1/2·25 of the figure for the compression strength perpendicular to the grain at the limit of proportionality.

The use of strength data in this way is an advance on former practice, which was too conservative as far as the strength properties of wood are concerned. On the other hand, data from tests on small, clear specimens must be used with understanding. For example, some commercial timbers are the product of several botanical species, and the strength properties of each, as determined by tests, may well show a range between the weakest and the strongest of more than 20 per cent. It would be wasteful always to use the figures for the weakest timber tested, and unwise to employ those for the strongest species. In most calculations it is probable that the mean figures of all the species tested could be used, but it is in such cases that a proper appreciation of the significance of the data is essential. It does not seem advisable to attempt to assess liability to fungal attack in service in a reduction factor, but this is what is done when the reduction factor is reduced from $\frac{1}{6}$th to $\frac{1}{7}$th or $\frac{1}{8}$th, depending on the 'site' conditions where the timber is to be used. The practice should be confined to temporary structures, and even with these it must be remembered that there is a tendency to retain them for appreciably longer periods than those originally planned. A margin is, however, secured by using figures for green material when these are applied to timber that will be at least partially seasoned before being used. The intention is, of course, to provide for strength in excess of the initial calculated requirements, on the grounds that such excess is discounted as decay starts; it is assumed that decay will be detected before the strength properties of individual pieces of wood have been reduced below the values used in the calculations. The Forest Products Division of the Department of Scientific and Industrial Research in Australia has taken a sound first step in the absence of detailed data. Australian timbers have been classified in four groups, and single values for certain strength properties have been selected for all the timbers in each group. These values, with or without reduction factors in different circumstances, may be used in the standard formulae applied in constructional design. It is obvious that the values selected must err on the side of caution, since the timbers in any one group will vary appreciably in strength properties.

Stress Grading

Various factors have been shown to influence the strength properties of wood, and the causes of variation in strength between different pieces of the same timber, in so far as they are understood today, have been discussed. The practical application of this knowledge rests on the availability of rapid means for separating timber into a few broad strength grades. Such means exist; they have been established by experimental procedure and the application of simple mathematical theory.

Research has established that, within certain limits, the strength of wood bears a relation to certain visible defects. Statistical analysis of test data discloses that these conform to simple mathematical laws; in particular, variation follows the **normal** or **Gaussian** frequency distribution.

Many measurable properties are influenced by several, often unrelated, causes, which it is frequently desirable to study singly. For example, if a random selection of 1000 men were made, and their heights recorded, it would be possible to construct a graph from these data: the resultant graph is a **frequency curve**. The heights in the example suggested might be expected to range from 5 ft to 6 ft 6 in. (1·524 to 1·9812 metres). By arranging the measurements in one-inch classes (25 mm classes if in metric units) it is found that the majority fall near the arithmetical mean of these extremes, and that progressively fewer fall in the classes further removed from the mean. In effect, height in adults follows a normal frequency distribution, which can be expressed graphically as in Fig. 29. If selection of men is restricted to those of one race, or to those between, say, 9 and 12 stone, different curves from those for men of all races, or men of all weights, would be obtained; the type of curve would be the same in each case, with the peaks for these different sets of data moved to the right or left. Taking one test at a time, similar frequency curves can be constructed from the test data for wood; an example is given in Fig. 30. Such a figure has been constructed from data for one species in one test. This shows a maximum stress ranging from about 14 N/mm² to 25·4 N/mm², with the greater number of pieces around 46 N/mm². In any random parcel of 100 pieces of such timber, it is to be expected that one piece might fail at about 14 N/mm², another between 14 and 16 N/mm², perhaps two around 18 N/mm², two more at 20 N/mm², and so on. The chances of picking the weakest pieces every time are obviously small, and therefore it would be very wasteful to fix the strength for the grade at the minimum of 14 N/mm². By accepting the possibility of a small number of pieces failing below a selected minimum, it would be possible to raise this minimum considerably and yet ensure that the bulk of the material was at or above the minimum selected. In constructional work generally, wood is used in built-up structures, the strength of which is considerably above that of the weakest member. Moreover, a factor of safety, or what the Americans not inaptly call a factor of ignorance, is provided in the calculation of stresses. Hence, the inclusion of a few pieces below a certain minimum is immaterial. Having decided on the number of pieces below the minimum that can be admitted,

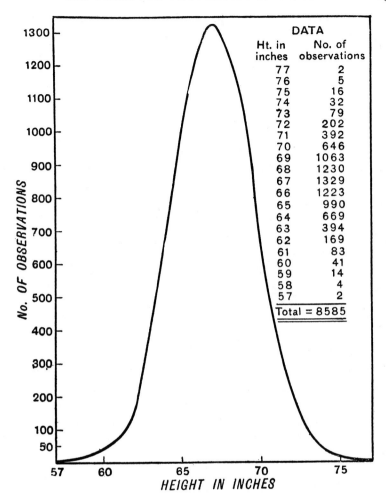

DATA	
Ht. in inches	No. of observations
77	2
76	5
75	16
74	32
73	79
72	202
71	392
70	646
69	1063
68	1230
67	1329
66	1223
65	990
64	669
63	394
62	169
61	83
60	41
59	14
58	4
57	2
Total = 8585	

FIG. 29. Normal or Gaussian frequency distribution. Height in inches of adult males in Great Britain, based on 8585 measurements.

Figures from final report of Anthropometric Committee, 1883

the allowable working stress is determined. The number of admissible potential failures will be governed by the degree of testing employed in the finished article: in aircraft construction, when a high degree of testing is carried out, very much higher working stresses can be admitted than in building construction where little or no testing is done.

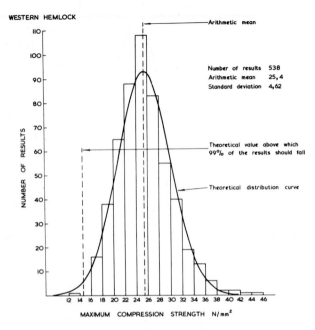

FIG. 30. Frequency distribution of maximum compression strength.

As with men and their heights, the peaks for test data can be influenced by restricting the data studied, and, provided the same percentage of potential failures to be admitted is not varied, the allowable working stress will be raised or lowered with different sets of data. For example, it is known that wood of low density has strength values below the mean for the species. It follows that if all pieces below a certain density are excluded, the peak of the frequency curve for, say, maximum strength in bending, for material above this minimum, will be higher than the peak of the curve that includes the whole range of material tested. Further, with the same reservations as to potential failures, the allowable working stress for the former could safely be fixed at a figure greater than for the latter.

Density is not a particularly convenient factor to work with in the timber yard, but such visible features as knots are. The influence of knots can be studied by testing material of the same lengths and sectional area, but with different percentages, measured by area, of knots, and the relevant frequency curves constructed. Separate curves would be provided for knot areas of 10, 20, and 30 per cent., or any other selected percentages. Admitting the same percentage of potential failures with each, three separate minimum allowable working stresses would be obtained. This is the theory behind stress-grading rules.

The influence of readily assessable features on strength properties are studied, and by determining the maximum knot area, or minimum number of rings per 25 mm, or a combination of such features, the allowable working stresses for different classes of material are determined. By visual inspection it is then possible to arrive at the grade of any piece of timber with assurance that 80, 90, 95, or any other previously selected percentage of such material will possess the minimum strength properties of its grade. Sufficient timber in structural sizes, and containing defects, has had to be tested for the influence of readily observable defects to be accurately determined. From these studies it has been possible to draw up simple rules that make it possible by rapid visual inspection to allocate any piece of timber of the species covered to its correct stress grade. It would have been wasteful to fix the strength of any grade at the minimum value for that grade, since this would mean that in any random parcel of x pieces, $x - 1$ could be expected to have a higher maximum stress value than the figure adopted for that grade. It was decided, therefore, to accept a probability, or mathematical chance, of 1 in 40 that no piece would fail below the failing stress fixed for that grade. A factor of safety was then applied, the factor selected being 27/64ths, to give the allowable working stress for the grade. The values that have been accepted for Northern European softwoods are a safe working stress in flexure respectively of $8 \cdot 274 f.$, $6 \cdot 895 f.$, and $5 \cdot 516 f. : f.$ is the stress per sq. in. in bending. This means that the minimum strength as determined by test for the $8 \cdot 274 f.$ grade is $\dfrac{8 \cdot 274 \times 64}{27}$ or 19·6 N. One piece in 40, or 25 in 1000, would fail at some figure below 19·6 N, but no piece would fail, or rather the chance is one in several thousand that no piece would fail, below the calculated allowable working stress of $8 \cdot 274 \text{ N/mm}^2$. The corresponding figure for the $6 \cdot 895 f.$ grade is 16·3 N, and for the $5 \cdot 516 f.$ grade 13·08 N. These stress-grading rules are further discussed in Chapter 17.

With such rules it is possible to use wood much more economically, with certain definite minimum guarantees. The much closer precision in the use of wood that is thereby gained places it in the field of engineering materials. At present only a very few stress grades have been worked out, but the potentialities of this approach to wood are obviously enormous, and the seemingly pointless testing of hundreds of thousands of pieces of wood to destruction takes on an altogether different significance. A further development, which would take matters much further, is the possibility of using a machine for stress-grading; this development is discussed on pages 338–9.

PLATE II

FIG. 1. Illustrating the testing machine, the test piece, and method of applying the load in impact-bending tests. The 50 lb. hammer is in the dropped position

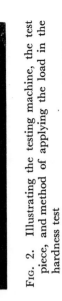

FIG. 2. Illustrating the testing machine, the test piece, and method of applying the load in the hardness test

Photos by F.P.R.L., Princes Risborough

PLATE 45

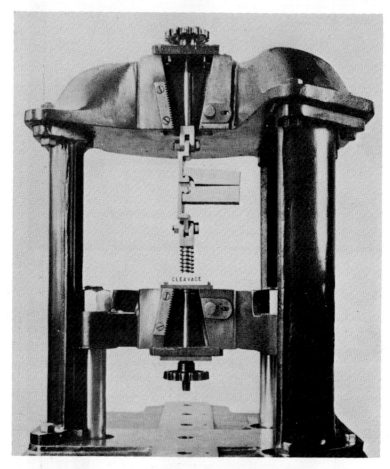

Illustrating the testing machine, the test piece, and the method of
applying the load in cleavage tests

Photo by F.P.R.L., Princes Risborough

CHAPTER 10

The Conductivity, Heat, and Energy Values of Wood

HEAT CONDUCTIVITY

IN many situations the ability of a substance to resist the passage of heat, electricity, or sound, is of the greatest importance. Dry wood is one of the poorest conductors of heat, and this characteristic renders it eminently suitable for many of the uses to which it is put every day, *e.g.*, as a building material, in the construction of refrigerators or fireless cookers, and as handles of cooking utensils. The handle of an all-metal teapot becomes as hot as the teapot itself in a relatively short space of time, but a wooden one remains comparatively cool. Good quality cooking pots, teapots, and other containers for hot liquids often have small buffers of wood inserted between the vessel and its handle to prevent the passage of heat by conduction to the handle.

The transmission or conduction of heat depends on two factors: (a) the **specific conductivity** and (b) the **specific heat** of the intervening material. Although the specific conductivity of dry wood substance is low, that of timber is even lower, for, as has been indicated, wood is a cellular substance, and in the dry state the cell cavities are filled with air, which is one of the poorest conductors known. The cellular structure of wood also partly explains why heat is conducted about two to three times as rapidly along, compared with across, the grain, and that heavy woods conduct heat more rapidly than light, porous ones.

The specific heat of a substance is the amount of heat required to raise the temperature of one gram of that substance 1° C. The specific heat of wood is about 50 per cent. higher than the specific heat of air, and four times as high as that of copper. It has already been inferred that the conduction of heat through wood is a matter of importance in the kiln drying of timber, since one of the aims of this process is to raise the temperature of the interior of planks and boards in the kiln. In these, and similar circumstances, the high specific heat of wood is a disadvantage. Fortunately, the movement of heat is more rapid in green timber and, as wood is usually more or less green when subjected to such treatments, the disadvantage of poor conductivity is less marked. Green timber conducts heat much more quickly than dry timber of the same species because of the water present, which is a much better conductor than air.

One effect of applying heat to a substance is to cause it to expand. Allowances to permit of expansion and contraction, with changes in atmospheric temperature, are made in all-metal structures such as bridges, rails, and steel-work structures generally. Woody tissue also expands with heat, but timber in use tends to shrink when heated. This apparent contradiction is easily explained. Timber in use always contains a varying quantity of moisture in the cell walls, which, on being heated, is lost to the atmosphere and, as has previously been explained, loss of moisture is accompanied by shrinkage. In consequence, although heat would cause the cell walls to expand, the loss of moisture from the walls results in shrinkage, which more than counteracts any increase in volume caused by the expansion of the woody tissue. In effect, it is not that the expansion of wood substance is low but that shrinkage is high; actually, in linear expansion along the grain, ash is almost identical with cast iron and steel, although the linear expansion of most other species is considerably less; across the grain, the linear expansion of beech is about six times as great as that of iron or steel.

The reaction of timber to heat has an important bearing on its suitability as a fire-resistant material. Because of the relatively high specific heat and poor conductivity of wood, wooden doors are often effective in preventing the spread of a fire for a considerable period. Wooden doors fail when shrinkage causes the different parts, e.g., panels, styles, and mouldings, to pull apart, leaving gaps through which flames can penetrate. Failure through shrinkage usually causes a breakdown of wooden doors long before the flames have been able to penetrate by combustion, or heat by conduction. Metal doors, on the other hand, conduct heat to the opposite side so quickly, and absorb so little heat themselves in the process, that they tend to pass on a fire from chamber to chamber with great rapidity unless constructed with an insulating core. In a light fire, papers filed in a steel cabinet will char and burn whereas under the same conditions those in a wooden cabinet might well come through unharmed. In spite of eventual failure, because of shrinkage, it has been shown that a well-constructed wooden structure is often efficient in *retarding* the progress of a fire: the distinction between *retarding* and *resisting* fire is important – wood is highly combustible, but is not highly inflammable.

ELECTRICAL CONDUCTIVITY

Absolutely dry wood offers practically complete resistance to the passage of an electric current, but the presence of contained moisture renders it a partial conductor. This phenomenon is the basic principle used in the design of electric moisture meters, described on pages 128 to 131. Besides the moisture content of a piece of wood, its density and the species influence its electrical conductivity. For example, lignum vitae, and certain other dense woods, have been used for insulation purposes; they are sometimes impregnated with wax to keep out moisture, thereby maintaining their insulating properties. The small differences in electrical conductivity of different woods of the same density can be explained by attributing such variations to the effect of differences in anatomical structure of different woods and the possible influence of certain inorganic extractives present in some woods.

ACOUSTIC PROPERTIES

The acoustic properties of wood are of importance in musical instruments and in building construction.

The laws of acoustics indicate that the power of conducting or cutting sound is linked with elasticity. Thus a piece of wood, so fixed as to be allowed to vibrate freely, will emit a sound when struck, the pitch of which will depend on the natural frequency of vibration of the piece. This in turn is governed by the density (since density affects elasticity) and dimensions of the piece. Wood, the elasticity of which has been destroyed, as for instance by fungal decay, will give a dull sound when tapped, in contrast to the clear ring of sound wood.

The property of resonance, or of vibrating in sympathy with sound waves, is also possessed by wood by virtue of its elastic properties. The special quality imparted to notes emitted by wood is very pleasing and causes wood to be extensively used in sounding-boards and other parts of musical instruments. Uniformity of texture, which comes from extreme regularity in growth throughout the life of the tree, and freedom from defects, are the essential properties of timber for sounding-boards. Slow-grown spruce, from restricted areas in Czechoslovakia, is perhaps the most famous source of supply of high-grade piano and violin sounding-boards. More recently, balsa has come into prominence for the construction of amplifying chambers of gramophones.

The ability of a material to absorb sound is dependent on its mass, the way in which it is fixed, and on the acoustic properties of the surface of that material, *i.e.*, whether the surface is capable of absorbing or reflecting sound. The cellular nature of wood is such that when timber is fixed so that it cannot easily vibrate, the surface has a deadening effect on sound waves; for this reason wood is valued as a flooring and paving material.

The Heat Value of Wood

Like other organic materials wood is combustible; under suitable conditions it will burn, and its constituents undergo oxidation with the liberation of energy in the form of heat. The fuel value of a timber depends largely on the amount of wood substance in a given volume, *i.e.*, on the density, and on the chemical composition of the wood substance, and on the state of dryness of the wood. As a general rule the denser the timber the higher its potential fuel value, but this may be modified by the presence in the wood of such substances as resin. The fuel value of resin is about twice that of wood substance, and, other things being equal, resinous woods have a higher fuel value than non-resinous woods. The influence of moisture content will readily be understood: wet wood has a much lower heating value than dry wood of the same species, because much heat is lost in transforming the contained moisture into steam. It is, therefore, anything but economical to use damp firewood: it lasts longer but gives out much less total heat than the same amount of seasoned wood, even on a dry-weight basis.

Another factor influencing the value of wood as a fuel is the ash content. Although this may not affect the heating value to an appreciable extent, for certain commercial purposes the amount of residual ash is an important consideration.

Dense woods burn more slowly, and with less flame, than light woods, which tend to flare up and burn away quickly. Decayed wood has a lower heating value than the same volume of sound wood of the same species. Comparison of relative fuel values of different species presents considerable difficulty, partly because subsidiary merits, such as even, regular burning qualities and low ash content, are not easily assessed, and partly because the volumetric unit of measurement of firewood – the cord – is capable of wide fluctuations in the actual amount of contained timber, or, in effect, of combustible material. The form of wood, *i.e.*, split billets, logs, or sawn waste, and the method of stacking, result in considerable differences in the fuel value of different cords of wood of the same species: not only may the volume of wood vary, but also its degree of dryness – split billets will dry more rapidly than logs. Comparison with other fuels should be on a dry-weight basis, but even this precaution does not take into consideration the legitimate charge against wood fuel of higher cost for storage and handling of the more bulky commodity. On a dry-weight basis, the heating value of coal is about 1·6 times as great as an equivalent weight of wood, and this figure may be of some value for approximate comparisons. Alternatively, a heat value of 8000 British Thermal Units per lb. of bone-dry wood may be used as a basis of calculations, suitable allowances being made in comparative costing figures for ease or otherwise of handling, increased cost of storage, ash percentage, and the like. In practice, it is advisable to assume that the efficiency of wood fuel is not 100 per cent. of its British Thermal Units value but some lesser percentage of this value.

Charcoal is wood fuel in an alternative form: it has a higher fuel value than ordinary wood, both on a volumetric and a weight basis. The advantages of charcoal over firewood as a fuel are largely economic ones, associated with low transport and handling charges per heat unit, the savings more than offsetting cost of manufacture.

No less than 53 per cent. of the total world consumption of wood is as fuel, the largest proportion of which is consumed in a most wasteful manner in primitive stoves or ovens, and often as open fires on the ground. Burned in this way, it is estimated that 90 per cent. of the fuel value of wood is lost, only 10 per cent. being utilized effectively. These figures take no account of the additional heat value lost by reason of burning green wood, when a considerable amount of the heat generated is absorbed in driving off excess of moisture in the wood. Admittedly, much of the 'wood' so consumed would be completely wasted were it not used as firewood, but the prodigious volume of material so involved intensifies the need for tackling this problem realistically. The natural resources of the world are not inexhaustible, and even a small improvement in the manner in which wood is used as fuel could make large quantities of potentially useful material available for other purposes. The development of burning wood in enclosed retorts, to yield producer gas, is discussed in the next section.

THE ENERGY VALUE OF WOOD IN INTERNAL-COMBUSTION ENGINES

Wood and charcoal can be used as a source of producer gas for internal-combustion engines. The choice of timbers for this purpose is a wide one, only mangrove species being less suitable than most, because of their high saline content which causes corrosion of cylinder walls. For economic reasons, however, the denser timbers are preferable to the less dense ones, and those with a small ash content are more suitable than those with a high ash content. Whenever possible only one species of timber should be used as fuel at a time, thereby ensuring uniform burning and output of producer gas.

In spite of the apparent low cost of the fuel in countries well stocked with forest, producer-gas engines cannot be said to rank as a serious competitor of petrol or Diesel oil engines if a high degree of efficiency in performance is required, and this position is likely to persist so long as the relative cost of the different fuels remains of the same order as at present. The inherent disadvantages of producer-gas engines are three in number: (1) they are of necessity heavier, on a power-weight basis, than petrol engines, (2) they require refuelling more frequently, and (3) the fuel, being bulky, is more expensive to transport and store than petrol or Diesel oil. In the tropics there is the added difficulty of obtaining firewood or charcoal of only one species in commercial quantities: the mixed composition of the forests makes such selection a practical impossibility, except in mangrove areas, and these timbers, as has been mentioned, are less suitable than most for producer-gas plants. On the other hand, when cheap transport is more important than efficiency in performance, producer-gas engines should not be too readily ignored.

The war years, which deprived many countries of their normal sources of supply of coal and petroleum products, once again focussed attention on producer-gas plants. Apart from generators for mobile vehicles, generators for stationary engines were designed, which undoubtedly functioned efficiently and economically. Even more interesting was the development of wood-gas stoves for cooking and the heating of domestic hot water. Stoves were produced on a commercial scale in Switzerland, Sweden, and Austria, capable of utilizing – in the form of producer gas – about 80 per cent. of the heat value of wood.

Another approach to the economical utilization of wood as fuel or motive power is the distillation of wood in special retorts to recover the alcohol products, which are similar in character to petrol. This is the specialized field of the wood chemist, involving commercial-size plants of high capital cost, and it must suffice to draw attention here to these derivatives of wood, the commercial exploitation of which could help to solve some of the pressing economic problems of the world today.

PART IV

Considerations Influencing the Utilization of Wood

The Seasoning of Wood

THE OBJECTS OF SEASONING TIMBER

SEASONED timber is admitted on all sides to be superior for practically all purposes to unseasoned timber, but the real reasons for the superiority are not always appreciated. It is generally realized that dryness has something to do with the superiority of seasoned timber, but it is also frequently supposed that seasoning is a maturing process that is closely dependent on the time factor. Next to dryness the most commonly acclaimed property of seasoned timber is its freedom from movement. This, we have seen from the discussion of variation in moisture content of seasoned timber, is not strictly true.

The primary aim in seasoning is to render timber as stable as possible, thereby ensuring that once it is made up into furniture, fittings, etc., movement will be negligible or for practical purposes non-existent; simultaneously, other advantages accrue. Most wood-rotting and all sap-stain fungi can grow in timber only if the moisture content of the wood is above 20 per cent.: hence, seasoning arrests the development of incipient decay in wood and removes the risk of infection of sound timber. Seasoning does not confer immunity from subsequent infection should the moisture content of previously dry wood be raised above the critical minimum, as a result, for example, of prolonged exposure to damp conditions. Several insect pests can live only in green timber, but others do not appear until wood is at least partially seasoned: those that require timber to be green cease their activity as the wood dries out, and in most cases cannot resume the attack even if the moisture content of the timber should subsequently be raised. Reduction in weight of wood accompanies loss of moisture; this is of practical importance as it reduces handling costs, and may effect economies in freight

AIR SEASONING OF TIMBER

Points illustrated in Plate 46

1. Level foundations which raise the pile well off the ground.
2. Stack not more than 6 ft wide.
3. Piling sticks (= stickers) in vertical lines and the support of short lengths and overhanging pieces.
4. Weather-tight sectional roof.
5. 2 in. oak with $\frac{1}{2}$ in. sticks to reduce rate of drying and prevent checking.
6. 2 in. beech with flush stickers, overhanging stickers and coatings to reduce end-checking.
7. Method of building sample board into pile.
8. Piling of sleepers and large-sectioned stock of free-drying timber for rapid seasoning.
9. Squares piled in stick and self-crossed.

---

charges. Reduction in freight charges applies particularly to inland transport, either rail or road, and has led, for example, to cedar shingles being sold in Canada on a moisture-content basis. Seasoning also prepares timber for various 'finishing' processes, *e.g.*, painting and polishing, and it is an essential preliminary if good penetration of wood preservatives is sought. Finally, most strength properties increase as timber dries and, although the increases may not in themselves justify the expense of seasoning, they are of more than academic significance.

Certain advantages may be secured by storing seasoned timber for very long periods before it is put into service; these periods are not the two to three years required for initial air seasoning, but periods of twenty to thirty years or more. In this time, dry timber will constantly absorb moisture and swell during wet spells, or lose moisture and shrink during dry periods. Each time there is a change in the moisture-equilibrium conditions the reactions of a piece of wood to such changes become progressively slower. In time it is to be expected that 'movement' or 'working' will be so retarded that timber exposed only to the slight moisture-equilibrium changes that occur in service indoors will be more stable than timber dried to a particular moisture content for the first time. This, however,

PLATE 46

Stack showing the various points of good piling technique. (Note that the surrounds of the stack are free from rubbish; compare this with the condition of the yard illustrated in Plate 47)

Photo by F.P.R.L., Princes Risborough

PLATE 47

Log piling. (Note the rubbish lying about; this is a bad feature of the shed illustrated)

is not what is ordinarily understood by air seasoning, and it is not an argument against kiln seasoning, which is superior to air seasoning if carried out intelligently. The special advantages accruing from prolonged storage are secured equally well whether the material is initially air- or kiln-dried.

Drying occurs because of differences in vapour pressure from the centre of a piece of wood outwards. As the surface layers dry, the vapour pressure in these layers falls below the vapour pressure in the wetter wood farther in, and a vapour-pressure gradient is built up that is conducive to the movement of moisture from centre to surface. Further drying ('seasoning') is dependent on maintaining a vapour-pressure gradient. This gradient is first established in a freshly sawn piece of wood as a result of loss of water vapour from the surface layers of the piece. The steeper the gradient the more rapidly does seasoning progress, but, in practice, too steep a gradient must be avoided.

Below the fibre saturation point drying is accompanied by shrinkage. The amount of shrinkage that will occur varies with the species and the degree of dryness attained; it will usually be greater tangentially than radially, and negligible longitudinally. If the tendency of a wood to shrink on drying is high the risk of stresses being set up in the outer layers is great with a steep moisture gradient: the outer layers want to shrink but are restrained by the wetter interior. The outer layers may set in a stretched condition, i.e., case-hardening occurs, or the tissues may be ruptured, i.e., surface checking results.

PRINCIPLES OF SEASONING

Experience has shown that the chief difficulty to be overcome in seasoning timber is the tendency of the outer layers of a piece of wood to dry out more rapidly than the interior. If these layers are allowed to dry much below the fibre saturation point, while the interior is still saturated, stresses are set up, because the shrinkage of the outer layers is restricted. The stresses may attain such magnitude that the tissues in the outer layers of the wood are actually ruptured, and surface splits or checks result. Rupture of the tissues results in fibres separating in the region of the middle lamella, whereas vessel walls break across where two separate members join.

The whole art of successful seasoning lies in maintaining a balance between the evaporation of water from the surface of timber and the movement of water from the interior of the wood to the surface. Three factors control water movements in wood: the humidity, the rate of circulation, and the temperature of the surrounding air. Temperature has a twofold effect: by influencing the relative humidity of the air, it governs the rate of evaporation of water from the surface of wood; and it also governs the rate of movement of water outwards in a piece of wood.

It is important to appreciate how these three factors interact. The rate of loss of moisture from wood depends on the humidity of the air in immediate contact with the surface layers, and on the dryness of the layers themselves. The rate of movement of water outwards in a piece of wood is dependent on the vapour pressure of the outer layers being lower than the vapour pressure further in, and on the differences in vapour pressure of successive layers not being excessive. If the outer layers are appreciably drier than the interior, greater resistance is offered to the movement of moisture outwards than when differences in vapour pressure, and consequently in moisture content, of successive layers are smaller; in extreme circumstances resistance may be such that diffusion of moisture from the inner layers outwards is brought to a standstill, the moisture in the interior of the wood being sealed in. Resumption of moisture movements in such cases can usually be achieved only by artificial means, e.g., steaming in a kiln. The relative humidity of the atmosphere, and its temperature, are all-important in the seasoning process: the lower the relative humidity of the air the better will it be able to take up moisture from the surface of a piece of wood, and, conversely, wood in contact with saturated air cannot dry at all; alternatively, high temperatures can explain the drying power of the atmosphere, although its relative humidity is high. At temperatures prevalent in the tropics, namely 80° to 90° F., comparatively high relative humidities, e.g., 70 to 80 per cent., still leave the air with appreciable drying powers because the amount of moisture that air at these high temperatures requires to take up in raising its relative humidity 1 per cent. is so much greater than the amount involved in raising the relative humidity by 1 per cent. at, say, 60° F. This factor in the temperature-humidity relationships of the atmosphere explains why it is possible to air-dry timbers in the humid climates of the tropics to as low moisture contents as those achieved in temperate regions, and in less time. In fact, unless the site conditions of tropical storage sheds are exceptionally unfavourable, the main problem is usually to retard the rate of drying to minimize checking and distortion. At the same time, surface drying is usually not sufficiently rapid to preclude sap-stain discoloration in timbers particularly susceptible to such infestation, *e.g.*, melawis, obeche.

Assuming no temperature changes, the relative humidity of the air increases as moisture is absorbed, and the affinity of the air for further moisture decreases; this, in turn, slows up the drying of the surface layers of wood exposed to such air. When air is absorbing moisture less rapidly, as a result of its relative humidity increasing, differences in moisture content of successive inner layers of wood exposed to such conditions will be less marked, the two factors thus combining to reduce seasoning stresses to a minimum. On the other hand, if the air in contact with the surface layers of a piece of wood is in constant circulation, its relative humidity may never become sufficiently high to retard the rate of absorption of moisture from these layers, while they, by drying out too quickly, offer greater resistance to the movement of moisture from the interior of the wood, and conditions of maximum stress in the outer layers result.

METHODS

Preliminary seasoning. Seasoning is sometimes begun before the tree is felled by **girdling** the trunk, *i.e.*, cutting away a strip of bark and wood completely encircling the stem. This is the general practice with teak in Burma, the main purpose being to reduce the weight of the timber so that logs will float. The girdle severs the supply of water from the roots, while, before they die, the leaves exhaust some of the water present in the trunk. The reduction in moisture content secured by girdling is very small, even over a period of a year, but six to twelve months is usually sufficient to ensure that logs of species that just sink when green will float after girdling; teak in Burma is girdled three years before felling.

Timber is sometimes purposely stored in log form to effect preliminary seasoning, although more often than not log storage is a matter of convenience in connection with the maintenance of timber supplies to a mill. In point of fact the loss of moisture from timber in the log is extremely slow, and for practical purposes seasoning may be said to begin only after conversion to boards or planks. There is, however, another aspect of log storage that may be of some practical significance. The parenchymatous tissue in the sapwood remains alive after the tree is felled, until the moisture content of the wood falls below the minimum necessary to sustain life, or until the food material in the cells is consumed. By remaining alive, the parenchymatous tissue uses up the food material essential to sap-stain fungi and certain insects, and the timber is thus rendered immune to infection from these sources.

Two methods of seasoning are in common use: **air**, sometimes called *natural*, and **kiln**, often called *artificial*, **seasoning**, although in commercial practice a combination of the two will often be more satisfactory and economical. So-called '**water seasoning**' is a misnomer: there can be no loss of moisture so long as timber remains waterlogged. The hygroscopicity of wood and its subsequent shrinkage may be reduced by 'water seasoning', but the benefits are not of sufficient magnitude to be of any commercial value as a method of seasoning. Prolonged storage in water, however, reduces the starch content of the sapwood of species rich in this substance, rendering such timbers less susceptible to powder-post beetle attack. There appears to be some doubt whether it is the reduction in starch content that is of importance, or whether it is the removal of other essential substances that occur in very small quantities in unsoaked wood. In the absence of these other substances timber still rich in starch appears to be immune to powder-post beetle infection. This has been demonstrated with known susceptible timbers: samples soaked in water for three months and subsequently exposed to attack remained immune although their starch content had not been appreciably reduced, whereas other samples from the same consignment that were not previously soaked were heavily infested when exposed to attack.

AIR SEASONING

Air seasoning aims at making the best use of prevailing winds and the sun, while protecting timber from rain. Wind, by circulating the air, prevents it from becoming saturated with moisture absorbed from seasoning timber, and the sun, by raising the temperature of the air, lowers its relative humidity. The combined effect of these two factors is to maintain the drying power of the air. Rain, on the other hand, increases the humidity of the atmosphere, and, as it is accompanied by lower temperatures, reduces the drying power of the air. If at the same time the timber is actually wetted, it may pick up appreciable quantities of moisture. As a general rule, the problem is to accelerate air circulation adequately, although – as explained on page 202 – in the tropics, and with timbers prone to develop seasoning defects, it may be necessary to reduce air circulation and thus slow up the rate of drying. For example, measures are taken to reduce air circulation with a timber such as oak when freshly converted in the warm summer months and, conversely, oak converted in the cool winter months can safely be exposed to greater air circulation.

Control of the climatic factors is best achieved in properly constructed, well-ventilated sheds, but with low quality timber such structures are impracticable on economic grounds. The most efficient shed is, moreover, only effective up to a point: even in weather-proof buildings the relative humidity of the air varies appreciably at different seasons of the year. Control of air circulation, whether in sheds or in the open, is effected by piling the timber in properly constructed stacks, the design of which is the most important consideration in air seasoning. Control of the movement of water in wood is more difficult. Water movement is, of course, affected indirectly by control of air circulation, but additional measures are advisable to compensate for the more rapid movement of moisture along the grain than takes place across it. If the loss of moisture from the ends of a piece of wood is not checked serious stresses are set up that result in bad end-splitting; to minimize this trouble some form of end covering should be adopted. It will be seen, then, that three factors are available for regulating air seasoning; namely, seasoning sheds, correct piling, and end protection of the individual pieces of wood in a stack.

Seasoning sheds. In its simplest form a seasoning shed may be nothing more elaborate than a large Dutch barn with temporary roofing. On the other hand, it may be a permanent building, consisting of a roof and four walls, the walls being louvred, so that air circulation through the building can be regulated with considerable precision. For softwoods, except of the higher grades, any form of seasoning shed was formerly regarded as prohibitive in cost, but for the more valuable hardwoods seasoning sheds have usually been regarded as essential, and a more or less permanent building with a corrugated-iron roof may often prove more economical in the long run than a purely temporary structure. Today, even the lowest grade of wood represents an appreciable capital investment, and sheds should be regarded as essential. In the tropics some form of shed is very necessary to protect timber against heavy rain and the very strong sun, but, because of the intense heat in the middle of the day, a corrugated-iron roof should be avoided: thatch, shingles, or rough boarding are better. Thatch was probably the commonest roofing material for seasoning sheds in the smaller Malaysian sawmills before the war, and no serious fires are known to have resulted.

Piling. Piling technique is the most important factor in air seasoning, because such points as the position and orientation of stacks, and their method of construction, largely govern air circulation.

The site of the seasoning yard is usually dictated by such circumstances as the necessity for proximity to the saw-mill, the land available, or the layout of existing buildings. Wherever possible, however, the site should be a naturally well-drained one, sufficiently removed from buildings to guard against the accumulation of stagnant air and/or the creation of air eddies.

The nature of the floor of seasoning sheds or yards is important. The most satisfactory is a good concrete floor, which will not hold moisture, and can be kept clean. A cheaper alternative is well-rammed earth (clay) or cinders. Sawdust is bad as it holds moisture, and results in the circulating air being damp, so that seasoning is retarded, and the development of wood-rotting fungi is encouraged. The floor must be kept clear of rubbish; wood waste left lying about provides opportunities for fungi and insects to breed and spread to sound timber, and such rubbish increases the fire hazard. All wood waste from the saw-mill, and that which inevitably collects in the seasoning shed or yard, should be collected and burned if it cannot be utilized as the raw material of some manufactured wood product, or be sold for some purpose or another. It is not sufficient to collect and dump wood waste in an unused corner of a yard, where it will be equally effective as a breeding-ground for fungi and insects, if not so great a fire hazard. The important points in stack-building are the orientation, foundations, spacing, and width of stacks, and the spacing and width of stickers.[1] Two alternative methods of orienting stacks with reference to the passage ways are possible: **endwise**, *i.e.*, with the timber at right angles to the passage ways, and **sidewise**, *i.e.*, with the timber parallel to the passages. Endwise piling makes for ease of inspection and tallying of the stock, but sidewise piling ensures better air circulation from the passage ways. In endwise piling the air is held up by the stickers and can only circulate by way of the narrow alleys between stacks. Economic considerations, and mill layout, however, usually determine the method selected, but where mechanical elevators are used for stack-building sidewise piling is obligatory. If several varieties of timber, requiring different seasoning periods, are dealt with in the same yard, sidewise piling is more convenient and economical: high handling costs result when one of a series of endwise-piled stacks is required out of turn, because turning space in the passages tends to be restricted, and timber coming out at right angles to the extraction ways may absorb much time in manoeuvring. By far the most common failing is to crowd sheds to their maximum capacity, the excuse being made that land values are so high that the fullest use must be made of a firm's storage capacity. This argument overlooks the fact that expelling moisture from wood is inevitably an expensive matter, but quicker air drying that is achieved by not over-filling seasoning sheds may well offset the higher rental costs per cubic metre of throughput because of the saving in fuel when such timber is finally kiln-dried just prior to use. The economics of this argument are worth investigating, although the general application of the theory is likely to have to be deferred until timber is regularly sold on a moisture-content basis.

For the foundations of stacks, baulks of timber are commonly used, but concrete, brick, or even wooden piers, are better, as they offer less resistance to the free circulation of air under a stack. If solid baulks are used they should be at right angles to the alleys. Wood in the foundations should be thoroughly sound and well-seasoned, and, if practicable, it should be treated with creosote or other wood preservative. If species susceptible to powder-post beetle attack are used for the foundations of hardwood stacks the timbers should be absolutely free from sapwood, otherwise infection

[1] See page 208.

in the foundations may spread, as drying progresses, to the timber in the stack. This precaution is not of the same importance in stacks of softwood timbers because all softwoods are immune to powder-post beetle attack. For permanent foundations consisting of baulks of wood, it is well worth while considering the possibility of a damp-proof course immediately under the baulks: an excellent method would be to provide concrete footings, and to bed the baulks to these with a bituminous mastic. The height of the foundations should be governed by the nature of the floor: a height of 20 to 30 cm is sufficient with concrete floors, but not less than 45 cm is desirable with earth floors. The foundations of open-air stacks should be sloped to permit of rain running off the top boards or planks in a stack, instead of soaking into the timbers below.

Unless solid baulks of timber, at right angles to the length of the stack, are used, a system of longitudinal members, or piers, with cross pieces or stringers, is necessary. For the cross pieces, metal rails are best as they interfere with air circulation least; they should be at right angles to the length of the stack, and should be covered with strips of wood to prevent the bottom layer of the stacked timber from coming in contact with the metal. If a stack is to be a large one, or if stickers are to be closer together than the spacing of the foundation piers or timber baulks, it is important to distribute the weight evenly over the foundations, and not over the bottom layers of timber in the stack. This can be achieved by a system of bearers and cross pieces, between the foundations and the bottom row of timbers in the stack, of sufficient strength to carry the weight above.

The dimensions of stacks must be kept within certain limits to secure rapid and uniform drying, and to avoid the risk of stagnant air accumulating in the centre of the pile, which can be a cause of unequal drying, and sometimes leads to fungal infection. Four metres is recommended as the maximum width for stacks, and one metre as the minimum width of passage ways. Excessive height is to be avoided for similar reasons, and there is the added disadvantage that tall stacks increase handling charges. Five metres is suggested as a reasonable maximum, but unless mechanical elevators are used, or the calls on yard space are particularly pressing, a stack should be under rather than over this height. Wide stacks are to be avoided in countries where termites (white ants) abound: there is a risk that termites may break through the foundations and attack the timber in the pile. Wide stacks make inspection beneath more difficult, and in such circumstances attack may go undetected for a long time.

In practice, the nature of the output of a mill often determines the size of stacks. Where the output is varied as to species, qualities, and sizes the stocks carried of any particular 'item' are probably sufficiently small to impose reasonable limitations on the size of stacks. On the other hand, a yard or store confining itself to a few timbers, in two or three sizes, will have to arrange to distribute the out-turn in stacks of suitable dimensions.

Circulation of air through a stack is secured by separating the successive layers of timber by strips of wood known as **stickers**, the thickness of which regulates the rate of air-flow. The stickers should be of sound, seasoned timber and, to avoid indentation of boards in the lower part of the pile, not of a harder type than the timber in the stack. The use of softwoods for this purpose is a safeguard against the introduction of powder-post beetles through infected stickers (*vide* also *Lyctus*, or powder-post, beetles, page 272). When a stack is dismantled the stickers should receive as much consideration as is given to the seasoned timber; they should be collected, bundled, and stored for further use. Proper sticker drill in the seasoning yard is well worth attention: the issue of stickers and their return to stock should be organized as for any other stores. It is common practice to use the lowest grades of wood for stickers, but this is a very short-sighted policy, since stickers are then always lightly regarded, and wastage is high. Twenty-five millimetre boards, piled with 25 mm stickers at 600 mm centres, give a sticker volume of about 4 per cent. of the timber in a stack: if stickers are used only once such wastage will be seen to represent an appreciable additional cost per cubic metre of timber handled at a time when even the lowest grades of wood are worth more than £17.50 per cubic metre sawn. Many years ago one large importer in this country purchased prime American black walnut for conversion to stickers, and found the outlay well worth while with a sticker life of fifteen years or more.

The most suitable thickness for stickers depends on the thickness of the timber to be seasoned, its drying qualities, and the season of stacking. For thin stock, of species not subject to serious degrade in seasoning, 37·5 mm stickers are suitable, but thick planks of species that are inclined to split or surface-check badly may require stickers as thin as 12·5 mm. The time of the year that stacks are built should also be taken into consideration: oak piled for the first time in autumn can safely be stacked with stickers of greater thickness than would be suitable for newly converted oak stacked for the first time in the late spring or early summer. To secure rapid drying, the thickest stickers that experience has shown can be used with safety should always be employed; it is doubtful, however, whether stickers more than 50 mm thick secure any further acceleration in the rate of air circulation. Stickers should be no wider than is absolutely necessary, as the area of timber in contact with them is hindered from drying at the same rate as the remainder, and, in certain timbers, such covered portions may become stained. Too narrow stickers, on the other hand, cause indentation and, for this reason, the width should never be less than the thickness, and for really soft timbers it may need to be greater. A maximum of 50 mm in width should, however, suffice for the most easily bruised timber.

The stickers at the ends of a stack should be wider than the remainder and should project about 12·5 mm beyond the ends of the stacks. By this means a buffer is provided against the rapid circulation of air over the ends of the timber, and a considerable amount of end-splitting is prevented. Furthermore, stickers should project slightly beyond the sides of sidewise-piled stacks to protect the timber from being bumped by traffic in the passage ways.

Stickers impede the circulation of air and, therefore, should not be unnecessarily numerous; on the other hand, an insufficiency of stickers results in the sagging of boards and planks. The distance between stickers depends on the thickness of the stock and its liability to warp; 12·5 mm boards require stickers 600 to 900 mm apart, but planks of 50 mm and upwards are usually sufficiently supported by stickers 1200–2400 mm apart, the spacing increasing with increase in thickness of the planks.

Plate 46 (see p. 200) illustrates the important details of stack construction; it is not suggested that timbers of such varying thicknesses and classes as those in the 'model' stack should be piled together, but the model was assembled to illustrate as many points as possible with the strictly limited amount of timber available to a research laboratory. It may be observed that the stickers are in vertical rows; this arrangement is essential to avoid unequal stresses on the lower layers of timber, which would inevitably result in a considerable amount of bowing. (See also Plate 47.) As far as possible, the timbers in a stack should be of uniform length, but when this is not practicable the longest pieces should be at the bottom; projecting ends must be supported as in Plate 46. Further, in any one row, all timbers must be of the same thickness, otherwise the thicker pieces carry the weight of the whole stack above them. Other points in stack construction are that the top layer of timber, and all projecting ends, should be covered with thin, dry boards, or, if in an open yard, by a raised roof. Thin-dimensional stock should be weighted by laying heavy baulks of seasoned timber on the top of the stacks, to prevent bowing and cupping. Timber should be stacked with stickers as soon after sawing as possible; **close piling**, *i.e.*, without stickers, even for a few days, is a fruitful cause of staining, and, if prolonged, it may result in serious losses from fungal decay.

A special form of piling, often adopted for hardwoods in England and in certain continental countries, is illustrated in Plate 47. In this method, each board is piled in sequence as cut, and the log is sold as a unit. The advantages claimed for this method of piling are that the merchant is not left with narrow widths and defective boards, and, with figured woods, 'matched' material is kept together. Incidentally, Plate 47 illustrates many of the 'mistakes' discussed in previous paragraphs: the baulks of timber as foundations effectively impede air circulation, the cleats nailed to individual boards will encourage end-splitting in such boards, failure to 'weight' the top of the stacks has resulted in distortion of the uppermost boards, and the rubbish lying about is increasing the fire hazard and providing a breeding-ground for insects and fungi.

Timbers liable to discoloration, *e.g.*, sycamore, are frequently seasoned by stacking on end, thereby avoiding the use of stickers; and baulks, sleepers, squares, etc., are usually 'self-piled' in various ways, the essential feature of which is that some of the pieces of timber act as stickers. A suitable method for rapid drying of sleepers, and short lengths of timber of similar dimensions, is illustrated in Plate 46, item 8; long baulks should not be stacked in this manner because of the risk of their bowing.

'**Self-piling**' is a common practice with softwood timbers in Scandinavian countries; in this method timbers of the same dimensions as the remainder of the stack are used as stickers, often with no greater distances between the 'stickers' than between the pieces of timber in the rows above and below. Unless the pieces used as stickers are short lengths, the stacks are too wide to secure a uniform rate of drying in the whole pile. When used for boards stacked flat, the method is obviously inferior to piling with proper stickers. But planks are frequently piled on their narrow face, the distance between the rows being the width of the planks, *i.e.*, one row of planks is laid flat, the next on their narrow faces, and the succeeding row flat, and so on. In this way the distance between the rows may be 150, 175, 200, or 225 mm, and, if the planks are only 50 to 75 mm thick, a relatively small area of timber in the stack is covered up by other green material. Such an arrangement ensures rapid drying of the surface of the timber, a very necessary condition if 'blue-stain' is to be avoided in European redwood, but it is too drastic for many timbers and might give rise to serious surface-checking.

If timber has to be stored on a building site for any but a very short period it should be built into temporary, roofed stacks. This is equally important, whether the timber is partially or fully seasoned when delivered. If partially seasoned, piling with stickers will ensure that some seasoning will occur on the site; that is, the best use will have been made of the interval between delivery and use. If the timber is seasoned when received it is imperative to keep it covered so that it will not become wetter before it is required; joinery should not only be under cover, it should be stored in a weather-proof shed.

It is common practice to close-pile softwoods on arrival in this country, and serious consequences have not resulted in the past from this practice, at least with European redwood. It has, however, been generally recognized that deck cargo that has become wet in transit cannot be so treated – it must be properly stacked until dry. Prior to 1932 close-piling was permissible because most of the imported timber came from northern Europe, and had been stacked for some time before shipment: it was more or less air dry. Piling prior to shipment is not usual on the Pacific Coast: the timber, unless kiln-dried, tends to be shipped as cut, which may be the day of shipment. Such green timber may escape harm in transit, but unless properly stickered on arrival in this country it may well develop dote over here, and it certainly will not dry out if close-piled (see also the discussion of dote on page 257).

End Protection. End protection is provided by coatings of various, more or less waterproof, substances, or by strips of wood or metal nailed to the ends of timber. Strips or cleats of wood are thoroughly bad for any timbers but thick planks, because the small longitudinal shrinkage of the strip is opposed to the much greater transverse shrinkage of the timber, with the result that shrinkage is restricted between the points of attachment of the strip, and end-splits frequently develop in the piece of wood. Splitting will almost invariably occur in timber up to 25 mm thick if wooden cleats are used, but planks 75 mm or more in thickness may not suffer any harm unless incipient splits are present before the cleats are attached; in these circumstances, however, the cleats are likely to be ineffective in preventing the splits from developing. If end protection is provided by a thin strip of metal, *e.g.*, hoop iron, the metal buckles concertina-fashion as the baulk shrinks and end-splitting may be avoided. S-shaped pieces of iron, driven into the ends of baulks of timber, are effective in preventing the development of splits that have already occurred, or in reducing the amount of splitting that would occur were no protective measures taken. More satisfactory than wooden cleats or iron ones, is the use of an end-coating, the essential qualities of which are impermeability to water and air, a semi-liquid state to permit of its being applied with a brush, and a capacity to harden on exposure so that it will set and not flake off when the timber is roughly handled. Many substances and mixtures fulfil these conditions so that choice is mainly one of expediency. Wax is an obvious possibility, but cost, and the necessity of having to apply it hot, make it unsuitable. The United States Forest Products Laboratory recommends a mixture of one part asbestine and one part barytes to two parts of hardened gloss oil; 5 litres of the preparation being sufficient for 10 m². A very successful mixture consists of finely powdered, unburnt brick clay and ground dammar (a resin compound soluble in petroleum spirits) in equal proportions, with sufficient paraffin to permit of spreading the mixture. The proportion of clay may be increased to about 55 per cent. to reduce the cost of the preparation. Clay or chalk, mixed with dung as a spreading medium, has also been used, but fine mud without a binding medium flakes off too easily to be of value. Various proprietary petroleum waxes are available on the market; they usually have the advantage over ordinary waxes, in that they can be applied cold, and they are more or less transparent. Sheets of plywood nailed over the ends of a stack to the projecting stickers are a cheap compromise, but they interfere with the circulation of air along the stack and may slow down the seasoning process too much for all but the more refractory timbers.

Kiln Seasoning

Kiln-drying is effected in a closed chamber, providing maximum control of air circulation, humidity, and temperature. In consequence, drying can be regulated so that shrinkage occurs with the minimum of degrade, and lower moisture contents can be reached than are possible with air seasoning. The great advantages of kiln seasoning are its rapidity, adaptability, and precision. It also ensures a dependable supply of seasoned timber at any season of the year; and it is the only way that timber can be conditioned for interior use requiring lower equilibrium moisture contents than those prevailing out-of-doors, or in unheated sheds.

There are other advantages gained by kiln drying. In properly operated kilns, every piece of timber in a kiln load can be dried to a uniform moisture content throughout. Moreover, the drying process also sterilizes the timber: the temperatures used, and the humidities maintained in a kiln, are lethal to any insect or fungus present in the timber when placed in a kiln. Such sterilization does not, of course, protect the dried timber against fresh infestation after removal from the kiln. Further, the resins or gums in certain woods are to a large extent set or hardened in kiln-drying, so that the risk of subsequent 'bleeding' from finished surfaces is reduced. On the other hand, errors in the technique of kiln operation may have serious consequences: seasoning degrade can be magnified, and whole consignments of timber hopelessly spoiled, by improper kiln-drying.

It is necessary to regulate kiln drying to suit circumstances: different timbers and dimensions of stock require drying at different rates. As a general rule, softwoods can withstand more drastic drying conditions than hardwoods, thin boards than thick planks, and partially dry stock than green timber. There are limitations as to the dimensions that can be kiln seasoned economically: unless timber is previously treated with certain chemicals, so-called **salt**

seasoning or **chemical seasoning** (*vide* page 235), kiln-drying is only suitable for material up to about 75 mm thick: above this figure the rate of drying would have to be so slow, to avoid serious seasoning degrade, that the method would be altogether too expensive. On the other hand, the occasions when large-sized timbers are required uniformly dried to low moisture contents of around 12 per cent. are extremely few so that restriction to the smaller maximum dimensions suggested does not really impose limitations of practical commercial importance.

In the past, standard drying schedules have often been employed, irrespective of the species, dimensions, or condition of the timber to be seasoned, and, too often, kilns have been little better than hot ovens. In such circumstances kiln-drying can be thoroughly unsatisfactory, resulting in serious damage to the timber. Such malpractices are undoubtedly at the root of many objections still levelled against kiln-drying, whereas extensive tests show that, if properly carried out, kiln seasoning is not only as successful as air seasoning, but in many respects is superior. For reasons of economy, it is common practice to air-dry timber initially, and to complete drying to the required final moisture content in a kiln. Provided air-drying is done properly, the combination of air- and kiln-drying is not open to any objections, and should prove much more economical than kiln-drying from green: 25 mm oak, kiln-dried from green, will have to occupy a kiln from 5 to 6 weeks if degrade in drying is to be avoided, and throughout this time consumption of fuel and power is incurred. The same timber, first air-dried to just below the fibre saturation point, should not require to be in a kiln for more than half the period.

A kiln consists of some form of more or less air-tight shed, fitted with heating apparatus, a supply of water or steam sprays, and, in some types of kilns, artificial means of accelerating air circulation. One of the greatest problems in kiln construction is the reduction of heat losses to a minimum. To this end cavity walls of brick or tile are generally used, and the interior surfaces are painted with some water-proofing substance. The doors are of wood or metal of several designs that aim at securing a tight fit.

The usual method of supplying heat to kilns is by a system of steam-heated coils over which the air passes before circulating through the stacks of timber. Other methods of heating could be used but steam is particularly suitable as it is easily regulated, and in many saw-mills it is available from the burning of wood waste. Special types of furnace-heated kilns were developed in the war years; these will be discussed after the more conventional types of kilns have been described.

The humidity of the air can be controlled by regulating the temperature, by admitting water or water vapour, or by changing the air through removal of saturated air from the kiln and replacing it with fresh air from the outside. In practice the manipulation of temperature alone is seldom sufficient, and a system of water or steam sprays, and inlet and outlet air ducts, are installed. The circulation of air is secured by '**natural-draught**' or mechanical means. The former method is dependent on temperature differences at different levels in the kiln, which cause air currents to be set up. **Forced circulation** is obtained by means of fans or blowers. Natural-draught circulation is sometimes further stimulated by the suitable arrangement of the steam or water sprays.

In theory the air in a kiln can be used indefinitely, if there is some means of de-humidifying it after it has passed through the timber. In practice some of the moisture is removed from the air by condensation on the walls of the kiln, but it is usual to arrange for a portion of the moisture-laden air to be drawn off and replaced by an equivalent amount of fresh air from the outside. The escape of used, humid air, and the introduction of fresh, relatively dry air, is usually designed to take place through special outlet and inlet channels, but a certain amount of interchange occurs as a result of natural leakage.

The rate of drying in different parts of a kiln varies, because the temperature of the air and its relative humidity vary at different levels, and arrangements have to be made to counteract this as far as possible. One method is to increase the rate of circulation, so that there is less opportunity for the air to become saturated before it has passed through the stack. Where artificial means of accelerating air circulation are not installed the same effect can be obtained by keeping down the size of the stacks of timber, or alternatively, the direction of circulation may be reversed by means of a double set of outlet and inlet ports, coils, and sprays.

It is important to follow the conditions of the air in a kiln closely during a run, and also the progress of drying in the timber. The first can be done by means of wet and dry bulb thermometers (**hygrometers**) (Fig. 31) suitably placed in the kiln, and illuminated, so that they can be read from outside the kiln. With readings of the two thermometers, the relative humidity of the air in a kiln is found by reference to appropriate tables or charts. Self-recording instruments are sometimes used in place of simple mercury thermometers; these incorporate inked pens, and charts (graph paper so ruled that the relative humidity is read direct). Such instruments are an essential part of a fully automatically operated kiln, and they provide a very useful record of conditions in a kiln throughout the entire run, but unless very carefully maintained – and they are delicate instruments – they have certain disadvantages in comparison with the simple mercury-in-glass thermometers. They are costly, relatively sluggish, and easily damaged by rough handling of the pens when changing the charts. Precautions are necessary in siting hygrometers in a kiln to ensure that they give a correct picture of the condition of the circulating air. A minimum of two instruments is essential, so placed that the state of the air, both as it enters and leaves the pile of timber, can be read. When only one instrument is used, control of the kiln is not infrequently determined from the condition of the air leaving the timber, when serious mistakes can be made. Such air will have a lower temperature and higher relative humidity than fresh air admitted into the kiln and heated before circulation. If the conditions of the moment call for air of the temperature and humidity of that leaving the timber, the actual air circulated is liable to be too warm and too dry, causing too rapid drying, with the risk of serious seasoning degrade. Instruments must be at least 15 cm from any wall so that they are in the path of the main air-flow. To enable hygrometers inside the kiln to be read from the outside port-holes should be provided in the end walls of a kiln, and electric lighting for illumination of the instruments: frequent opening of kilns to read instruments would be extremely wasteful of steam. Low-powered field glasses or opera glasses should be used for taking readings. Thermometers graduated in degrees centigrade have been found rather easier to read than Fahrenheit thermometers. All hygrometers require maintaining in good order; with the simple mercury-in-glass type, this merely involves maintaining the water level in the receptacle, and the syphoning wick, in perfect condition. Distilled water should be used, and the calibration of the instruments should be checked annually.

Progress of drying in the timber should be followed by means of **'test'** or **'sample boards'**. At the commencement of a run, the moisture content of the parcel as a whole should be determined from a sufficient number of samples by the oven-dry method described

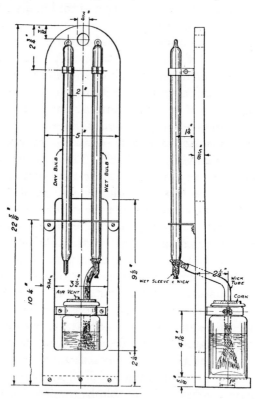

FIG. 31. A standard type of wet and dry bulb thermometer
By courtesy of the Director, F.P.R.L., Princes Risborough

on pages 123–126. The boards or planks from which the samples are taken also serve as 'sample boards' for following the progress of drying. After the moisture-content samples have been cut from either end the remaining length of each board should be at least 1·5 metres. The sample boards should be immediately weighed and returned to the pile, but so distributed that they will provide a picture of the progress of drying in the whole consignment. The stickers above the boards are notched so that the boards can easily be withdrawn for periodical weighing, and then be replaced. The weights of the boards, at the selected final moisture content, are arrived at by calculation, as explained on page 126. Intermediate weighings give, by calculation, the moisture contents of the moment, and the progress of drying is, therefore, followed closely. The drying schedule can be modified according to whether it is revealed that drying is too rapid or too slow.

In spite of reasonable care it is possible for drying stresses to be set up in the course of a kiln run, sufficiently serious to cause case-hardening or honey-combing,[1] if not actual visible splits or checks. When such stresses are suspected, and also as part of the routine study of drying progress, the consignment should be tested for such stresses. For this purpose, strips or prongs from test pieces cut from the sample boards are used, *vide*, Fig. 32. Cross sections 12 mm thick, cut 22·5 cm or more from the ends of the sample boards, provide the test pieces. For the 'strip' test pieces the cross sections are cut into four equal strips, parallel to the original surfaces of the boards, as in Fig. 32, *a*. Alternatively, the outer strips can provide the prong-shaped test pieces by cutting away the middle portion to within 25 mm of one end, as in Fig. 32, *b*. A study of the behaviour of the strips or prongs will indicate the nature and extent of the drying stresses. In the early stages of drying, tension stresses tend to be set up in the outer layers of the wood. At this stage the strips and prongs will immediately curve outwards when cut, as in Fig. 32 (ii). When the samples are allowed to dry for 12 to 24 hours, to a uniform moisture content, the slightly wetter inner faces of the strips or prongs will shrink more than the outer faces, and, if permanent stresses have occurred, the strips or prongs will eventually bend inwards, as in Fig. 32 (iii). At a later stage in drying, the behaviour of the test pieces is different. By this time the stresses become reversed in the wood, and the core is in tension while the outer layers are in compression. Test pieces cut from boards in this condition will immediately curve inwards, and, when allowed to dry to a uniform moisture content, the extent of curvature will be increased. When no permanent set has occurred the strips or prongs will become parallel, as in Fig. 32 (i), on reaching a uniform moisture content, that is, after 12 to 24 hours' exposure to normal atmospheric conditions. If tests reveal that serious drying stresses have developed, the

[1] For definitions of these terms see page 251.

drying schedule must be modified to relieve these stresses; this aspect is discussed in the section on kiln operation.

It was necessary to dilate on the importance of correct piling technique in air seasoning; correct and careful piling is of no less importance in kiln seasoning, and the trouble taken is always justified because the appearance and quality of the load at the end of a run will be superior, compared with a poorly piled load. With very few exceptions, piles must be built with stickers as for air

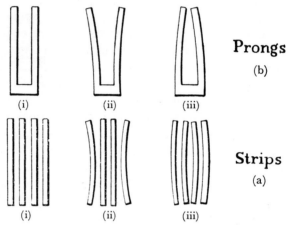

Fig. 32.   Prongs or strips used in testing for drying stresses

By courtesy of the Director, F.P.R.L., Princes Risborough

seasoning – the exceptions are certain classes of dimension stock. The stickers should be selected from clean, dry timber, and finished 25 mm by 25 mm, or for thin boards 25 mm by 19 mm. Softwoods, except larch, and hardwoods not inclined to warp of 50 mm and upwards in thickness, should have stickers spaced at 90 cm centres; for boards, *i.e.*, material less than 50 mm thick, the spacing should be reduced to 60 cm. 50 mm material of larch, beech, birch, and other timbers that tend to warp appreciably in drying, should have stickers spaced at 60 cm centres, and for boards of these species the spacing should be reduced to 45 cm centres. Still more refractory timbers, *e.g.*, elm, should be piled with stickers at 30 cm centres.

The same precautions must be taken regarding the alignment of stickers in vertical rows as in air seasoning, and loads must be distributed to the foundations by means of a system of cross stringers and bearers when the stickers are spaced closer together than the distance between the main supports of a stack. Special supports must be provided for long boards that overhang the ends of a stack. Stacks of boards may be weighted to advantage, to reduce distortion in the upper rows of a stack; concrete slabs are suitable, but ferrous metal weights must not be used with timbers such as oak and chestnut because the tannin in these woods may lead to serious staining when brought in contact with iron in a humid atmosphere.

Construction of kilns. Concrete floors and footings, with provision of drainage for the floor, are recommended. 275 mm brick cavity-wall construction is probably the most economical form of walling; slight ventilation of the cavity is desirable. Reinforced concrete, with provision for thermal expansion, is suggested for the roof of the kiln. The roof of a double-stack kiln will require supporting on steel joists. An independent light roof to protect the kiln, air outlets, motor, etc., from the weather, and also to permit of additional insulation of the kiln roof with sand or sawdust, is strongly advised.

Doors can be a serious source of heat loss; side-hung doors, with heavy stiles and rails, have been proved unsatisfactory. The authorities at Princes Risborough recommend a centre-hung door, constructed with a timber frame, an inner sheet-metal face, protected by paint from corrosion, with an air gap for insulation between this and the outer face of resin-bonded or other waterproof plywood or matchboarding. Further provision against heat loss is secured by 19 mm wide felt strips between doors and jambs, with clamps around the edges so that the door can be pulled up hard against the jamb.

Of the two types of forced-draught kilns, the external-fan kiln is more compact than the internal-fan type, and the regulating apparatus is particularly accessible, but the air is admitted and withdrawn at the same point, so that unless baffles and dampers are carefully arranged there is a risk of the air circulation short-circuiting.

Type of kiln. There are two main types of timber-drying kilns, namely, **progressive** and **compartment** kilns: the former are almost always operated by the 'natural-draught' method, but the latter may be natural- or 'forced-draught' operated.

Progressive kilns. In progressive kilns green timber is admitted at one end and moved gradually to the other, where it emerges dry. The air flows in the opposite direction to the movement of the timber, so that the material that has been longest in the kiln receives the hottest and driest air. In passing through the piles of timber the air absorbs moisture, which increases its relative humidity and lowers its temperature, so that at the loading end the wet timber comes in contact with relatively cool, humid air. The severest drying conditions are, therefore, at the exit end, where the timber is

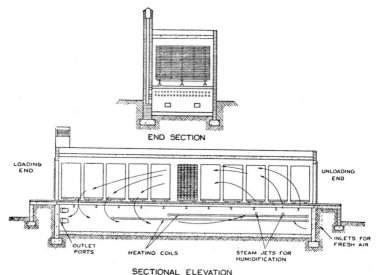

END SECTION

SECTIONAL ELEVATION

FIG. 33. Progressive kiln

By courtesy of the Director, F.P.R.L., Princes Risborough

best able to accommodate itself to them, and the mildest at the loading end, where the timber is least able to stand up to rapid drying. After circulation through the length of the kiln, part of the air is discharged into the atmosphere, and the remainder returns below the floor of the kiln to be re-circulated. Fresh air is admitted at suitable openings to compensate for the amount discharged, and this, mixed with the returning cool air, is heated prior to circulation through the timber. Fig. 33 illustrates the principal points in the design and mode of operation of a natural-draught, progressive kiln.

The uses of progressive kilns are limited since their successful operation depends on a steady supply of timber of the same species and dimensions. The reason for this is that the drying conditions cannot be modified as each new load is added, and while the kiln still contains partially dried loads of a particular type. In addition to lack of flexibility, progressive kilns cannot be regulated with great precision, and this renders them unsuitable for timbers that are difficult to season. Further, as the main air-flow is longitudinal, the timber should be piled at right angles to the length of the kiln (to ensure uniform drying) and this restricts progressive kilns to short stock unless kilns of great width are provided. The advantages of progressive kilns are held to be that, once placed in efficient operation, they require less skill to run, and the output is more or less continuous, compared with compartment kilns.

Compartment kilns. A compartment kiln, like a progressive kiln, consists of a closed chamber, but it differs in operation in that the timber remains in the same position in the kiln throughout the drying period, the temperature and humidity of the circulating air being constantly changed. Such kilns may be operated by the natural-draught or the forced-draught method. The air may be circulated cross-wise from top to bottom of the kiln, or *vice versa*, or from end to end; and the circulation may be reversible. The timber may be stacked on edge, when the main circulation is usually up through the stack and down the sides of the kiln; or flat-piled, when the circulation is up one side, across the stack, and down the other side. Flat piling, the more common method, is sometimes arranged with a central flue, the air travelling up the flue and circulating to either side.

Conditions at the commencement of a run are mild; that is, the relative humidity of the air in circulation is high and its temperature is low. As drying proceeds the temperature is raised and the humidity is reduced, and, as a result, the drying conditions become more severe, but are kept within bounds to prevent degrade of the timber.

The great merit of compartment kilns is their flexibility in operation, coupled with the fact that they can be designed to give precision of control. Flexibility is desirable when the out-turn of a mill is constantly changing, both in species and dimensions of stock; and maximum control of drying conditions is essential for the successful drying of difficult timbers. In effect, compartment kilns are to be preferred to progressive kilns in all but special circumstances; in practice the special circumstances are mills producing continuous supplies of a single timber, of one thickness.

910111213141516171819202122232425262728293031323334353637383940

pipes condense the moisture absorbed from the timber by the circulating air and, since removal of this moisture is continuous, more rapid circulation of air is secured than is possible when de-humidifying of the air in a kiln depends on admission of fresh air from the outside by means of hand-operated vents. Another type of kiln is the **water-spray compartment kiln** designed by Mr H. D. Tiemann of the U.S. Forest Products Laboratory. In this type, water sprays are arranged in rows along the side walls of the kiln and heated air passes through them, and is cooled in the process until it reaches a state of saturation. The saturated air is then re-heated, to reduce its relative humidity to any required figure, before circulation through the timber; by this means the condition of the air in circulation can be precisely controlled. Water-spray compartment kilns are, however, complicated in design, and in operation, and they consume appreciable supplies of water and heat; greater precision in control can be secured with a good type of forced-draught kiln, and such kilns can be quite simple in design, besides being easier to operate than the water-spray type.

Forced-draught kilns may be of the **external-fan** or **internal-fan** type. In the external-fan kiln the air is heated, humidified, and set in motion by apparatus located outside the kiln, whereas in the internal-fan type the heating, humidifying, and circulating apparatus is situated inside the kiln, either in the roof or basement. Figs 35 and 36 illustrate an internal-fan kiln (double-stack pattern), the design of which was evolved at the Forest Products Research Laboratory, Princes Risborough. The fans, heating coils, steam sprays, and used-air outlets, are situated in the roof, and the fresh air is admitted at floor level. This arrangement makes for economy in construction over basement-type kilns, while providing reasonable accessibility, immunity from flooding, and ease of loading. The kiln illustrated in Figs 35 and 36 is undoubtedly one of the most efficient and economical types of kiln at present available; it embodies the results of extensive research and practical experience in operation of overhead, internal-fan kilns, having been developed at the laboratory from earlier prototypes. It probably offers more uniform air circulation than any kiln of this size yet designed, and it provides the maximum timber accommodation for a minimum of standard engineering components.

The width of stacks in the double-stack kiln should be about

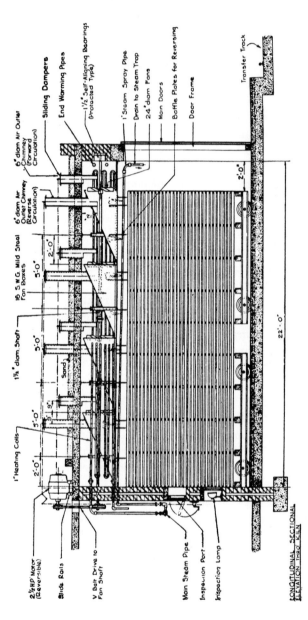

FIG. 35. Internal-fan kiln (double-stack pattern)

By courtesy of the Director, F.P.R.L., Princes Risborough

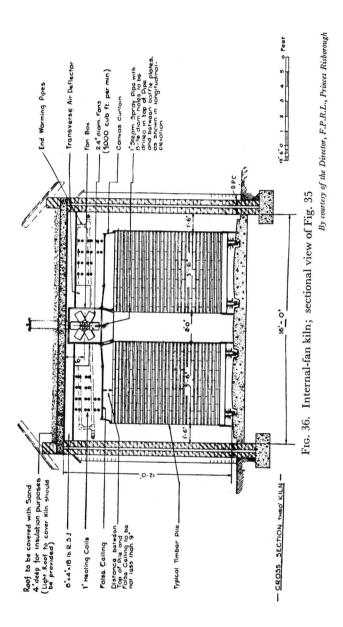

Roof to be covered with Sand
4' deep for insulation purposes
(Light Roof to cover Kiln should
be provided)

End Warming Pipes

Transverse Air Deflector

Fan Box

24" diam. Fans
(5000 cub ft. per min)

Canvas Curtain

1" Steam Spray Pipe with
6-¹⁄₁₆" diam. holes to be
drilled in top of pipe
and between baffle plates
as shown in longitudinal-
elevation

6"x4"x18 lb. R.S.J

1' Heating Coils

False. Ceiling

Distance between
Top of Pile and
False Ceiling to be
not less than 9"

Typical Timber Pile

DPC

1'-6' 5'-6' 2'-0' 3'-6' 1'-6'

16'-0'

12'-0'

CROSS SECTION THRO' KILN

2'-6'·0 1 2 3 4 5 6 Feet

FIG. 36. Internal-fan kiln; sectional view of Fig. 35

By courtesy of the Director, F.P.R.L., Princes Risborough

1·7 metres and should not exceed 1·8 metres. Allowing space for two stacks, a 0·6 metre centre corridor (the diameter of the fans), and inlet and outlet ducts, the maximum width of the kiln is 5·2 metres. The height from the bottom of the timber stack to the false ceiling of the kiln should not exceed 2·4 metres. If the volume of timber to be dried does not fill the kiln the unused space must be blanketed off with curtains to prevent the circulating air from short-circuiting over the tops of the stacks.

Fans are required at intervals of 1·5 metres, and 0·6 metre is required between the ends of the stacks and the kiln doors. These requirements impose a minimum economical length of 3·6 metres, and where this dimension provides too large a kiln for a mill's requirements some other pattern is likely to be more economical.

For mills requiring kilns of smaller capacity than the one described above – and in this country, where many thicknesses and species are handled by one mill, this will often be the case – a smaller, cross-shaft, overhead fan kiln has been designed at the Forest Products Research Laboratory, *vide* Fig. 37. The aim has been to provide a highly efficient kiln with the minimum of metal constructional parts, as these call for skilled labour in erection. By substituting 0·91 metre diameter fans for the 0·6 metre diameter fans used in the double-stack kiln, and a completely reversible system of air circulation, it has been possible to increase the width of the pile to 2·13 metres, and yet ensure uniform drying of the whole load. Adding the 0·45 metre for inlet and outlet ducts on either side, the maximum width of the kiln is 3·05 metres. The height from the bottom of the stack to the false ceiling is 2·6 metres, and a clearance of 0·23 metre between the top of the pile and the ceiling is recommended. The whole length of the kiln may be utilized for timber, but in this case access doors to the side ducts should be provided. The length of the kiln can vary within limits to meet a mill's needs, so long as adequate provision is made for drainage.

Considerable latitude in the selection and design of the heating arrangements is possible; ordinary 25 mm steam piping and fittings have been found quite satisfactory. Details depend on the timbers to be dried and the maximum temperatures likely to be required: these and other points are set out in *Forest Products Research Laboratory* leaflet No. 18.

Forced-draught compartment kilns have the great merit of

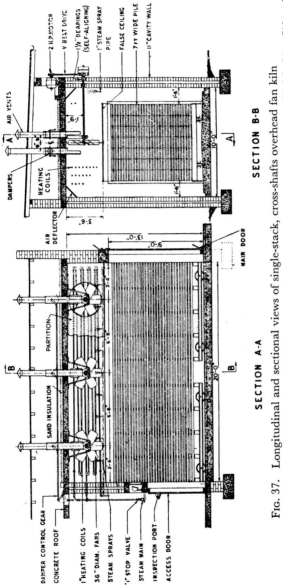

FIG. 37. Longitudinal and sectional views of single-stack, cross-shafts overhead fan kiln

By courtesy of the Director, F.P.R.L., Princes Risborough

adaptability for drying any kind of timber in any condition, and they can be regulated with precision to suit all circumstances. Such kilns, however, are more costly to build, and complicated to construct and operate, than natural-draught kilns.

Furnace-heated kilns. The progressive and compartment kilns so far described depend on steam for heating: several attempts have been made to utilize the heat from burning wood-waste direct to dry timber, and **furnace-heated kilns**, as they may conveniently be called, of proprietary manufacture have been on the market for some years. In their simplest form, reliance is placed on moisture expelled from the drying wood to provide suitable humidity conditions in the kiln, and not unnaturally this has often proved unsatisfactory for drying refractory timbers. The use of wood-waste for fuel, and the elimination of a boiler and steam piping, are, however, obviously attractive points in a simple, inexpensive kiln. For these reasons attention was paid to the design of furnace-heated kilns in the war years both at the Forest Research Institute, Dehra Dun, and at the Forest Products Research Laboratory, Princes Risborough. A kiln evolved at the latter laboratory is illustrated in Figs. 38 and 39. This follows closely the design of the single-stack compartment kiln described above; it is 6·1 metres long, and accommodates a trolley-loaded pile of timber 1·8 metres wide. Air circulation is secured with 0·91 metre diameter fans, reversible in operation, and driven at a speed of 500 r.p.m. by a 2 h.p. electric motor. Heating of the kiln is effected with 125 mm diameter wrought-iron flue pipes and headers. The flue gases from the sawdust-burning furnace enter the heating system through a manifold, situated half-way along the length of the kiln, travel to the headers at either end, and return through the pipes to an exhaust manifold.

Attention has been given to design of the furnace to ensure automatic feeding, steady burning, and constant heat output, and with a little experience on the part of the operator these ends are achieved. Unlike earlier patterns of furnace-heated kilns, reliance is not placed entirely on moisture expelled from the drying timber to maintain suitable humidity conditions; high humidities are provided by means of a system of water drips. Hence, humidity control is not effected solely by manipulation of the air inlet and outlet ducts. A distributor tank feeds water uniformly through pipes to convenient points along the length of the kiln, where it flows into horizontal

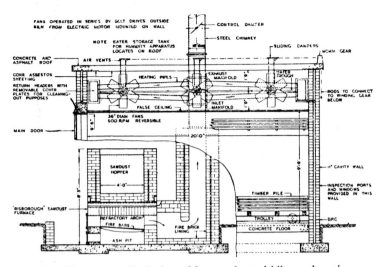

FIG. 38. Longitudinal view of furnace-heated kiln, and section
through furnace

By courtesy of the Director, F.P.R.L., Princes Risborough

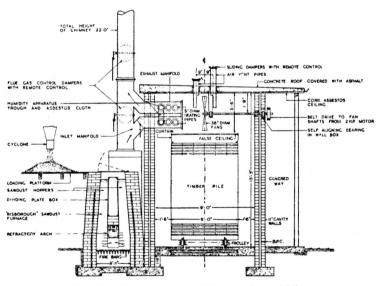

FIG. 39. Sectional view of furnace-heated kiln

By courtesy of the Director, F.P.R.L., Princes Risborough

troughs. Arrangements are made, by means of V-shaped pieces of copper wire, bound with cotton bandage, for drips of water to fall directly on to the top surface of the flue pipe below.

Separate consignments of 50 mm Douglas fir, 25 mm oak, and 75 mm Scots pine have been successfully dried in a kiln of this design; the Douglas fir from 50 to 17 per cent. moisture content in 9 days, the oak from 85 to 14 per cent. in 36 days, and the Scots pine from 70 to a 15 to 20 per cent. range in 15 days. In all cases it was not found possible to increase the humidity at the end of the run sufficiently to relieve the case-hardening stresses completely. In spite of this limitation, a furnace-heated kiln of this type obviously has considerable possibilities, and is a great advance on kilns similarly heated, but without provision for forced-air circulation or augmenting of the humidity by water drips. The kiln was fully described in *Wood*, September 1944.

Choice of kiln. The primary conditions for determining the choice of the type of kiln to be installed depends on such points as first cost, the volume of timber to be handled, the space available for the kiln, the condition of the timber to be dried (*i.e.*, whether green or partially or fully air-dry), and the kind of timber. The special merits of the different types of kiln have already been indicated.

Kiln installation involves considerable capital expenditure, so that first cost should not be allowed to weigh too heavily in the selection of a particular type of kiln: a higher initial cost may justify itself by proving more economical in the long run. For example, the more expensive forced-draught kiln provides a greater out-turn per cubic metre of kiln space than the natural-draught type, and it ensures the maximum control of drying conditions, thereby reducing degrade in seasoning to a minimum. Either of these factors may well justify the higher initial cost of installation of a forced-draught kiln. Of the two classes of forced-draught compartment kilns, the external-fan type is the more compact, and the operating apparatus is especially accessible, but the internal-fan kiln, particularly if of the overhead variety, provides more uniform drying conditions and fewer problems in securing proper air circulation. The last point is all-important; a kiln constructed in accordance with a standard set of plans will present problems of its own before air circulation under working conditions is correctly adjusted. The conclusions reached, as a result of extensive research, indicate that the overhead internal-fan kiln has much to recommend it, and, when starting from scratch, as opposed to modernizing an old-type kiln, it would seem inadvisable to ignore these findings. The double-stack and single-stack kilns described and illustrated above have been proved satisfactory in commercial operation; they are not unduly costly, and they are relatively simple both in construction and operation.

Kiln operation. Successful kiln-drying is very largely dependent on the skill of the kiln operator: a poorly designed kiln will give better results in the hands of a good man than the most up-to-date and efficient kiln in unskilled hands. It is essential to confine any run to one species and one dimension of stock; it is equally important to make proper use of kiln instruments, by siting them correctly and maintaining them in good condition; and progress of drying must be followed by means of sample boards. Given these essentials, it is also advisable to use a drying schedule that provides a margin for error: typical schedules for different timbers, published by the Government research laboratories, have been evolved from experimental practice and provide such margins. Modifications may still be called for: good-quality material will tolerate more severe drying conditions than poor-quality timber; quarter-sawn boards will dry more slowly, but with less degrade, than flat-sawn material of the same species.

Examples of typical drying schedules are given in Tables 5 and 6. In both schedules it will be seen that the initial temperature is lower than the final temperature, and the initial relative humidity is higher than the final relative humidity. In effect, conditions in the kiln are made progressively more severe as drying proceeds.

Temperature and humidity control in the kiln is effected by manipulation of heating coil and spray valves, with occasional adjustments of air inlet and outlet dampers. Economy in operation of kilns is dependent on making the maximum use of the moisture extracted from the timber for maintaining the required humidity of the circulating air. This is effected by allowing only very slightly more moisture to escape *via* the outlet ducts than is being extracted from the timber, the deficit being made good by comparatively small amounts of steam from the steam spray system.

Warming up must not be too rapid because of the time lag in heating the wood to the recorded temperature of the air. Too rapid heating of the air in a kiln may result in condensation of

TABLE 5.

Kiln Schedule A

Suitable for timbers which must not darken in drying and for those which have a pronounced tendency to warp but are not particularly liable to check.[1]

Moisture content (%) of the wettest timber on the air-inlet side at which changes are to be made	Temperature (Dry bulb)		Temperature (Wet bulb)		Relative Humidity % (Approx.)
	° F.	° C.	° F.	° C.	
Green	95	35	87	30·5	70
60	95	35	83	28·5	60
40	100	38	84	29	50
30	110	43·5	88	31·5	40
20	120	48·5	92	34	35
15	140	60	105	40·5	30

TABLE 6. *Drying schedule for timbers that dry very slowly, but which are not particularly prone to warping*[2]

Moisture content (%) of the wettest timber on the air-inlet side at which changes are to be made	Temperature (Dry bulb)		Temperature (Wet bulb)		Relative Humidity % (Approx.)
	° F.	° C.	° F.	° C.	
Green	120	48·5	115	46	85
60	120	48·5	113	45	80
40	130	54·5	123	50·5	80
30	140	60	131	55	75
25	160	71	146	63·5	70
20	170	76·5	147	64	55
15	180	82	144	62·5	40

[1] Schedule A of Appendix III, *A Handbook of Hardwoods*, 1956, H.M. Stationery Office, London. Reproduced by courtesy of the Director, Princes Risborough Laboratory, Princes Risborough, Aylesbury, Bucks.

[2] Schedule G of Appendix III, *A Handbook of Hardwoods*, 1956, H.M. Stationery Office, London. Reproduced by courtesy of the Director, Princes Risborough Laboratory, Princes Risborough, Aylesbury, Bucks.

moisture on the surface of the timber, which takes appreciably longer to warm up. Too high temperatures in the course of a kiln run are to be avoided, as high temperatures may darken the whole consignment. Rapid cooling at the conclusion of the run is frequently possible, but there is a risk that the hot timber will heat the cool air entering the kiln, making it appreciably drier, and this may lead to a renewal of case-hardening stresses or even splitting of the wood. It is suggested that a difference of 9° F. (5° C.) between wet and dry bulb readings should be maintained in the initial warming period, until the desired dry bulb reading is attained, and a similar difference during cooling, until the dry bulb reading has dropped to about 80° F.

The behaviour of different timbers in kiln-drying varies enormously; in general, softwoods are much less refractory than hardwoods. With the latter, it is common practice partially to air-dry the stock first, as otherwise the kiln run is too long to be economical. Notwithstanding preliminary air seasoning, commercial kiln-drying schedules may occupy anything from a few days to three weeks or more.

CHEMICAL SEASONING

Air and kiln seasoning are methods of drying wood, aimed at making the material more suitable for use. **Chemical**, or **salt seasoning** as it is sometimes called, has exactly the same object.

Chemical seasoning relies on the principle that aqueous solutions of certain chemical substances have lower vapour pressures than pure water. By treating the outer layers of wood with certain salt solutions the vapour pressure of the contained moisture in these layers is reduced, which establishes a vapour-pressure gradient in the piece; there is no immediate drying in the surface layers. Further, the equilibrium moisture contents of such chemically impregnated timber are higher than the equilibrium moisture contents of untreated wood, *vide* Fig. 40. In consequence, it is possible to maintain a vapour-pressure gradient, and therefore movement of moisture outwards in the piece, while the equilibrium moisture content of the surface layers is above fibre saturation point. The risk of setting up drying stresses in these layers is thereby eliminated. In theory, the careful adjustment of temperatures and humidities subsequently ensures continued, uniform drying to the final moisture content required, at a faster rate than is possible in safety with untreated wood.

Chemical seasoning involves treating the surface layers of green timber with a suitable chemical salt before the seasoning process is commenced.

Many chemicals and combinations of chemicals have been tried, and such different methods of application of the chemicals as dry-spreading, soaking, dipping, and spraying. The outstandingly cheap chemical for chemical seasoning is common salt (sodium chloride), which is, however, bad from the corrosive standpoint, and because of the liability of the treated timber to sweat subsequently. Urea has neither of these objections; it has been employed on a considerable scale, both experimentally and commercially.

The depth of penetration of the chemicals in chemical seasoning is usually not great: wet sapwood will take up the chemicals very quickly, but only the outer layers of the heartwood are affected. This is because absorption occurs by diffusion, which is rapid only if the moisture content of the wood is high enough to provide a continuous film of water within the wood into which the chemical can diffuse.

When drying treated timber, as in a kiln run, a moisture gradient is set up as in untreated wood, but with the all-important difference that the equilibrium moisture contents of the treated outer layers are different from those of untreated wood. Movement of moisture from the centre to the surface of the wood occurs while the equilibrium moisture content of the outer layers is well above fibre saturation point. This condition can be maintained for some time by constant adjustment of the relative humidity and temperature of the air in the kiln. By eliminating drying in the surface layers in the early stages, the whole consignment in a kiln can be dried in relative humidities that would cause untreated wood to check. Large-size timbers, 150 mm by 150 mm and up, of Douglas fir, for example, will surface-check when dried with relative humidities as high as 90 per cent., whereas the same timbers, after chemical treatment, have been safely dried with relative humidities of 75 per cent.

Chemical seasoning is most effective when the vapour pressure of the drying air is kept in equilibrium with the vapour pressure of the solution in the wood. This can be arranged, and yet allow the untreated interior to attain a moisture content in equilibrium with the drying conditions while the treated layers remain above fibre saturation point. It will be seen from Fig. 40 that the equilibrium moisture content of treated wood is at fibre saturation point until the relative vapour pressure of air at 70° F. falls below about 0·77, and for such atmospheric conditions the equilibrium moisture content of untreated wood is around 17 per cent. The curves for other temperatures are, of course, different, but they show a similar relationship. It will be apparent that, by selecting a suitable

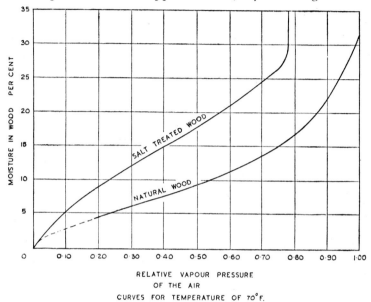

Fig. 40. Equilibrium moisture contents for salt-treated and natural wood for different relative vapour pressures (temperature of 70° F.)

By courtesy of the Director, F.P.L., Madison, U.S.A.

schedule for a kiln run, chemically treated wood theoretically can be dried almost to the final stage before the outer layers are dried below fibre saturation point, and seasoning stresses in the outer layers could, therefore, be eliminated.

Were wood to be dried so that the outer layers were still above fibre saturation point while the moisture content of the interior fell to 17 per cent., stresses would result in the interior sufficient to cause honey-combing. This has actually occurred in experiments with salt seasoning. It follows that care must be exercised to ensure that, while eliminating drying stresses in the outer layers, stresses in the interior are also avoided.

Apart from the seasoning of large-size timbers, the practical outcome of chemical seasoning rests on this: can a suitable schedule be selected that is sufficiently fast to outweigh the cost of the chemical used and the double handling that the chemical treatment involves? Kiln-drying schedules that give perfectly satisfactory results have been worked out for untreated refractory timbers. Chemical seasoning must, in the circumstances, reduce the time of the run in terms of money, by at least as much as will be expended on the chemical treatment, for it to have any practical advantages. It is in this direction that the claims of chemical seasoning have yet to be proved. For example, with suitable drying schedules, urea is undoubtedly satisfactory, but with such refractory timbers as English oak, it has not been possible to reduce the drying schedules sufficiently to justify its use on economic grounds. Complete success with English oak has only been achieved with schedules no more drastic than those suitable for untreated material of this species. The occasions when really large-size timbers are required fully seasoned must be comparatively rare, and the fact that they can be dried only with the aid of chemical treatments does not, in itself, bring chemical seasoning within the sphere of a practical seasoning method in ordinary circumstances.

There is another aspect to chemical seasoning than merely changing the moisture retention qualities of the surface layers of treated wood to permit of accelerating the drying process. This is the possible influence of the chemicals used on the hygroscopic properties of the cell wall structure. This aspect has been referred to as anti-shrinkage treatments. The theory is that during the process of treatment, and subsequent drying of the treated wood, some of the chemicals used may diffuse into the fine structure of the cell wall. As drying progresses moisture is given up, but the chemical absorbed is deposited in the walls and prevents their normal contraction. It will be apparent that appreciable quantities of a suitable chemical must be used to have a marked influence on total shrinkage – once more it is an economic question. At the same time, it is obviously desirable to assess both the moisture retention qualities and the anti-shrink properties of salts that may be used in chemical seasoning.

But cost, moisture retention, and anti-shrink are not the sole criteria. The chemical selected should be soluble in water, non-poisonous and harmless to handle, and with good storing qualities;

it should not discolour the wood, nor affect such subsequent processes as painting, varnishing, and gluing. It must be non-corrosive to metals, as this would otherwise restrict the usefulness of the timber when dried, besides damaging the kiln in which the drying was done. Moreover, the dried, treated timber must not be (a) more recalcitrant to work with tools, (b) inclined to sweat when exposed to more humid conditions, (c) more inflammable, (d) more liable to insect or fungal attack, and (e) the electrical conductivity should not be increased. Some chemicals have fire-retardant and toxic properties, which enhance their usefulness.

In conclusion, it is not inappropriate to quote from an American publication, *A primer on the chemical seasoning of Douglas fir*,[1] that appeared as far back as November 1938: 'In appraising the commercial significance of chemical seasoning a distinction must be made between what the process will do under ideal conditions and what it will profitably do under average commercial conditions. To say that timbers of a given size can be chemically seasoned without surface checking is quite different from saying that treated timber of the same size can safely or profitably be kiln dried in the average run of commercial kilns in a practical length of time. . . . In kiln-drying the more refractory items after chemical treatment, a hair-trigger control of drying conditions is required which is not often attained in commercial kilns.' These truisms are no less applicable today, in spite of the knowledge and experience that has been accumulated in the intervening years. Chemical seasoning undoubtedly has wide possibilities, but it is not the panacea of every seasoning problem. Further, when comparing drying schedules for untreated timber with those recommended for chemical seasoning, it is imperative to know whether the same latitude has been allowed in both schedules. Drying schedules that have been published from time to time by Government research laboratories give the operator a margin for error. Such margins can by no means always be counted upon in schedules proposed by the enthusiasts for chemical seasoning.

DRYING OF WOOD ELECTRICALLY

The resistance of wood to the flow of electrical currents has been adapted for many years for measuring the moisture content of

[1] Publication No. R 1278 of the Forest Products Laboratory, Madison, Wisconsin.

wood, *vide* pages 128 to 131. A development from this is the use of electricity for the drying and seasoning of timber. Initial experimental work, involving the use of low-frequency currents, has not given encouraging results: excessive splitting has been the common experience, suggesting the impracticability of this method for drying. It is not difficult to explain such excessive splitting: as wood dries it develops a very high resistance to the flow of an electric current, and consequently there is a considerable output of heat. This heat still further accelerates drying in the immediate vicinity. In consequence, drying tends to be anything but uniform throughout the piece: the layers of wood immediately in contact with the plate or electrode dry out rapidly, while those further in are still wet. Such unequal drying sets up seasoning stresses, which are so pronounced that rupture of the tissues – or serious splitting – is inevitable. This, in point of fact, is the experience to date in the restricted amount of experimental work that has been done in the drying of wood by low-frequency electrical currents.

A development of electrical drying of wood has been the application of high-frequency alternating currents for drying purposes, and in plywood manufacture, for example, it may be said that high-frequency heating – also called radio-frequency heating or di-electric heating – has passed beyond the experimental stage and is a proven, practicable commercial method for the setting of glues. The application of the method to dimension timber, *i.e.*, boards, planks, and scantlings, has, however, been on a comparatively small scale, and under laboratory conditions. The results to date have, however, been highly encouraging: there seems to be no question that timber in ordinary commercial sizes could be so dried, both rapidly and with the minimum of degrade, but it cannot be regarded as economically practicable on a commercial scale for timber, except in veneer form, in the foreseeable future, because of the very high capital cost of the electrical apparatus required, and the cost of electricity, compared with steam, as a source of heat.

The theory behind the application of a high-frequency alternating current to the drying of wood is interesting, and it explains the freedom from seasoning degrade that the method secures. Certain substances are poor conductors of electricity, and when placed in the field of an alternating current of high frequency, they become hot. This is explained by the frequent re-alignment of the randomly arranged molecules of the poor-conducting substance, induced by the constant change in direction of flow of the electrical current: such changes in position, rapidly repeated, cause the generation of heat. Expressed diagrammatically, we may assume that the complex molecules of wood substance are arranged as in Fig. 41 (*A*). The molecules are then placed in the field of an alternating current of high frequency by means of two plates, or electrodes, placed on either side of the piece of timber, and connected up in a high-frequency field. When the current is switched on the molecules proceed to rearrange themselves, taking up positions as in Fig. 41 (*B*).

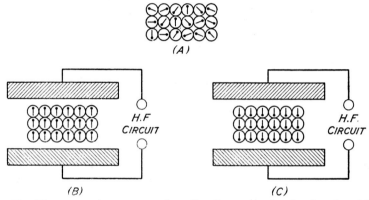

FIG. 41. Diagrammatic representation of realignment of molecules placed in a high-frequency field. (Drawn by T. W. Paddon, Esq.)

By courtesy of the Cleaver-Hume Press

Alternation of the current, that is, reversal in direction of the flow, results in realignment of the molecules as in Fig. 41 (*C*). The constant change in direction of the flow of the electrical current, causing constant change in position of the molecules, results in the generation of heat. This heat in turn induces water movements in the wood: in fact, drying or seasoning proceeds. The great advantage of this method of drying is that, from the commencement, all the molecules making up the wood substance of the piece, enclosed in the high-frequency field, are involved: heat is generated equally in the centre and in the outside layers of the wood. In consequence, moisture movements set up are not confined to the outer layers, as with low-frequency currents, and drying of the wood proceeds uniformly throughout. It follows that this reduces seasoning stresses to a minimum, and, consequently, degrade is eliminated.

SEASONING OF TIMBER WITH
DEHUMIDIFYING EQUIPMENT

An innovation in 1964 was the application of dehumidifying equipment to the seasoning of timber. By 1970 Westair Dynamics Limited had evolved a specially designed plant for drying timber. The principle of this method is to draw dry air through stacked timber in a 'drying' chamber, so that it picks up moisture. The warm wet air is then drawn through a refrigeration circuit, where the air is cooled to its dew point, securing condensation of the moisture absorbed from the timber. The condensed water is taken away *via* an outlet drain tube and the air is then passed over heating coils prior to being expelled from the unit for re-circulation, using forced-draught circulation with high-level fans as in standard steam-heated, forced-draught circulation kilns. The three most commercially relevant facts are:

1. Drying times are longer than with conventional systems.
2. The quality of timber after drying is excellent.
3. The capital and running costs of drying are extremely low.

Diagrammatic examples of Westair Standard and High Capacity Timber Seasoners are illustrated in Fig. 42 (i) to (iii). Low-capital outlay is an unquestionable advantage of this method of drying timber, it being possible to adapt an existing building as a suitable drying chamber. Those who have developed this equipment claim that its advantages are absolute safety in operation, minimum degrade of timber, and totally efficient use of energy, resulting in low operating costs.

FIG. 42. Typical drying room layouts showing examples of Westair standard and high capacity timber seasoners.

By Courtesy of Westair Dynamics Ltd.

(i) The most commonly used layout for a Westair kiln. In this figure one Standard Timber Seasoner is shown drying approximately 7 m³ (250 ft³) timber (capacity will vary according to thickness). Particular note should be taken of the fractional horse-power circulating fans positioned in the dropped beam above the timber stack. These fans create the primary air circulation in the kiln and the Timber Seasoner draws air from this primary circulation. (a) walls and roof max. 'U' value 0·25 (b) drain to suitable point (c) standard timber seasoner (d) doors to suit application (e) air flow (f) standard timber seasoner (g) air deflectors fixed to unit (h) fractional horse-power circulating fan in dropped beam (j) flexible baffle of canvas or heavy gauge polythene to stop short circuiting of air (k) floor level.

(ii) Where floor area is at an absolute premium the Standard Timber Seasoners can be positioned with the circulating fans in the dropped beam above the timber stack. If one compares (ii) with (i) it can be readily appreciated that a much larger timber stack can be accommodated in the same sized building with the arrangement as in (ii). This arrangement does, however, require that the roof is 'load bearing' and also means that maintenance of the units is difficult. (a) walls and roof max. 'U' value 0·25 (b) drains to suitable point (c) standard timber seasoner (d) doors to suit application (e) standard timber seasoner (f) drain to suitable point (g) fractional horse-power circulating fan in dropped beam (h) dropped beam supporting circulating fans (j) flexible baffle of canvas or heavy gauge polythene to stop short circuiting of air (k) floor level.

(iii) The arrangement incorporating the High Capacity Timber Seasoner. The unit is shown positioned external to the kiln in a chamber of its own. The controls of the unit are positioned in such a way to be adjusted from within the external chamber which allows the operator completely free access to the unit during the drying operation. (a) walls and roof max. 'U' value 0·25 (b) drain to suitable point (c) high capacity timber seasoner on base (d) doors to suit application (e) high capacity timber seasoner (f) base (g) dropped beam supporting circulating fans (h) flexible baffle of canvas or heavy gauge polythene to stop short circuiting of air (j) floor level.

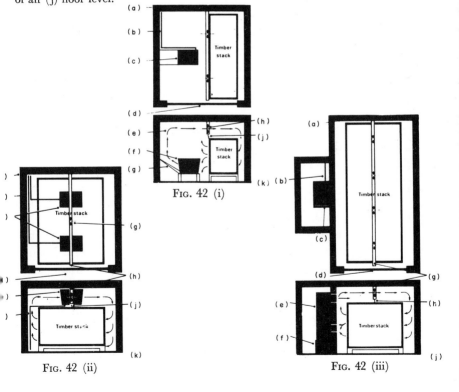

FIG. 42 (i)

FIG. 42 (ii)

FIG. 42 (iii)

CHAPTER 12

Defects in Timber

TIMBER, being a natural product, is seldom entirely free from blemishes and other imperfections that tend to lower its economic value; these are spoken of collectively as **defects.** A feature that in some circumstances is considered a blemish may, in different circumstances, be held to enhance the appearance of a piece of wood, when it would not of course be classified as a defect.

Defects may be classified under two broad heads: natural defects, that is, defects resulting from factors influencing the growing tissues of the living tree, and defects resulting from the activity of external agents or the subsequent treatment of felled timber. Defects caused by fungi and insects are discussed in separate chapters (Chapters 13 and 14).

NATURAL DEFECTS

Knots are, perhaps, the commonest defect in timber. As a tree increases in diameter it gradually envelops the bases of branches; the portions of the branches enclosed within the wood of the trunk are called knots. If the branches are alive at the time of their inclusion their tissues are continuous with those of the main stem and the knots so formed are said to be **live** or **tight knots.** When a branch dies a stump remains, which is gradually surrounded by the tissues of the trunk, but, being dead, its tissues are not connected with enveloping tissues of the main stem, and a **loose** or **dead knot** results; such knots fall out, either when timber is converted, or after it is seasoned and when it is being worked up. The broken stubs of dead knots provide ready access for decay and, consequently, dead knots are frequently unsound.

Knots vary in size from little more than a pin-head to many milli-metres in diameter. They also vary in shape, according to the angle at which they are cut through in conversion. They are defined in CP 112 (1967) under five heads: **splay, arris, edge, margin,** and **face knots.** A splay knot is one cut more or less parallel to its long axis, appearing cone-shaped on one face and semi-circular on the adjacent face or edge. Such knots have also been called **spike knots.** An arris knot, as the name suggests, is one that appears on two adjacent edges, the piece of included branch being inside the piece, instead of appearing on one face as in a splay knot. An edge knot is one appearing, either round or elliptical, on the edge of a board or plank, whereas a face knot is a similar knot, but on the face as opposed to the edge of a board or plank. A margin knot is defined as one "appearing on a face outside the middle half of the depth of the face near to, or breaking through an edge" (CP 112:1967). Knots have an important bearing on the utilization of timber; in many species they are the primary cause of degrade, *i.e.*, a lowering of quality below the best or **prime grade.** Knots may spoil the appearance of boards, although in "knotty pine" the abundance of knots is regarded by some as a decorative feature, and such timber is specifically selected for panelling. The effect of knots on strength properties is discussed on pages 64 to 66; they may also adversely affect seasoning, machining, and ultimate painting.

Bark pockets. Pockets of bark are sometimes enclosed in the wood of the main stem. They result from injury to the cambium. Growth ceases locally until the adjacent cambium has completed the occlusion of the damaged area, resulting in portions of bark becoming embedded in the wood. Such pockets obviously constitute a defect, the seriousness of which depends on the size of the pocket and the extent to which decay may have developed in the vicinity. A bark pocket in a plank of chengal is illustrated in Plate 49, fig. 2; this pocket undoubtedly originated from an old tapping cut.

Pith flecks. Patches of abnormal parenchymatous tissue, called **pith flecks,** occur in some timbers, as a result of the tunnelling of the cambium by the larvae of certain insects. Pith flecks are usually wider tangentially than radially, and extend considerable distances vertically; their inner faces follow the outline of the cambial sheath and their outer faces are irregular in outline (Plate 48). Pith flecks are a common feature of some timbers, *e.g.*, alder, birch, maple, sycamore, but they are not sufficiently constant in occurrence to be of more than subsidiary value in identification.

Included phloem, although a normal feature of some timbers, is usually considered a defect as far as utilization is concerned. (See pages 46 and 47 and Plates 25 and 23, fig. 4.)

Pitch pockets are described on page 31, and illustrated in Fig. 7. In the Canadian literature they are classified according to size as *small, medium,* or *large*. When a pitch pocket is cut through at its widest part, so as to appear as a shallow opening on the longitudinal face of a piece of timber, it is called a **pitch blister. Pitch seams** or **shakes** are openings along the grain that follow the outline of the growth rings. Pitch pockets, blisters, seams, or shakes are for the most part defects of softwood species, but similar defects also occur in one hardwood family, the *Dipterocarpaceae*. In this family, however, defects of the types referred to are much smaller than typical, corresponding defects in softwoods, and their contents are usually solidified dammars or oleo-resins. A resin pocket in a plank of balau is illustrated in Plate 49, fig. 1.

Gum veins are traumatic canals that occur in some woods (Plate 22, fig. 4); they are usually filled with dark-coloured deposits. In some timbers, *e.g.*, jarrah, they are usually infrequent in occurrence, but timber from fire-swept forest may contain gum veins in considerable numbers that constitute definite defects; in other timbers, *e.g.*, African walnut, they are so frequently present as to constitute a characteristic feature of the wood, which may enhance its appearance.

Mineral streaks are defined as 'localised discoloration of timber, in the form of streaks or patches usually darker than the natural colour, which does not impair the strength of the piece'.[1] Mineral streaks have been found in sycamore and wych-elm. The term has also been applied to the light-coloured streaks occurring in timbers of the family *Dipterocarpaceae*, *e.g.*, lauan, meranti, seraya, keruing, gurjun. These streaks are really the resin canals in longitudinal section, which, because of their white or yellow contents, show up against the red or brown background of the wood.

[1] *British Standard Terms and Definitions Applicable to Softwoods*, British Standards Institution, No. 505, 1934.

Resin streaks or **pitch streaks** are narrow brown streaks extending along the grain, and fading out gradually, that occur in spruce, Douglas fir, and other softwoods (Plate 54, fig. 4). They are caused by local accumulations of resin in the tracheids. Resin streaks may sometimes be confused with discoloration caused by incipient decay, but they can be distinguished because the darkened wood is not soft or otherwise affected; the strength properties are not influenced in any way.

Strawberry mark is a minor defect sometimes encountered in sitka spruce. The discoloration is caused by accumulation of resin, and consists of a red-brown zone up to 2·5 cm wide and 3·75 cm high, running radially through the wood. In truly radial faces the defect appears as a bar of darker-coloured wood running across the piece, but, if the cut surface is oblique, the discoloration usually appears more as in Plate 49, fig. 4. Unless the accumulation of resin is accompanied by enlargement of the rays to many times their normal size, the strength properties of the wood are unaffected, and the defect is no more than a minor blemish.

Latex canals are described on page 46, and illustrated in Plate 24. The large canals occur in relatively few woods, *e.g.*, those of the family *Apocynaceae*; in grading they cannot be treated as ordinary defects as they are a natural feature of the structure of such timbers. For some purposes, *e.g.*, as cores for face veneers, and uses in plywood manufacture, requiring timber in short lengths, the presence of latex canals is immaterial.

Compression failures are actual horizontal ruptures in fibre or tracheid walls, which arise 'naturally' in some timbers from wind action bending the bole in the early years of a tree's life. Compression failures may also be induced by careless felling, or as a result of over-stressing of timber in service. The ruptures or fractures in the cell wall, called **slip planes,** are conspicuous in specially prepared sections, examined under polarised light. In the actual plane of failure slip planes are very numerous, and it is usually impossible to determine whether the timber has been overstressed previously. If, however, only partial failure has occurred, producing **compression creases,** *i.e.*, puckering of the surface of the wood, only apparent on careful examination, the extent of slip planes are an indication of the severity of over-stressing.

Stresses that develop as converted timber dries may cause ruptures in adjacent fibres to link up, producing zigzag hair cracks across the grain (Plate 49, fig. 3, and Fig. 43, *g*). The hair cracks are not easy to detect, except in planed material, but their presence should be suspected if the cross-cut ends of boards or planks are at all carroty in appearance, or if there are numerous horizontally-broken fibres across the swage marks left by a circular saw on longitudinal faces.

Compression failures are often associated with the presence of **'spongy'** or **'punky heart'**, now more correctly called **brittle heart,** which is particularly common in logs of lauan and meranti, but it also occurs in African mahogany and other tropical timbers, *e.g.,* afara. Compression failures usually extend considerably beyond the conspicuous limits of brittle heart, which explains why defective boards or planks may find their way into otherwise well-graded parcels. Compression failures are a serious defect if present in large numbers; their influence on strength properties is discussed on page 180. Affected boards may even break in two when lifted, which gave rise before the war to the expressive term *'three-men-boards'* for a low-grade of packing case stock shipped from Singapore to Mauritius, the third man being required to support the board in the middle!

Compression failures have also been called *thunder-shakes, lightning-shakes*, and *cross-breaks*.

DEFECTS ARISING FROM OTHER THAN NATURAL CAUSES

SEASONING DEFECTS

Next to knots the commonest causes of degrade in timber are defects resulting from faulty seasoning technique. It has been explained that as wood dries it shrinks, and that the shrinkage is not uniform in all directions. Moreover, the outer layers of a piece of wood tend to dry out more rapidly than the interior, resulting in temporary or permanent distortion of the timber, and even in the separation or rupture of the tissues. The different forms of permanent distortion of timber and of ruptures of tissues are defects; separately and together they are referred to as **degrade in seasoning** or **seasoning degrade.** Permanent distortion gives rise to various forms of **warping**, and ruptures of tissues to **checks**, **splits**, and **shakes.**

Cupping is a warping across the width of a board (Fig. 43, *b*). In flat-sawn material one surface is more nearly radial than the other, and, since radial shrinkage is less than tangential, the side towards the pith tends to shrink less than the opposite face; if this side of a board does shrink less than the opposite face, the board will be distorted or warped when dry; it becomes bowed or **cupped** in cross section. Cupping can be reduced by proper piling, *vide* last sentence of next paragraph. The defect does not occur in truly quarter-sawn material.

Twisting is the spiral or corkscrew twisting of a board or plank in a longitudinal direction as it dries, which, in extreme cases, may render the timber valueless (Fig. 43, *a*). Twisting can usually be traced to spiral or interlocked grain, although it may also result from unequal shrinkage brought about by variation in density within the board or plank; it can usually be minimized, if not completely eliminated, by weighting stacks with heavy baulks immediately the pile is built and before drying has commenced.

Bowing is a warping (or sagging) from end to end of a piece of timber (Fig. 43, *d*); that is, it is similar to cupping except that it occurs along the length of the piece and not across its width. Bowing results from too wide spacing of stickers, which causes the timber to sag under its own weight.

Spring is distortion in the longitudinal plane, the board or plank remaining flat (Fig. 43, *i*). **Spring** is not uncommon in boards from near the 'core' or 'heart' of a log, and is the result of the sudden release of internal stresses when the log is sawn through. All timber is more or less subject to spring, but that of certain species is more susceptible than others. In extreme cases, spring can be so serious as to make conversion uneconomic: some kempas from swamp areas springs to such an extent that distortion of both ends of a scantling is measurable in centimetres rather than in millimetres.

Checks and *splits* are separations or ruptures of the wood-tissues in the longitudinal plane and are distinct from the horizontal fractures, 'thunder-shakes' or compression failures, that have their origin in other factors than drying stresses. A **check** is a separation of the fibres that does not extend through the timber from one face to another (Fig. 43, *e*), end **splits** are separations extending from face to face. An end split is one that occurs at the end of a log or piece of timber (Fig. 43, *f*).

Checks and splits may close up if the dry timber is subsequently exposed to damp conditions, but once the fibres have separated they cannot actually join together again, and the checks and splits are present although they may not be visible.

Shakes. Serious splits are often called '**shakes**', but it is better to confine the use of this term to separations of the fibres in timber of large size or in the log; shakes may originate from other causes than drying stresses, *e.g.*, from careless felling, internal stresses existing in the living tree that are released when the tree is felled.

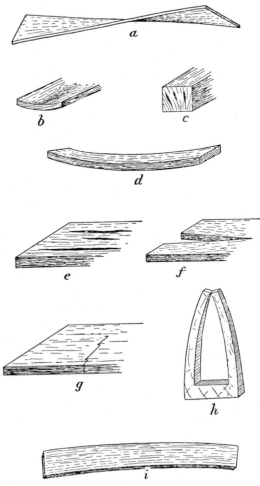

FIG. 43. Various defects in timber: *a*, twist; *b*, cupping; *c*, honey-comb checks; *d*, bowing; *e*, checks; *f*, end split; *g*, compression failure; *h*, behaviour of test sample from a case-hardened board; *i*, spring

Shakes are of several types, *e.g.*, **ring-shake**, where the separation follows a growth ring, **star-shake** where the ruptures radiate outwards from the pith.

Case-hardening. When timber is dried so rapidly that the outer layers want to shrink while the interior is still saturated a stress is set up, because the outer layers are restrained from shrinking normally; they may eventually shrink the full amount, when checks will result, or they may set in a distended or stretched condition (**tension set**); pieces of timber in which this latter condition occurs are said to be **case-hardened**. Case-hardening can be removed by steaming, to restore moisture to the outer layers. If steaming is applied in time, and subsequent re-drying is properly controlled, previously case-hardened timber that has not surface-checked or become honey-combed is in no way inferior to timber that has never been case-hardened. Case-hardened timber may cup and develop other forms of distortion when it is subsequently re-sawn or worked up (see also discussion on pages 135 and 136, and Fig. 32).

Honey-combing. If case-hardening is not relieved by steaming, the outer layers set without shrinking the normal amount, and when the interior dries below the fibre saturation point it, too, is restrained from shrinking, and interior checks may result. This condition is known as **honey-combing**. The stresses set up are greatest tangentially, because shrinkage is greatest in this direction, and the resulting separation of the fibres is always initiated where the tissues are weakest: that is, along the rays (Fig. 43, *c*).

Checks caused by conditions leading to case-hardening may close, and honey-combing checks may not extend to the surface, so that the defects cannot always be detected before the timber is worked up. A simple test for case-hardening is described on page 219–20, and illustrated in Figs. 32 and 43, *h*. Honey-combing can be detected by cutting a plank through about 30 cm from the end and noting whether there are any internal checks along the rays on the freshly exposed end.

COLLAPSE

Some timbers are liable to a defect known as **collapse** if kiln-dried slowly at too high humidities, or at too high temperatures. The use of high humidities while timber is high in moisture content, and the type of timber, may also appreciably affect the amount of collapse occurring. Collapse can also occur in rapid air-drying of very green timber of a few species, *e.g.*, Tasmanian oak, western red cedar, cypress, and hemlock: the cells are flattened in drying, which is manifest in the more porous early wood, producing, in extreme cases, a corrugated surface – hence **washboarding** – of quarter-sawn faces (Plate 50). Collapse results in excessive and often irregular shrinkage, and may lead to appreciable distortion, *vide* Plates 51 and 52; in some extreme cases severe internal checking may occur. The defect is serious because, in addition to the loss of timber in trimming mis-shapen pieces of wood, and the abnormally high shrinkage already referred to, the strength properties of the wood may also be reduced. Collapsed air-dry wood is, however, usually stronger than non-collapsed or reconditioned wood, primarily because it is more dense.

No uniform behaviour can be assigned to the effect of position in the tree in relation to the occurrence of collapse. As regards the locality factor, Australian experience suggests that timber from moist or swampy places, and fast-grown, immature trees, is more prone to collapse than material from other sources.

The cause of collapse has given rise to divergent theories. There are indications that some collapse may ordinarily occur in the drying of any timber, but this is not what is normally meant by the term. Until recently it was supposed that pronounced collapse required the cells to be completely filled with water, and the cell walls to be almost completely impervious. Collapse was then explained by the effect of internal tension forces set up during drying in the water occupying the whole of the cell cavities of affected cells. This theory called for no bubbles in the cell cavity water larger than a certain minute minimum, and supposed that if water were lost by diffusion through the cell walls faster than air or water vapour entering the cells, a state of tension would arise in the water remaining in the cell cavity. The forces of cohesion between the water molecules and faces of contact between the water and the cell walls would result in a state of tension of the water sufficient to contract and draw in the cell walls. A recent mathematical analysis by W. H. Banks and W. W. Barkas attributes collapse to surface tension phenomena, and explains the occurrence of collapse when some air is present. Collapse is naturally more severe with kiln-

PLATE 48

Pith flecks in birch. The upper portion of the figure shows the appearance of pith flecks on cross-section, and the lower portion their appearance on a longitudinal face

Photo by F.P.R.L., Princes Risborough

PLATE 49

FIG. 1. A resin pocket in
a plank of balau

Photo by F.R.I., Kepong, Malaysia

FIG. 2. A bark pocket in a
plank of chengal

*Photo by Timber Research Laboritory,
Kepong, Malaysia*

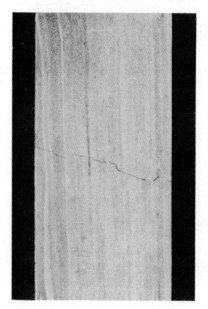

FIG. 3. Typical compression failure
in a seasoned board; such failures
open as the timber dries and
become more conspicuous

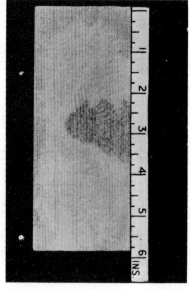

FIG. 4. Strawberry mark in Sitka
spruce: appearance on quarter-
sawn surface

Photos for Figs. 3 and 4 by F.P.R.L., Princes Risborough

PLATE 50

'Collapse' producing 'wash-boarding' in a board of Tasmanian oak

Photo by Division of Forest Products, C.S.I.R., Australia, by courtesy of Dr H. E. Dadswell

PLATE 51

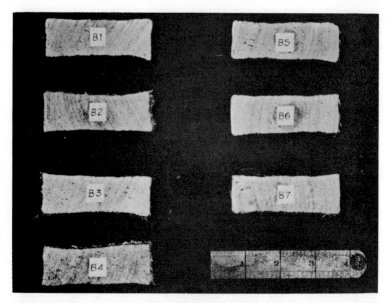

Fig. 1. Before treatment

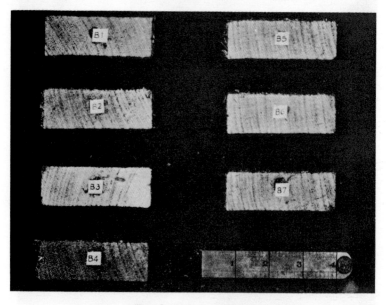

Fig. 2. After treatment
Re-conditioning of collapsed Tasmanian oak

Photos by F.P.R.L., Princes Risborough

drying, because the temperatures used increase the rate of drying and also the plasticity of the cell walls.

A method of **reconditioning** has been evolved for removing collapse with complete success. It consists in heating collapsed timber in a chamber in saturated air to a temperature of about 210° F., and maintaining these conditions for some hours before allowing the timber to cool. Plate 51, fig. 2, shows the effect of re-conditioning the timber illustrated in Plate 51, fig. 1, for a period of four hours; the final moisture content was increased only by ½ per cent., and there was an appreciable gain in cross-sectional area. Slight checking of no commercial importance occurred on the edges; the strength properties of the re-conditioned timber were not lowered in comparison with uncollapsed material of the same moisture content. Machining qualities and working properties generally were improved. Equally successful results have been obtained with other species.

Timber that is badly warped or cupped, without apparently being collapsed, may also be successfully re-conditioned: distortion may be sufficiently removed to secure a marked increase in the final cross-sectional area. Even apparently 'normal' shrinkage has been reduced by a re-conditioning process, *vide* Plate 51. This consignment had cupped slightly, but suffered little distortion, and re-conditioning effected an increase of 23 per cent. in available cross-sectional area over the original kiln-dry dimensions. Australian work suggests that when reduction in 'normal' shrinkage is achieved, it is because unsuspected collapse has been removed.

Re-conditioning, then, besides removing collapse, may be employed to remove distortion resulting from other causes, and it is also effective in reducing normal shrinkage. It is not suggested that re-conditioning should be a complement to kiln seasoning in every case, but the results obtained have been so striking that its possibilities are not without practical significance; they deserve to be more fully investigated on a commercial scale.

CHAPTER 13

Decay and Sap-Stain Fungi

GENERAL PRINCIPLES

Most forms of decay and sap-stain in timber are caused by certain plants, called **fungi**, that feed either on the cell tissue or cell contents of woody plants. It is important to distinguish between **wood-rotting fungi**, responsible for *decay* in timber, and those that feed on the cell contents, causing *stains*. The former consume certain constituents of the cell wall, and lead to the disintegration of woody tissue, whereas the latter remove only certain stored plant food material in the cell cavities, leaving the cellular structure intact. Wood-rotting fungi seriously weaken timber, ultimately rendering it valueless, whereas **sap-stain fungi** spoil the appearance of wood, but do not affect most strength properties. Sap-stain is, in effect, not a preliminary stage of decay, but such stained timber, exposed to suitable conditions, may later be attacked by wood-rotting fungi.

Wood-rotting and sap-stain fungi belong to a large group of plants that includes edible mushrooms and toadstools. The visible mushroom or toadstool is the **fruit body** or **fructification** of the fungus, the vegetative parts of the plant being out of sight, in the feeding medium. The fruit bodies of wood-rotting fungi are frequently flat, fleshy or woody, plates, the undersides of which bear **spores** or seeds. The destructive part of a fungus is its vegetative system, or **mycelium**, made up of numerous exceedingly fine tubes called **hyphae**; these may become matted together to form a felt-like mass. Hyphae grow by elongating at their tips, passing from cell to cell of the host plant, feeding on the walls or cell contents in their path. The complete life cycle of a fungus is, therefore, (1) spore, (2) hyphae, (3) mycelium, (4) fruit body or fructification, and (5) spore.

All fungi feed on organic material of either plant or animal origin.

Those of interest to the user of timber attack *either* living (probably unhealthy) trees, *or* felled timber. One condition essential to the development of all fungi is the presence of sufficient moisture: initial infection will not occur in any timber below 20 per cent. moisture content. Moreover, reduction of moisture below the critical minimum causes all fungi to cease growing. When growing vigorously, some fungi are, however, capable of extending their attack to adjacent timber near the critical moisture content; they will not ordinarily initiate attack unless the moisture content is about the fibre saturation point, and many fungi require appreciably higher moisture contents to initiate attack. Some fungi transport moisture from an outside source, or produce their water requirements from the break-down of cell-wall substance in their path, enabling them to raise the moisture content of wood just below the 20 per cent. level to this level. Reduction in moisture content is not sufficient to bring about the immediate death of a fungus: hyphae are capable of remaining in a dormant condition for several months, but even those of the 'dry rot' fungus will not survive for a year in normally dry conditions.

The conditions essential to fungal growth are four in number:

(1) Food supplies   (2) Adequate moisture
(3) Suitable temperature   (4) Air (oxygen) supplies

In most circumstances growth of fungi in wood under service conditions is dependent solely on the presence of adequate moisture. The wood itself, or the cell contents of the sapwood of certain species, constitutes the necessary food supply, and oxygen can always be obtained, except in a vacuum or gas-tight chamber, or under completely water-logged conditions. Some timbers, however, contain extractives of an oily or solid nature that are poisonous to fungi; these substances render the wood unsuitable as food, thus explaining why some timbers are 'naturally durable'. That is, they are resistant to attack, but not immune. Practically all fungal growth ceases at or below freezing point (32° F.), and is very slow at temperatures below 40° F. Hence, in temperate regions with really cold winters, fungi may be quiescent in the winter months, and temperatures are a limiting factor in polar regions; optimum temperature conditions vary with different fungi, but are in the region of 65° to 85° F. In apparent contradiction to the foregoing, the wood-work of cold-store rooms and refrigerators is not infrequently attacked by fungi: this is because the wood is not at the temperature of the artificially cooled space, being part of the surrounding insulation and, therefore, at some higher temperature, often not much lower than that of the outside atmosphere; further, such timber is liable to become damp as a result of condensation from warm air striking the cooler timber of the cold-store.

The presence of decay is often visible from the spongy appearance of the ends of logs, compared with the more fibrous end surfaces of sound logs. Actual decay is the final stage of attack; by this time the colour and texture of the infected wood have been changed, and most strength properties have been appreciably reduced. Beyond the decayed area, and in the early stages of infection, a state of **incipient decay** exists: no visible changes in structure may be apparent, and only slight colour changes or softening may occur; strength properties are likely to be only very slightly reduced. Sterilization at this stage will arrest decay, when, provided attack was very slight, such timber is likely to be as good as sound stock, and it will be perfectly safe after sterilization, as far as the risk of spreading infection is concerned. On the other hand, in carrying out repairs necessitated by fungal activity, it is not sufficient merely to cut out all visible signs of 'decay', a margin of safety should be secured by cutting well back beyond the last traces of any incipient decay. In addition, adjacent surfaces, whether of brickwork, masonry, concrete, or metal, should be heat-sterilized, so long as the fire-hazard permits, or the surfaces should be treated with a fungicide. The limitations of heat-sterilization are discussed later, *vide* page 317; the special precautions to be observed in selection of timber for repairs is discussed on pages 54 to 55.

WOOD-ROTTING FUNGI

Actual decay in timber may be detected by the abnormal colour of the wood, by the transverse fractures of the fibres on longitudinal sawn faces, and by lifting the fibres with the point of a penknife, when, if the timber is decayed, the fibres will snap, instead of pulling out in long splinters. The important distinction between the wood-rotting fungi, and other fungi, is that the former live on certain constituents of the woody cells of plants.

It has already been stated that the two main constituents of wood substance are cellulose and lignin. The **brown rots** feed mainly on the cellulose, and the **white rots** feed both on cellulose and lignin, but to a varying extent, depending on the particular fungus. The different wood-rotting fungi can be further subdivided, according to the form decay takes, into cubical, spongy, pocket, stringy rots, and so on. The terms **dry rot** and **wet rot** are misleading: 20 per cent. moisture content is a critical minimum moisture content for active growth of all fungi. Most fungi require the moisture content of wood to be between 35 and 50 per cent. for optimum growth, and some appear to prefer still wetter conditions. Precise figures for different fungi are virtually impossible to arrive at; it is known, however, that the minimum moisture content required for spores to germinate is higher than the figure for infection of wood adjacent to actively growing mycelium, or for fungi already present to continue growing. In the final stages of decay, caused by those fungi that continue active in wood near the critical moisture content, the wood may be dry and friable, and this has led to the use of the term 'dry rot', whereas, in the final stages of decay caused by fungi that require wood to be comparatively wet for them to initiate or continue attack, the affected area is often itself wet, hence the term 'wet rot'. The term 'wet rot', however, is sometimes mistakenly applied to the slow disintegration of wood exposed to the weather, *i.e.*, **weathering**, which may be quite independent of fungal activity. Constant exposure results in a softening of the surface of wood, and repeated wetting by rain, followed by rapid drying in the sun, leads to the development of numerous surface checks that split the surface layers into small cubes or rectangles. Once disintegration has commenced, conditions are, of course, favourable for fungal infection.

Apart from the confused use of the term 'wet rot', the attempted distinction from 'dry rot' is unsound; fungal attack is dependent on damp conditions. Equally, 'dry rot' is an unsatisfactory term: in the tropics it is often applied to damage caused by dry-wood termites, *vide* pages 279 to 282.

Pocket rots take the form of apparently localized areas of infection, scattered over the surface of a board or plank; areas are frequently discoloured, or they may be white; they are easily dented with a thumb-nail or the point of a penknife. In the timber trade this type of infection is frequently referred to as **dote**.

Many fungi are essentially forest problems: there are fungi that attack logs that have been lying in the forest until thoroughly saturated, and there are others that attack standing but unhealthy or over-mature trees, or freshly felled logs. Fungi that attack timber in service may belong to neither of these classes; they are relatively of much greater economic importance, but they can be kept within bounds provided simple precautions are observed.

Forest fungi. Fungi that attack old logs are unimportant, since such logs would never be converted for timber. Fungi that attack standing trees, or freshly felled logs, on the other hand, are responsible for losses, principally to owners of forest; they need not be a problem to the consumer of wood. These fungi belong to the class that attack decidedly wet wood, which, once seasoned, is, however, safe from development of further decay. Moreover, for indoor purposes, such seasoned wood is unlikely to encounter conditions that would lead to renewal of attack by 'forest' fungi that require relatively high moisture contents for their active growth. During conversion of timber known or suspected of having been attacked by 'forest' fungi, all wood containing visible areas of infection should, and probably would, be discarded, when, so long as the material recovered is properly piled to ensure rapid drying, or, preferably is kiln-dried after conversion, it is perfectly safe to use. On the other hand, if such timber is close-piled when green, any incipient decay present is likely to develop, and fresh infection from spores of the same, or similar, fungi, may occur. Most dote in timber can be traced to bad practices at the mill, or to close-piling of imported timber shipped green (see also discussion on page 212 and below).

Trametes serialis Fr. A brown pocket rot responsible for complaints of dote in Douglas fir, has led to the assertion in some quarters that this timber is less durable than European redwood. Decay takes the form of small, spindle-shaped pockets, of soft, discoloured, dark-brown wood (Plate 54, fig. 1); thin white threads of hyphae may or may not be visible. When the wood has had time to dry out, attack of this nature is often more apparent, because of the development of shrinkage cracks in the affected zones. Timber containing such visible decay will have lost mechanical strength, and should obviously not be used. The fungus most frequently responsible for this type of infection is *Trametes serialis*, a species common in America, but of rare occurrence in the United Kingdom, unless present in imported timber.

Douglas fir is no more susceptible than European redwood to fungal attack, and the dense form is rather more resistant, but much Douglas fir is imported in a green condition, practically straight off the saw, whereas the Baltic timber is usually stick-piled immediately after conversion, and for some time before shipment. Any fungus present in green timber, and spores that may have alighted on the surface of wet boards, are favourably placed if the material is close-piled on arrival overseas. On the other hand, if properly stacked on landing, the timber will dry out before any fresh decay can occur, and that already present, if not visible when the timber was green, would become apparent during drying, allowing affected wood to be rejected instead of being put into service. *Trametes serialis* is also the most usual cause of dote in Sitka spruce.

Trametes pini (Thore) Fr. A **white pocket rot** is sometimes found in Douglas fir; it occurs as small elliptical areas 3 to 12 mm. in length, with pointed ends (Plate 54, fig. 2). Such infected timber is un-suitable for use, but the fungus responsible, *Trametes pini*, is essentially one that originates in the standing tree, and does not spread to any extent after felling.

An unidentified fungus has been responsible for incipient decay in imported ash, beech, and birch; it usually appears as a white flecking of the surface. When consisting only of much lighter-coloured, small, round or oval, areas that can be removed by planing to a depth of 3 to 4 mm, the sound wood beneath is perfectly safe to use. In a more advanced stage, attacked timber is, of course, worthless, *vide* Plate 54, fig. 3. In a still more advanced stage, fine black lines may be visible; the wood is then very brittle.

There is nothing that the importer or ultimate consumer of wood can do to combat forest fungi other than arresting development of such infection by proper stacking of green timber on receipt; it is presumed that all reasonable care will be taken at all times to destroy any timber found to be infected. Provided these two points are observed, forest fungi will not be a problem in timber yards. Eradication in the forest is more difficult: sound silvicultural technique and good forest management are the key to minimizing such infection.

Fungi that attack wood in service. Fungi that attack wood in service may be either 'dry rots' or 'wet rots'. In buildings the former are usually much more serious than the latter, but in coal-mines wet rots are mainly responsible for decay. By far the most important fungus destructive to wood in service is *Merulius lacrymans* (Wulf.) Fr., which will attack drier wood than most fungi, although not wood below about 20 per cent. moisture content. Because of the necessity for adopting appropriate remedial measures promptly, if *Merulius lacrymans* is the causal agent of decay, it is important for all who are responsible for the care of buildings to be able to identify the fungus correctly. It must not be confused with the cellar fungus, *Coniophora cerebella* Pers., which is capable of attacking only definitely wet wood, and is therefore usually not a serious problem. Nor should *Merulius lacrymans* be confused with mould growths that frequently appear as tiny green or black tufts on damp wood. Such moulds do not cause decay, but their presence is indicative of damp conditions, favourable to the attack of wood-rotting fungi.

The most effective method of eliminating wood-rotting fungi from interior wood-work is to use sound, seasoned timber, free from fungal infection in the first place, and to provide and maintain efficient ventilation so that the timber will not be exposed to damp conditions subsequently. Basement conditions call for effective damp-proof courses, both horizontally and vertically. Linoleum-covered floors should be sparingly washed with water, and such floors should be adequately ventilated beneath. If timber is to be laid direct on concrete it should be laid in a bituminous mastic. This is standard practice with wood-block floors, but strip floors are frequently nailed to fillets let into the concrete, leaving a 12 to 19 mm clearance between the underside of the floorboards and the top of the concrete. This is thoroughly bad practice, and a fruitful source of 'dry rot' infection, even with supposedly naturally durable hardwood floors of such timbers as oak or teak. When this type of floor is required, pressure-treated softwood fillets should be used, and the air space between floor and concrete should be filled with a bituminous substance. Such floors tend to be less satisfactory than wood-block floors: there is a risk of action between the wood preservative and bituminous material, and it is exceedingly difficult to ensure that the air space between the floor and the concrete is completely filled.

PLATE 52

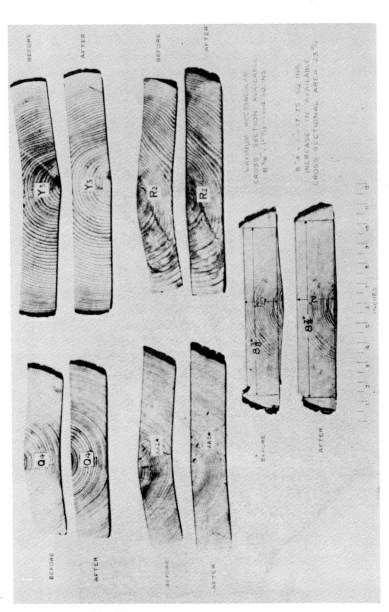

Cross-sections of ash planks before and after re-conditioning

Photos by F.P.R.L., Princes Risborough

PLATE 53

Neglect of maintenance as a cause of decay in wood. (a) Stopping up of rain water head and a leak in the lead-lined gutter behind produced conditions leading to (b) vigorous growth of *Merulius* on the gutter bearer and (c) serious decay of the floor joists immediately below

PLATE 54

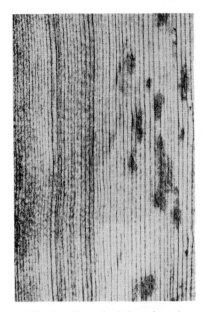

FIG. 1. Dote (incipient decay)
in Sitka spruce

FIG. 2. White pocket rot in
Douglas fir

FIG. 3. White pocket rot in
Canadian birch

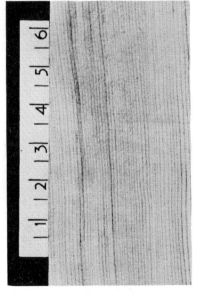

FIG. 4. Resin streaks in
Sitka spruce

Photo by F.P.R.L., Princes Risborough

PLATE 55

FIG. 3. A board of Scots pine showing sap-stain fungal infection of the sapwood

Photos by F.P.R.L., Princes Risborough

FIG. 2. *Coniophora cerebella*: portion of a decayed joist showing dark strands of mycelium, and longitudinal cracks in the wood

FIG. 1. *Merulius lacrymans*: portion of a decayed joist showing two fruiting bodies, mycelium, and deep cracks along and across the grain

Any timber built into brickwork, masonry, or concrete should be recognized as subject to the decay hazard. Unless it can be ventilated all round, only pressure-treated material is really suitable in these circumstances; an open-tank treatment of the ends of long timbers, *e.g.*, joists and bearers, that can be done in a steel drum on the building site, is a possible and practicable compromise. Once rot makes its appearance, it is imperative to take active measures immediately. The causal agent must be identified, and all decayed and infected wood removed and burned. Replacements should be with sound, preferably kiln-dried, material; if recurrence of attack is at all probable, wood preservatives should be employed prophylactically. Lastly, but of the utmost importance, steps must be taken to discover what gave rise to the unsatisfactory conditions culminating in fungal decay, so that these conditions can be corrected before repairs are put in hand.

Wood in contact with the ground, exposed to the weather, or inevitably exposed to fungal infection as in coal-mines, must be recognized as having a relatively short life. It is then a simple matter of economics to decide whether the use of more resistant timbers is the correct answer, or whether the use of wood preservatives is the better solution. If the latter course is adopted, then adequate treatments are essential, *vide* Chapter 15. In general, when the decay hazard is unavoidable, adequately pressure-treated non-durable timbers are usually more economical, and give a longer service life, than the most durable timbers untreated. The descriptions of the different fungi, given in the notes on different species, have been culled from leaflets and bulletins, published by the Princes Risborough Laboratory, to which reference should be made for more detailed information.

Merulius lacrymans (Wulf.) Fr., which is the commonest fungus responsible for 'dry rot', is a brown cubical rot. The appearance, both of the fungus and of infected wood, depends on the stage attack has reached, and on the growth conditions for the fungus. In damp conditions the fungus develops as white, fluffy, cotton-wool-like masses spreading over the surface of attacked wood. In drier conditions the mycelium forms a grey-white felt over the wood, usually with small patches of bright yellow or lilac. Branching strands may develop from the felt, varying in thickness from coarse threads to strands as thick as a lead pencil. These strands are made

FIG. 44. Top: typical cubical breakdown following *Merulius* attack; bottom similar breakdown in a piece of timber caused by *Coniophora cerebella*.

up of hyphae that conduct water; they can penetrate the mortar of a brick wall, and cross steel-work and concrete to reach new feeding grounds, *i.e.*, as yet uninfected wood. The fructifications are soft, fleshy plates, with white margins (Plates. 53 (*b*) and 55, Fig. 1). Numerous folds or shallow pores occur on the surface of a fructification, and contain the rust-red spores. These are microscopic, and so light that they are easily blown about; they are sometimes produced in such quantities that a whole room may be covered with a rust-red layer of spores. The fructifications sometimes grow vertically, in the form of a thick bracket, when the pore-bearing surfaces become elongated like small stalactites. Water may be exuded in drops by the fructifications – hence the name *lacrymans* or weeping.

The fruit bodies, which grow out into the air and light, are frequently the first indication of dry rot in a building, but unventilated rooms or shut-up houses that have been infected usually have a characteristic musty odour. Slight waviness on the surface of panelling, or the sinking of a floor, may be the first warning of extensive damage. Infected wood is soft when tested with the blade of a penknife, and it will not 'ring' when struck. Wood beneath a coating of mycelium is wet and slimy to the touch, but in the final stages of attack it is dry and friable, brown in colour, and breaks up into cube-shaped pieces.

Poria vaillantii (D.C.) Fr., formerly called *Poria vaporaria* Pers., is a brown cubical 'dry-rot', responsible for decay in buildings, but also occurring in coal-mines; it requires the presence of more moisture in wood than *Merulius* to initiate infection. The final stages of attack are similar in their effect on wood to the action of *Merulius lacrymans*. The hyphae and mycelium remain white and soft; the fruiting body is plate-shaped, covered with fine pores, and also white. The hyphae penetrate brickwork, but not deeply, and attack is therefore usually more localized than with *Merulius*.

Coniophra cerebella Pers., the cellar fungus, is also a brown rot, which was formerly not regarded as a cubical rot because, in the final stages of attack, the decayed wood usually develops characteristic longitudinal splits or cracks, *vide* Plate 55, fig. 2. Cubical breakdown, virtually indistinguishable from *Merulius* attack, may occur, *vide* Fig. 44; the timber decayed by *Coniophora* differs in not being permeated with mycelium. Ultimately the wood is extremely brittle, and can be powdered in the fingers. The hyphae are always fine; they rapidly turn brown or almost black. The fructification, which is rarely seen, is a thin plate, olive-green in colour. *Coniophora* favours decidedly wet conditions; it is very liable to occur where there is persistent water leakage or condensation. It is the commonest fungus found attacking modern, mass-produced, external joinery that has received no preservative treatment.

Paxillus panuoides Fr. is a brown rot requiring very moist conditions. The hyphae are paler than those of *Coniophora cerebella*, the mycelium is rather fibrous, and yellow or violet. The fruit bodies, which are often bell-shaped, are olive-green, with deep gills on the under surface.

Phellinus megaloporus (Pers.) Heim., syn. *Phellinus cryptarum* Karst., is a white rot, only recorded as attacking oak; it has occasionally been found active in building timbers. The hyphae are white and fibrous; the fructification is a thick, leathery plate or bracket, buff-coloured, with darker brown pores. In the final stages of attack, the wood is reduced to a soft white mass in which the hyphae are embedded.

Lentinus lepideus Fr. is a brown cubical rot, requiring moist conditions. It attacks timber out-of-doors, *e.g.*, telegraph poles, railway sleepers, and paving blocks. The fructification is a brown, woody mushroom. The fungus and decayed wood have a characteristic aromatic odour. Cartwright and Findlay ('The Decay of Timber and its Prevention', H.M. Stationery Office, 1958) record that this fungus 'occurs quite frequently on worked timber which has been imperfectly creosoted'.

Poria xantha Lind. non Fr. is also a brown cubical rot, requiring moist conditions. It is commonly found in greenhouses. The fructification is a thin plate, yellow in colour.

Of the foregoing, *Merulius lacrymans* and *Coniophora cerebella* are the most common fungi likely to be encountered in wood in service, and the first named is by far the most serious of all wood-rotting fungi, being responsible for untold damage annually. Good constructional design is the best prophylactic measure against ultimate fungal attack, but, where the decay hazard is unavoidable, the choice should be made between the naturally resistant timbers, or non-resistant timbers adequately treated with suitable wood preservatives, depending on which of these two alternatives is the more economical in the particular circumstances.

The wood-rotting fungi described above are *Basidiomycetes*, one of the groups of so-called 'higher fungi'. More recently it has been found that another group of fungi – the *Ascomycetes* – previously regarded as more important as a cause of staining, include several species responsible for decay in timber in exceptional locations. *Chaetomium globosum* Kunze has been found to be the cause of decay of framing and louvers in cooling towers. Decay takes a different form from that caused by the more usual wood-rotting fungi, completely decaying the surface of the timber attacked while the wood beneath appears to be still perfectly sound. Microscopic examination of the wood immediately beneath the completely decayed surface reveals the presence of large numbers of hyphae boring longitudinally in the cell walls. Although *Chaetomium* and related species can cause damage of considerable economic importance, the very wet conditions necessary for attack explain why these fungi, which have been called 'soft rot', are not a problem in ordinary buildings.

Mention should also be made of two fungi, unimportant in themselves, which are nevertheless an indication that dangerously damp conditions exist. Elf cups (*Peziza* sp.), which are yellow-brown cups about 25 mm in diameter, not infrequently develop on plaster ceilings following flooding from defective plumbing or frost damage. Unless steps are taken to secure rapid drying out of the affected areas there is a risk of subsequent 'wet rot' or 'dry rot' infection. Another fungus that should similarly be regarded as a warning that dangerously wet conditions have become established is a species of inky cap (*Coprinus* sp.). This fungus produces small soft toadstools that dissolve into an inky fluid. Findlay (*Dry rot and other timber troubles*) describes this species as often growing 'on damp cellar walls', but they may also appear on the underside of ceilings saturated by persistent plumbing leaks or defective internal gutters.

The Sap-stain Fungi

Sap-stain or **blue-stain** in timber is caused by several species of fungi of the mould type. These fungi are distinct from those that cause decay; hence, 'blue-stain' is not an incipient stage of decay, but its presence may be an indication of conditions favourable for the attack of wood-rotting fungi. Moreover, badly blued timber should be suspected of possibly also containing incipient decay or dote.

All staining of wood is not necessarily 'blue-stain'. Green timber rich in tannins that comes in contact with iron, as in sawing, may become stained blue-black. This is the result of chemical action, and the stained areas are usually superficial and easily planed off. Similarly coloured stains, caused by sap-stain fungi, penetrate wood deeply and rapidly (Plate 55, fig. 3).

Some wood-rotting fungi cause discoloration, but staining from this cause is accompanied by softening of the wood, whereas blue-stain fungi have little or no effect on strength properties, other than reducing resistance to impact bending, sometimes by as much as 40 per cent., which is of material importance in timber for tool handles and athletic goods. On the other hand, 'blue-stain' is responsible for degrading large quantities of susceptible timbers, because their value is reduced for decorative purposes, and, if heavily stained, they may be unsatisfactory for paint finishes.

Besides the fungi responsible for 'blue-stain', there are several other mould fungi that stain wood green, pink, purple, and, more rarely, brown; the majority of these produce a powdery or downy growth of mould that is easily brushed or planed off.

'Blue-stain' in softwoods is caused by several species of the genus *Ceratostomella*; attack is confined to the sapwood. 'Blue-stain' in the light-weight, light-coloured tropical hardwoods, such as obeche and ramin, is usually the result of *Diplodia* infection; attack is not confined to the sapwood, but may extend right through a log. Several mould fungi attack the light-coloured temperate hard-woods, not necessarily confining their attack to the sapwood. Ash and poplar are liable to be discoloured a dark brown, and oak a pale yellow.

The discoloration caused by mould fungi is not a stain in the true sense of the word: it is the presence of numerous dark-coloured hyphae in the translucent cells of the wood that produces the tinting visible on the surface. The fructifications are small, black, flask-shaped **perithecia**, as large as a pin's head, often with long necks, and containing numerous spores.

As with wood-rotting fungi, four conditions are necessary for mould fungi to grow: (1) sufficient moisture (actually appreciably more than is necessary for the more important wood-rotting fungi),

(2) food supplies, in the form of starch and sugars stored in the cell cavities, but not the wood-substance of which the walls are composed, (3) suitable temperatures, and (4) oxygen (obtained from the air). The right type of food material in sufficient quantities is a limiting factor: in most softwoods these requirements are only found in the sapwood, but in Sitka spruce attack may spread to the heartwood. The presence of bark saturated with moisture also inhibits mould growths from want of air. Relatively high temperatures are necessary for active growth: the optimum is between 70° and 80° F.; below the optimum, growth is very slow. In temperate regions favourable temperature conditions only exist in the summer months. Reduction in moisture content in the surface layers of wood can rapidly become a limiting factor: the fungi require moisture contents above the fibre saturation point of wood to initiate attack.

Unless infection has occurred in the forest, rapid reduction of surface moisture in converted timber is the simplest method of inhibiting mould growths. Kiln-drying immediately after conversion is the surest safeguard. Piling in properly built stacks, with stickers of maximum thickness, does not always ensure sufficiently rapid drying to prevent staining of particularly susceptible timbers. With these timbers the use of chemical dips is more or less essential, and is standard commercial practice in parts of northern Europe and North America, *vide* pages 310 and 311.

Susceptible timbers will often become infected by sap-stain in the forest after felling, either from the ends of logs or through places where the bark has been removed or damaged in felling. Avoidance of bark injury, and the use of end-coatings, are effective temporary measures, which should be regarded only as auxiliary to rapid extraction from the forest, followed by immediate conversion at the mill. Suitable end-coating materials are hardened gloss oil, containing 10 per cent. of cresylic acid, creosote, tar, or even lead paint. Any areas where the bark has been removed should also be dressed with the material used for end-coating.

Once timber has been dried below the critical stage for 'blue-stain' infection, these fungi should not constitute any problem to the consumer. If conditions in service become such that fresh blue-stain infection could occur, conditions favouring the attack of wood-rotting fungi would also exist, presenting a much more serious problem, and one requiring drastic and immediate remedial measures.

CHAPTER 14

Worm in Timber

THE damage referred to as **worm** in timber is the result of insect activity, but in salt water, teredo or ship-worm, and a form of wood-louse belonging to the crustacean family, are responsible for damage of this type. Insects tunnel in timber, spoiling the appearance of exposed faces, and, if the tunnels are numerous, they may so reduce strength properties as to make the wood valueless. Some insects only attack living trees or newly-felled logs, some only seasoned wood, and others only the sapwood of certain species. In consequence, the presence of insect damage is not in itself necessarily a cause for alarm: the damage may be of the first type and therefore of no consequence in seasoned timber, beyond the disfigurement caused. Moreover, some insects and crustaceans commonly associated with timber are of no importance because they do not attack it. For example, the land form of wood-louse is to be found under any piece of wood that has been left in contact with the ground, in sheds, or in the open for any length of time: these are the small oval-shaped crustaceans that roll up into balls when touched. Although probably the most familiar creature associated with timber, wood-lice are of no practical importance, as they do not attack wood. At most they are an indication that storage conditions are not good, and may lead to infection by wood-rotting fungi. On the other hand, dangerous pests are often overlooked because their insignificant appearance results in their escaping notice.

Everyone is familiar with the stages in the development, called a **life cycle**, of moths and butterflies from egg to caterpillar or **larva**, followed by a resting period or **pupation** stage as a **chrysalis**, until the emergence of the adult moth or butterfly. Few moths attack timber, although some are serious forest pests of standing

PLATE 56

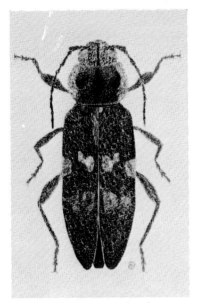

FIG. 1. The house longhorn beetle (×5)

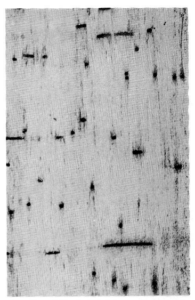

FIG. 2. 'Pin-hole' borer damage in African mahogany. (Note direction of the galleries at right angles to the grain of the wood)

FIG. 3. 'Shot-holes' in a meranti board (compare for scale with Fig. 2)

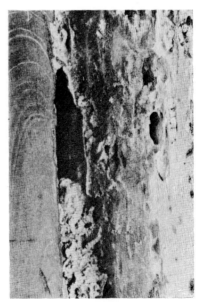

FIG. 4. Subterranean termite damage: wood hollowed out and filled with mud

Figs. 1, 2, and 3, *photos by F.P.R.L., Princes Risborough*

PLATE 57

FIG. 1. Portion of an oak plank showing *Lyctus* attack confined to the sapwood

FIG. 2. *Lyctus brunneus* Steph.
(×12)

FIG. 3. *Lyctus* frass

Figs. 1 and 2, *Crown copyright reserved*
Fig. 3, *photo by F.P.R.L., Princes Risborough*

PLATE 58

Figs. 1 and 2. The common furniture beetle, *Anobium punctatum* De G.
(× 14 approximately)

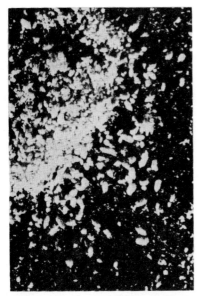 

Fig. 3. Frass of the common furniture beetle. (Note typical elongate pellets)

Fig. 4. Damage by the common furniture beetle in a structural timber of Scots pine

Figs. 1, 2, and 4, *Crown copyright reserved*
Fig. 3, *photo by F.P.R.L., Princes Risborough*

PLATE 59

Fig. 1. The death-watch beetle (×3 approximately). (Note characteristic bun-shaped pellets in bottom right-hand corner)

Fig. 2. Portion of an oak wall plate decayed by fungal activity and attacked by the death-watch beetle

trees, but other insects that have similar complex life cycles are responsible for heavy losses to timber producers and users.

By far the most important of these insects in temperate climates are beetles, a knowledge of the life cycles of which has indicated the most effective stage for introducing measures of control. Some beetles are destructive in the larval, or feeding, stage, and others do more damage as adult beetles. With the former, the larvae feed on wood, either the substances of which the cell walls are made, or the contents of these cells. With the latter class of beetles the larvae feed on a fungus, introduced by the adult beetle at the time of egg-laying; the fungus, in turn, obtains its nourishment from the wood. Tunnelling, which is the destructive work of 'beetles', is done by the larvae in the first type of pest, and by the adult beetle in the second. With the first, control must aim at killing the larvae before they can get into timber, or, failing this, wood liable to attack should be rendered unfit or unattractive as food for such larvae. With the second type, the adult beetles must be denied access to potentially attractive egg-laying grounds, *i.e.*, usually freshly felled logs, but also standing trees of some species; failing this, control should aim at destroying the fungus, which will at least ensure that any attack in progress will cease. The duration of the life cycle is another important factor in deciding the most suitable method of control; with some insects the life cycle is completed in a few months, but with others it may last two or three to several years. If the life cycle is a long one infestation may have occurred, and the destructive agent have been at work, for a considerable time before it is discovered; in these circumstances methods of control can seldom be applied at a sufficiently early stage to be of practical value. If sterilizing treatments are selected as the control measure, repetition of the treatment must be the more frequent the shorter the life cycle of the pest, because sterilization does not confer immunity from reinfestation. All control measures, but especially sterilization treatments, should be applied when the pest is in its most vulnerable stage, *i.e.*, egg, larva, pupa, or adult, depending on the particular pest involved.

The different types of insects, the means of their identification, and the methods of control, are discussed fully in various pamphlets and bulletins issued by the Princes Risborough Laboratory, Princes Risborough, the Forestry Commission, and other Government organizations.   Below, the groups of insects of importance to timber users are briefly described.

I. FOREST AND MILL-YARD PESTS

(1) *Longhorn beetles and certain moths.* The eggs of these insects are laid in crevices, or just under the bark, of living but usually un-healthy trees, or newly felled logs.   The adults do no tunnelling, the damage being done by the larvae, which feed on wood substance. The galleries are 3 to 25 mm in diameter, and oval in cross section; they are packed with coarse **'bore dust'** or **frass**.   The damage done is considerable, but, except for the house longhorn borer, infestation occurs only in green timber;   the larvae may continue feeding in relatively dry wood, but they will not migrate to adjacent seasoned stock.   Rapid extraction of felled logs, immediate removal of bark of susceptible timbers, and heat-sterilization of infested wood, will secure adequate protection against most longhorn borers.

The house longhorn (*Hylotrupes bajulus* L.), Plate 56, fig. 1, & Plate 60, fig. 3, is a serious pest in parts of northern Europe : in some districts infestation has to be notified, and appropriate remedial measures are compulsory.   The pest has long been known here, and has become of sufficient importance in some parts of the pine country of Surrey to necessitate by-laws requiring timber used in repairs or new work in roofs to be treated with approved wood preserva-tives.   The pest attacks the sapwood of softwoods, and usually only timbers in the roof space.   As the life cycle is upwards of 10 years, serious damage may result before the first flight holes bring the attack to light.   The frass contains pellets that resemble flints of petrol lighters when examined with a pocket lens.   The finding of such frass or oval flight holes is not proof of house longhorn infesta-tion;   still less that there is continuing activity.   Other longhorn borers produce similar frass and oval flight holes.   Active house longhorn borer infestation here has only been found in Surrey, and more recently, in Essex.   Even in these areas no *active* infestation has been found in any building more than 50 years of age unless new timber has recently been used for repair.

Rhagium bifasciatum F. is one of the commonest longhorns in Britain, but it is of little economic importance; it attacks decayed softwoods. *Tetropium gabriele* Weise is the larch longhorn, which con-fines its attack almost entirely to larch. It is of common occurrence in England and Wales, and has been responsible for appreciable damage when simple precautions have been omitted. Oviposition may occur in unhealthy trees, but is more usual in sound, felled logs that have been on the ground throughout a summer. The larvae feed beneath the bark, and enter the sapwood for pupation. Oviposition can be entirely eliminated by barking logs when felled.

(2) *Pin-hole borers.* These pests belong to the families *Scolytidae* and *Platypodidae*; they feed on a mould fungus, introduced by the adult beetles, and not on wood; the fungus grows on the walls of the galleries, which are wholly constructed by the adults and not by the feeding larvae. Little is known of the fungus, which continues to be called **ambrosia fungus** (hence ambrosia beetles), the name first coined by Schmidberger in 1836. The adult beetles tunnel spirally, at right angles to the grain, into living trees and newly felled logs, and lay their eggs in specially constructed 'egg-chambers'. At the time of egg-laying they also introduce a fungus into their galleries on which the larvae feed when hatched. The galleries of different species vary from 0·5 to 3 mm in diameter; they are usually oriented at right angles to the grain of the wood; that is, at right angles to the vertical axis of the tree. The galleries are usually empty, but they may become plugged with resin or other compounds; the walls of the galleries are stained black, and the tunnels themselves may be surrounded by an elongate-oval area of tissue discoloured by the fungus. Pin-hole borers ruin the appearance of considerable quantities of timber, and the galleries may be so numerous as to reduce strength properties appreciably, but attack does not continue in, nor can it spread to, seasoned timber, because the fungus on which the larvae feed requires moisture. Moreover, the whole of the gallery system of these borers is constructed by the adult beetle, so that the damage is usually done before the timber gets to the mill, often before the tree is felled. In the circumstances, the timber merchant can purchase infested timber without fear of the attack becoming any worse than it is at the time he makes his purchase. Wormy grades of mahogany, lauan, seraya, and meranti contain damage of this type. The 'shot-holes' referred to in the Empire grading rules[1] are caused by the same type of beetle, referred to as 'pin-hole' borers here, and are distinct from 'shot-hole'

[1] *Grading rules and standard sizes for Empire hardwoods,* prepared by the Advisory Committee on Timbers, Imperial Institute. See also page 334.

borers of trade terminology in this country and America. Typical 'pin-hole' borer damage in a tropical hardwood is illustrated in Plate 56, fig. 2; Plate 56, fig. 3 gives an idea of the difference in size of the galleries of different species of 'pin-hole' borers.

II. PESTS OF SEASONING YARDS

The serious pests of timber yards in temperate regions are the so-called **powder-post beetles**, belonging to the families *Bostrychidae* and *Lyctidae*. The larvae of these two families do not live on cell-wall substance, but on the starch content of the sapwood of certain timbers. In this respect, the food requirements of powder-post beetles resemble those of the sap-stain fungi, but, unlike fungi, the larvae have to devour the cells to obtain the starch they seek. The egg-laying habits of the two families of powder-post beetles differ, as do their demands in regard to the degree of dryness of wood favoured for egg-laying.

(1) *Lyctus, or powder-post, beetles.* These beetles lay eggs in the vessels of wood, and the larvae tunnel about, feeding on the starch contained in the storage cells. The attack is confined to the sap-wood of certain hardwoods;[1] it does not occur in the heartwood of any species, although, in emerging, adults may tunnel through heartwood immediately adjacent to sapwood, *vide* Plate 57, fig. 1. The size of the vessels is a limiting factor: they must be large enough to admit the **ovipositor** (egg-laying tube) of the adult female beetle, since it is in the vessels that the eggs are almost invariably laid. The fine-textured timbers, such as beech, with vessels below 0·1 mm. in diameter, are ordinarily immune.[2] Further, the starch content of the sapwood must be sufficiently high for the powder-post beetle to select the timber for ovipositing purposes. Several timbers, with large enough vessels, fail in this respect and, consequently, are immune to attack. Infestation occurs in partially- or fully-seasoned timber.

The life cycle of *Lyctus* from egg to adult beetle is normally about one year, but the period may be as short as ten months, and, where

---

[1] *Lyctus* attack has been reported in the sapwood of a softwood, *Pinus canariensis* C. Sm., grown in South Africa.

[2] I have observed *Lyctus* infestation in *Ilex* sp., the vessels of which are not more than 0·05 mm. in diameter, but this is exceptional. Attack unquestionably resulted from the *Ilex* timber being in physical contact with a susceptible timber containing active attack.

food supplies are deficient, the life cycle may be considerably extended: to two, or even three to four, years. Adults normally emerge from April to September, appearing in largest numbers in June, July, and August. Immediately on emerging, the adults mate, and the female begins egg-laying, being most fastidious in regard to the suitability of the particular piece of wood selected for egg-laying: it must be rich in starch.

There are several species of *Lyctus*, and the related genus, *Minthea*, the former being cosmopolitan, and the latter tropical. There are four common species of *Lyctus* in Great Britain: *L. brunneus* Steph. and *L. linearis* Goeze being the most abundant. *Minthea rugicollis* Walk. is the commonest species of *Minthea*; adult beetles have emerged in this country, but the species is not known to have bred here.

Soaking of logs in water is known to be an effective means of rendering susceptible timbers immune to powder-post beetle attack, but the soaking period is too long for normal commercial use. Bamboos and thatching materials are, however, regularly soaked in the East prior to use. Experiments have established that the starch content of the storage tissue is reduced by prolonged soaking, and this explains subsequent immunity of soaked materials from powder-post beetle attack. Starch depletion is, however, not the sole factor: laboratory-scale experiments with yellow meranti, a timber with sapwood normally rich in starch, showed that soaking periods too short to remove noticeable quantities of starch (as determined by the iodine test) were effective in rendering test blocks immune when subsequently exposed to attack. These, and similar experiments, suggest that the presence of some other, and presumably more soluble, substance than starch is essential to induce powder-post beetle attack; the substance or substances have not, as yet, been isolated or identified. Starch depletion can be achieved in this country by storing logs in the shade, with the bark on; the parenchyma cells remain alive sufficiently long to exhaust the stored starch during respiration. Exposed to the hot sun, or in the tropics, parenchyma cells die too quickly from lack of moisture to ensure exhaustion of starch, and there is the added risk during storage of still more serious losses from other forms of insect attack.

The galleries of powder-post beetles are similar in diameter to those of the smaller 'pin-hole' borers. They run along the grain;

that is, parallel with the vertical axis of the tree, but, as attack progresses, the separate galleries become merged, and all the attacked wood, with the exception of a thin outer skin, is eventually reduced to a flour-like powder (Plate 57, fig. 1). Powder-post beetles are responsible for enormous damage to the sapwood of susceptible timbers, *e.g.*, ash, oak, and agba, probably causing heavier financial losses in British yards than any other insect pest, and they may become a still more serious problem in the future as more and more of the light-weight tropical hardwoods are used as softwood substitutes.

Control of powder-post beetle attack presents many difficulties. The total exclusion of sapwood of susceptible species will secure 100 per cent. immunity from attack; this course is probably only economically justified in first-class joinery, flooring, panelling, and furniture. With structural timbers, the amount of sapwood to be tolerated in susceptible timbers should be specified, *e.g.*, sapwood not to exceed $\frac{1}{8}$ of the width of any face, and not to occur on more than two faces of any one piece of timber.

Prolonged storage of logs in water, as stated above, is effective in securing immunity from *Lyctus* attack, but the storage period for timber in log form is too long for the method to be practicable. Alternatively, prophylactic measures may be resorted to in an endeavour to secure immunity from infestation during the seasoning and storage period, and prior to manufacture, since it is in these stages that infestation usually occurs. To reduce the risk of infestation in timber yards and factories it is recommended that stocks be inspected twice yearly in March and October. Yards should be kept clean and free from accumulating sapwood waste, and only softwood or heartwood piling sticks should be used. Chemical control, by dipping of green timber immediately after conversion, is a practicable solution of the *Lyctus* problem in the United Kingdom up to the manufacturing stage; that is, while timber is in stack for seasoning, or in store awaiting manufacture. The immersion period ordinarily recommended may be sufficient to confer complete immunity from attack of timber dipped in veneer form, but is not effective for dimension timber once the latter is dressed or re-sawn, because surface penetration of the toxic chemicals alone is secured. The transitory nature of the protection is, however, of high commercial importance, where all the evidence points to infestation occurring prior to manufacture: if timber could reach the joinery and wood-working shops immune from infestation, losses from *Lyctus* attack would become of negligible importance in the United Kingdom. Long immersion periods, with heavy absorption of chemicals, are not called for in this country, but in Australia they have been found both necessary and economic; the chemicals used and the sterilization treatments recommended are discussed in Chapter 16.

(2) *Bostrychid powder-post beetles.* Except for differences in size of galleries, the damage done by these insects is similar to that of the previous group. That is, the larvae tunnel in partially- or recently-seasoned sapwood of certain hardwoods, reducing the wood to a flour-like dust, which is slightly coarser in texture than that of the *Lyctus* group; the galleries and exit holes are up to 3 mm in diameter. As with *Lyctus* attack, damage is confined to the sapwood.

Bostrychid beetles are typically larger than those of *Lyctus* species, and they are pests of tropical rather than of temperate regions, although there are a few species in temperate climates, including *Apate capucina* L., found in European oak. The adults are characterized by a hooded, roughened thorax covering the head, with a three-jointed club at the end of the antennae. The life cycles of the different species have not been the subject of such critical study as those of *Lyctus*, but it is apparent that the adults will infest timber in an appreciably wetter condition than that favoured by *Lyctus*: I have observed stacks of mersawa in Malaysia heavily infested by a *Bostrychid* beetle within a few days of the timber being sawn, whereas *Minthea* (the commonest of the *Lyctus* group in Malaysia) would be unlikely to infest timber until it had been in stick for several weeks.

Bostrychid beetles also differ from the *Lyctidae* in their egg-laying habits: the adults bore into the wood, constructing a Y-shaped egg-tunnel, which is kept free from dust, and in which the female lays her eggs. When the eggs hatch the larvae continue burrowing, but longitudinally, as do *Lyctus* larvae, packing the gallery system with fine, flour-like dust. The methods of control of *Bostrychid* beetles are identical with those for *Lyctus* beetles.

III. PESTS OF WELL-SEASONED, OLD WOOD

Furniture beetles, and the death-watch beetle, belong to the family *Anobiidae*, which are important pests of seasoned wood, although, paradoxically, occurring naturally in decaying stumps out-of-doors. There are several species of furniture beetles, belonging to more than one genus, but the most frequently encountered 'indoor' species is *Anobium punctatum* De G.   The death-watch beetle is *Xestobium rufovillosum* De G.

(1) *The common furniture beetle.* The natural home of this pest is out-of-doors, in decayed trees and posts, but it is better known as a pest of well-seasoned softwoods and hardwoods. The damage is done by the larvae, which hatch from eggs laid in cracks in the wood, in joints of made-up wood-work, and, more rarely, in old flight holes. The larvae travel along the grain, but as they feed and grow they tunnel in all directions, filling their galleries with loosely-packed, granular frass, which feels gritty when rubbed between the fingers (Plate 58, figs. 3 and 4); the pellets are appreciably thinner than those in longhorn borer frass.

The life cycle and biology of the common furniture beetle have not yet been fully investigated. The adults emerge in May, June, July, and August, and mate, when the females lay their eggs in suitable places; they will not lay on smooth surfaces. The eggs hatch shortly after they are laid, and the larvae commence tunnelling into the wood, on which they feed. The length of the life cycle is known to be as little as one year, but it is now thought that the life cycle, in this country, is more usually three years. The life cycle may well be considerably extended when food supplies are not entirely suited to the pest's requirements.

The common furniture beetle is widely known as a pest of old furniture, and of hardwood constructional timbers in period houses, but in the last 30 years it has been recognized as a common pest in the sapwood of softwoods in buildings of all ages. It was formerly thought that initial attack did not occur until the timber had been in service for several years, and that it was necessary for the timber to have 'matured' in some way for it to become attractive to the beetle. Entomologists now conclude that initial infestation may occur as soon as the timber has become seasoned, but the presence of attack may not be discovered until several life cycles have been completed, and flight holes are quite numerous.

In small articles of furniture and wooden ware the damage is not confined to sapwood, but in beams and constructional timbers generally, it usually is. In furniture and small wooden articles the damage done may be quite serious, as, for example, attack in the leg of a chair, but in structural timbers attack is mainly confined to the sapwood, and structural damage is only serious when the amount of sapwood is abnormally high. Attack in softwood timbers may develop to the extent of causing some timbers to collapse, if sufficient sapwood is present, but the presence of a beetle population is probably of more importance because of the risk of subsequent infestation of furniture.

Control can be effected by repeated applications of suitable preservatives, particularly those of the solvent type, and by sterilization or fumigation; neither of the last two mentioned methods confers immunity from fresh infestation, but they are more certain in eradicating existing infestation. To reduce the risk of infestation, it is recommended that care be exercised in the purchase of second-hand wooden articles as these may well be infested, thereby constituting a source of infestation for the spread of furniture beetles to sound timber. Fuel logs, and garden woodwork in close proximity to the house, are likely breeding grounds of furniture beetles from which infestation can spread indoors. The galleries of some furniture beetles may be distinguished from the flight tunnels of dry-wood termites (*vide* pages 279 to 282) by their being plugged with a black substance.

(2) *The death-watch beetle.* Death watch beetles lay eggs in crevices, cracks, or old exit holes, and the larvae do the damage by tunnelling in, and feeding on, the wood. Attack is usually confined to old timbers of several species of hardwoods, but it has been known to spread to adjacent softwood timbers.[1] Attack is not confined to the sapwood, but it is more likely to begin in sapwood than in heartwood. Adequate moisture, and the presence of fungal decay, are conditions favourable for infestation. The galleries made by the larvae are about 3 mm in diameter; they are filled with coarse frass, containing bun-shaped pellets (Plate 60, fig. 1). Removal of decayed wood, and the causes of decay, are the first essential steps in eradicating death-watch-beetle infestation. All too frequently, however, major structural damage has occurred before the infestation is discovered, when replacement of the attacked timbers, rather than *in situ*

---

[1] I have once encountered widespread and quite heavy death-watch beetle infestation in softwoods in an old house at Surbiton that contained no structural hardwoods, joinery, or flooring.

chemical treatments, is the only practicable course. It is essential to check the construction of attacked buildings, as, although large-sized timbers may have been used initially, subsequent shrinkage may have loosened vital carpentry joints. Moreover, the earlier craftsmen frequently used timbers the wong way round, so that apparently generous timber sections have been loaded from the commencement almost to the limit of their safe working stresses. Wherever possible, it is preferable to use pressure-treated softwood timber in repairs, but, if for aesthetic reasons hardwoods, and usually oak, must be used, such timber must be well seasoned and free from sapwood. Not infrequently, the elimination of dampness responsible for the initial fungal decay, the replacement of decayed and heavily attacked timbers, and the introduction of steel straps and the like to restore structural stability, will suffice, because any continuing attack will die out when conditions adverse to the rapid development of the beetle are established.

The life cycle and biology of the death-watch beetle have been exhaustively studied by the late Dr R. C. Fisher, who concluded that the length of the life cycle 'is dependent upon the moisture content of the timber, the presence and extent of fungal decay, and also upon temperature'. Under optimum conditions the life cycle may be only one year, but in less favourable circumstances it is prolonged over two to several years. Heavy infestation is often accompanied by *Corycorinetes coeruleus* Deg, a steely blue, hairy beetle, which is predatory on the death-watch beetle.

IV. OTHER TIMBER BEETLES

Several other beetles may sometimes be responsible for causing damage to timber in service, but the only two likely to be encountered at all frequently in this country are the wood-boring weevils, family *Cossonidae*, and *Ernobius mollis* L. The wood-boring weevils are essentially secondary infestation, following on fungal decay, and, in tackling this, the secondary pest is also eliminated. *Pentarthrum huttoni* Wallaston is probably the commonest wood-boring weevil, and the damage done resembles that caused by the common furniture beetle. The frass or dust is rather finer, usually round, and the conspicuous 'snout' of the weevil is a final clue to the pest at work (Plate 60, fig. 2).

Ernobius mollis L. is of no economic importance; it has not as yet acquired a popular name. It is a reddish or chestnut-brown beetle, up to 6 mm long, which leaves flight holes resembling those of the furniture beetle. The frass consists of small bun-shaped pellets, resembling those of the death-watch beetle but appreciably smaller; the frass is characteristically a mixture of red-brown and white particles because the larvae feed on bark of softwoods, often just penetrating the outer sapwood. Attack is quite common when the inner bark has been left on the waney edges of carcassing timbers. Removal of the bark brings an attack to an end, and there is no need for any additional chemical treatment. Damage is only likely to result if grounds with bark still adhering have been used as fixings for panelling. In boring their way out, the adults may leave flight holes that disfigure the panelling, besides giving rise to alarm.

It is important to stress that mere discovery of a flight hole, or even the finding of bore dust, is not necessarily a cause for alarm. It is important to identify the dust before having resort to expensive *in situ* chemical treatments, which, so often, are quite unnecessary because attack has ceased, or, alternatively, substantial structural repairs may be called for and *in situ* chemical treatments will be totally inadequate and probably unnecessary.

The wharf borer, *Nacerda malanura* L., should also be mentioned; it is a large beetle, superficially similar to some longhorn beetles, which leaves a flight hole about 3 mm in diameter; attack is confined to decayed timber.

V. TERMITES

The insect pests discussed in the preceding pages are those that commonly occur in temperate regions; the same species, or close relatives, and other insects unknown in cooler regions, are pests of timber in the sub-tropics and tropics. In these regions **termites** or **white ants** are the most serious insect pests; they probably cause more damage to timber annually than do all other insect pests together. Termites also destroy many other commodities, accounting for losses running into hundreds of thousands of pounds annually. Simple control measures exist, which would reduce the termite problem in the tropics to negligible proportions but, paradoxically, effective precautions of any kind are rarely taken.

There are many species of termites, but those that attack timber may be classified into one or other of two broad groups: **subterranean termites** and **dry-wood termites**. The former live in large colonies in the ground, and must retain an unbroken covered earthway from the soil to their feeding grounds. There is no risk of attack commencing from the swarms of termites that fly into houses at night; these individuals shed their wings and die by the morning, because they fail to regain a nest before their water requirements can be met. Dry-wood termites live in small colonies in thoroughly seasoned wood on which they feed; they require no access to the soil.

Termites do not make definite tunnels in timber, but they do tend to feed in restricted zones that become packed with mud as attack progresses (Plate 56, fig. 4). Attack on timber in the ground is somewhat different from attack above ground: naturally resistant timbers tend to be gnawed by termites, but the soft, non-resistant species may be completely hollowed out, except for an outer skin of wood. Contrary to popular statements, no timber is immune to subterranean termite attack, but the range in resistance of different timbers is appreciable: exposed to conditions of equal intensity of attack one timber may last less than six months and another more than ten years, with many other timbers with a variable serviceable life between these extremes. Moreover, resistance to fungal decay is not necessarily an indication of resistance to subterranean termite attack. For example, such naturally decay-resistant timber as oak heartwood does not show up well if exposed to termite attack. Conversely, some tropical timbers that are not regarded as particularly durable in their countries of origin, where termites are a more serious problem than fungi, may prove exceptionally resistant in temperate regions, where fungal decay is the serious hazard – examples are kempas and kapur. Hardness of a timber is no criterion of its powers of resistance.

Dry-wood termites invariably feed just below the surface of wood, and in most timbers attack is more or less confined to the sapwood; they produce granular dust, appreciably coarser than that of furniture beetles, which showers out when the sound skin of wood left on the surface of an attacked piece of timber is broken. The feeding termites also push out granular dust through their exit holes, the mounds of this dust often being the only evidence of attack in progress. The flight holes resemble those of some furniture beetles, but they are not plugged with black solid deposits.

Termites need not be nearly so serious a timber problem as they are generally supposed to be, and, of the two types, dry-wood termites are more troublesome than subterranean termites, because of difficulties of control: differences in the habits of the two groups result in precautions for rendering buildings proof against sub-terranean termites being ineffective against dry-wood termites. Effective control against the former is secured by proper design and construction of buildings: ant barriers must be provided at ground level. The simplest barrier in raised buildings is an oil channel around the feet of posts where they emerge from their foundations; where walls are contiguous with the foundations a strip of metal, extending 50–75 mm from the wall face, and projecting downwards at an angle of 45°, let into the damp-proof course below the ground-floor floor joists, is effective. With solid floors it is necessary to provide an impervious floor all over the site: at least 150 mm of con-crete, laid on foundations, proof against settlements, and extending through all walls, is recommended. Where these barriers cannot be provided, as, for example, with much timber used in contact with the ground (railways sleepers, fence posts, and poles), no timber will last indefinitely, and the choice is between the naturally resistant timbers or the less resistant ones adequately treated with wood preservatives. Chemical control of subterranean termites, usually involving the introduction of white arsenic powder into the runways, is effective in keeping down the termite population, but it is a palliative and not a curative measure, in spite of claims to the contrary. Extermination of termites by bacteriological methods is claimed for some proprietary products, but investigation of these has failed to establish that they work in any other way than as direct arsenious poisons. Brush applications of wood preservatives, unless repeated at frequent intervals, i.e., six to twelve months, are not effective in rendering wood immune from subterranean termite attack. Pressure processes are in a different category: with adequate absorptions of suitable preservatives, wood can be made to outlast its mechanical life, i.e., a properly treated railway sleeper will fail from rail-cutting or spike-killing rather than from termite attack or, for that matter, fungal decay.

Dry-wood termites cannot be controlled in the same way as

subterranean termites: given a susceptible wood, or exposure to attack, no economic methods exist for conferring immunity from dry-wood termite activity. The pests can be positively eliminated from building timbers and other indoor wood-work by screening buildings with fine metal gauze, a course that is extremely expensive and usually impracticable. Alternatively, reasonable precautions can be adopted that will minimize the risk of dry-wood termite attack, with dependence on curative measures when attack occurs. An American authority recommends the painting of wooden surfaces as an effective method of denying dry-wood termites entry into timber, and where this is practicable the course should be adopted, *e.g.*, for joinery. Planing has been suggested, but this is not an economical measure for carcassing timbers. It seems probable that the total exclusion of sapwood may appreciably delay dry-wood termite attack, even if it does not confer complete immunity.

The nature of the damage caused by dry-wood termites is apt to be misleading; at first sight the damage appears devastating, but closer inspection usually reveals the destruction to be less severe than was thought. In furniture, panelling, and high-class joinery, even a small amount of damage may be serious, because the appearance is spoiled, but in carcassing timbers it is necessary for the damage to be sufficiently serious to weaken the structure before alarm need arise. In practice, dry-wood termite attack is localized – some wooden members may be attacked while adjacent ones are quite free. Moreover, it is usually only parts of such members that are infested, and then often only to a depth of about 12 mm. If the infested zone is removed, and the remainder of the timber is liberally dressed with an oil-solvent wood preservative, or even a volatile toxic substance such as orthodichlorobenzene, attack will often cease, and the reduced member is usually still strong enough to carry the load required of it. It is probable that preparations suitable for eradicating furniture beetle attack will also prove effective against dry-wood termites.

VI. MARINE BORERS

Although not insects, several marine organisms, of which the teredo or ship-worm is probably the best known, are responsible for heavy losses with timber used in salt water. Intensity of attack varies in different regions, but is generally much more severe in tropical than in temperate climates: even the naturally resistant species such as greenheart and billian may have a very short service life in some tropical waters. The damage done takes the form of tunnelling, either vertically or horizontally, in the wood, which may be so extensive as to destroy the strength properties of a timber member completely.

Any wood used in brackish water[1] is liable to attack, and only in situations where infestation is known to be slight is it economical to depend on naturally resistant timbers. Pressure treatments with capacity absorptions of creosote or other good preservatives have been found effective in temperate waters, but metal sheathing, or studding the timber with nails, is likely to prove more economical wherever marine borers are particularly active. These aspects are discussed in Chapter 15. Many, but not all, of the timbers that are resistant to a greater or less degree to marine borer attack have been found to contain silica deposits in their storage tissue. Some timbers containing appreciable quantities of silica, however, have not revealed any particular resistance when exposed to attack.

[1] There is some evidence to suggest that water can be too salt for optimum development of the teredo, whereas very low concentrations have been found to be associated with exceptionally heavy infestation.

CHAPTER 15

The Preservation of Wood

GENERAL PRINCIPLES

ALTHOUGH no timber is immune to deterioration and ultimate disintegration if exposed for a sufficiently long period to ordinary atmospheric conditions, the serviceable life of individual pieces of wood varies considerably, depending on the species concerned, the amount of sapwood present, the use to which the timber is put, and the situation and atmospheric conditions to which it is exposed. For example, sound wood has been recovered from the Egyptian tombs and from piles driven into mud hundreds of years ago; in these, and similar instances, preservation is to be attributed to protection from the atmosphere rather than to inherent durability of the timber used. There is no doubt that many species generally considered non-durable would last indefinitely under such conditions. The persistence in the forest for several centuries of sound stumps of western red cedar is, however, only explained by the natural durability of that species when exposed to ordinary atmospheric conditions. On the other hand, even the most durable timbers may last only a few years if placed in a warm, damp, and badly-ventilated position.

The principal causes of deterioration of wood in service, as distinct from deterioration during seasoning, are fungal infection, termite and other insect or marine-borer attack, mechanical failure, and fire. The resistance of a timber to these agents of destruction may frequently be increased by the use of a suitable chemical, applied as a preservative. One or two substances were used for this purpose as long ago as Roman times, but the extensive use of wood preservatives is a development of the last hundred years. In practice, preservatives are usually applied to non-durable timbers so as to render the treated wood sufficiently resistant to the agents of

PLATE 60

FIG. 1. Bun-shaped pellets of death-watch beetle frass

FIG. 2. The common wood-boring weevil, *Pentarthrum huttoni*, Wollaston (× 10)

FIG. 3. Softwood roofing timber attacked by the house longhorn

PLATE 61

FIG. 1. Open-tank treatment of poles

FIG. 2. Pressure-treating plant

Plate by F.P.R.L., Princes Risborough

deterioration to warrant its replacing a naturally durable, but more expensive, timber. The chemicals used are legion, and several methods of application are advocated. The selection of the most suitable chemical and method of treatment are of the utmost importance and must be based on a thorough understanding of the scope and limitations of preservative treatments; none confers complete immunity, and a treatment suitable for one particular set of conditions may be useless for others.

FIRE-PROOFING

Wood is highly combustible but non-inflammable; that is, although in favourable circumstances any wood may be burnt to ash, timber is, comparatively speaking, not readily ignited. This statement applies equally to the most resinous species and to those furnishing the poorest firewood. Certain timbers are classed as fire-resistant under the London County Council bye-laws; such timbers have withstood a standard flame test, or have shown themselves capable, under certain conditions, of resisting the passage of flame during a definite, arbitrary period.

Wood heated to a temperature of about 250° C. will decompose, producing inflammable gases and charcoal. This is what happens when the surface of wood is exposed to a flame or to radiant heat. If the inflammable gases are produced in sufficient quantities and ignited, their combustion raises the temperature of wood further in, and, in consequence, the fire is kept going until all the wood is ultimately completely consumed or burned.

Timber above a certain critical thickness, which is of the order of 12 mm, cannot support self-maintained combustion, and continued burning is only possible so long as the surface receives an additional heat output from some radiant source such as the flames of a neighbouring fire. The formation of charcoal on the outside of a piece of wood probably acts as a screen against radiant or conducted heat, thereby retarding distillation of inflammable gases from within. In consequence, the rate of burning is much reduced, and, when the layer of charcoal is sufficiently thick, burning may become so slow that insufficient heat is produced to continue the decomposition of wood further in, and the fire goes out. This is what happens with timbers of large dimensions, and explains why, in quite large fires, heavy timber posts and beams often survive when a building is otherwise completely gutted by fire.

The low thermal conductivity of wood has an important bearing on the way burning wood transmits a fire. Some of the heat produced is immediately radiated outwards, some is absorbed in raising the temperature of wood just inside the burning area, and the remainder is conducted to the opposite face, whence it is radiated. Further, the behaviour of flames is important. They rise upwards, and therefore transmit more heat to wood above the source of the flames than to that below or to the sides. Hence, wood held vertically and ignited at the bottom will burn more readily than the same wood ignited at the top or held horizontally. This explains why doors, panelling, and other vertically disposed timbers constitute a greater fire hazard than beams and floors (the greater size of beams and floors is also in their favour).

So-called 'fire-proofing' processes are effective in so far as they prevent flaming of the inflammable gases, or combustion of the charcoal: they do not prevent chemical decomposition of the wood, but they alter the form that this decomposition takes. They are, therefore, **fire-retarding** rather than **fire-proofing processes**. Gay-Lussac postulated certain theories regarding the action of 'fire-proofing' salts as long ago as 1821, and these theories, enlarged upon by subsequent workers, were generally accepted until modern fire research replaced earlier theories. It was held that suitable chemicals acted in one or more of the following ways:

(1) The chemical melts at a temperature below that at which wood decomposes, forming a glaze over the surface and preventing access of oxygen to the wood.

(2) The chemical decomposes under heat, yielding non-inflammable gases that dilute the inflammable gases from the decomposing wood sufficiently to produce a non-inflammable mixture.

(3) The chemical vaporizes at relatively low temperatures, absorbing sufficient heat to prevent the temperature of the wood rising to the critical decomposition point. This is the action of water on a fire: it requires more than six times as much heat to turn boiling water into steam than is required to heat the same volume of water from 60° F. to boiling point. Heat dissipated in this way makes less available for continuing the chemical decomposition by burning of whatever else is 'on fire'.

Modern research indicates that the efficient fire-retardant chemicals for the treatment of wood are those that, under heat, increase the yield of solid charcoal and water vapour at the expense of inflammable vapours responsible for flaming. The inflammable gases are produced in insufficient quantities to provide an inflammable mixture, so that the treated timber does not flame, and the denser charcoal is so modified that it does not glow under normal conditions. The action, being a chemical one, depends for its efficiency on intimate contact between the fire-retardant chemical and the wood. In practice, this calls for impregnation of the wood, usually under pressure, with a solution of the chemical: brush applications are altogether too superficial to be of any value. Large-size timbers cannot be completely impregnated, but an average penetration to a depth of 25 to 37 mm is desirable. The alternative, of copious brush applications, essential for timber *in situ*, is much less effective than treatments under pressure.

Several chemicals could be expected to act in the way mentioned above, but for reasons of cost, and such other factors as corrosion, hygroscopicity, and toxicity, selection of a fire-retardant for wood is narrowed to two or three: monammonium phosphate, boric acid, borax, mixtures of these, and diammonium phosphate.

For timber *in situ*, fire-retardant paints confer some degree of protection, and are probably to be preferred to brush applications of aqueous solutions of fire-retardant chemicals. These 'paints', which are really plasters, function as an insulating and reflecting layer; they do not control the chemical breakdown of wood under heat. The paints may consist of a thin mixture of calcium sulphate plaster, or sodium or potassium silicate with an inert filler. The following formula for a fire-retardant paint has been proposed in B.S./A.R.P. 33:

Sodium silicate	50·8 kg
Water	45·4 kg
Kaolin	68·0 kg

Such paint should be applied either by two brush coats or by a spray, to give a covering of 1·86 to 2·3 m² to 4·55 litres. This, and other fire-retardant paints, have a limited practical application: they offer practically no protection in the event of an intense fire, they are not durable under exposed conditions, and they cannot be applied to timber already painted with an oil paint.

Notwithstanding its combustibility, wood is, and always will be, an essential constructional material. Its specific heat is relatively high, and its conductivity is low: by comparison, most metals require less heat than wood to raise their temperature any given amount, and they conduct heat more rapidly. Hence, wood is often effective in retarding the rapid spread of a fire. Moreover, initial failure of constructional wood-work in a fire can usually be traced to the phenomenon of shrinkage. Wood, like other materials, expands when heated, but this expansion is more than offset by shrinkage of the wood consequent upon the loss of moisture from the fine cell-wall structure. Shrinkage may result in built-up timber, e.g., panels and styles in a door, pulling apart, thereby providing gaps through which flames can pass before the wood itself is consumed. On the other hand, the delay, compared with the rapid transference of heat by conduction through metal barriers, is sometimes sufficient to enable a fire to be got under control before serious damage has occurred.

A combination of different factors results in all-timber dwellings actually constituting a smaller fire risk in America, from the insurance standpoint, than dwellings constructed of alternative materials. The timber house is naturally better insulated than a brick one, and, in consequence, does not call for excessive strain on the heating system to maintain comfortable conditions indoors, and over-straining of heating systems, to make good heat losses through walls and roof, is a fruitful cause of fires. Another source of fires in dwellings is a defective flue. With any reasonable standard of construction, chimney flues are completely isolated from the timber frame in an all-wood house, and, consequently, defects developing in flues are less likely to give rise to fires in timber dwellings, compared with brick structures, where the same precautions are not taken, and the brickwork supporting joists, etc., is of necessity bonded to the brickwork around flues.

Because of the tendency for wood to char, rather than burn rapidly itself, 'heavy-timber' construction is often preferred in America to unprotected steel framing, when the fire hazard is unavoidably high; the burning contents of a building produces heat sufficient to cause steel framing to buckle, resulting in collapse of the building and a total loss, whereas charring of heavy timbers delays collapse, often enabling the fire to be got under control before the timber members have been so weakened that they are liable to fail, and the building, if not its contents, is saved. 'Heavy-timber' construction can seldom be considered in this country because of the relatively high cost of large-sized timbers.

The scope for fire-retarding treatments in non-timber producing countries is, for economic reasons, strictly limited. Ordinarily, sound constructional technique, the use of fire-stops (pieces of wood or incombustible material) to seal off floors and partitions, and insistence on seasoned timber, properly framed together, for joinery and interior finishings generally, will meet all normal requirements. By-laws require the use of so-called fire-resistant timbers for certain purposes. Selection of such timbers in other circumstances is, however, seldom justified: if timber is suitable, other factors than the fire hazard should govern choice of species. Precautions against incendiary bomb attack introduced abnormal factors, and fire-retardant treatments might have had very real application. Joinery and finishings in ships, exhibition wood-work, and stage scenery are in a different category, and even where the regulations do not call for fire-retardant treatments, their use is generally to be recommended. Wherever laminated-wood construction is an economically sound method of construction, as it is today in many timber-producing countries, the extended use of fire-retardant chemicals under pressure would appear to be fully justified.

MECHANICAL WEAR OR FAILURE

In many situations the life of timber is limited by mechanical wear, against which ordinary preservative treatments are ineffective. The serviceable life can sometimes be extended, however, by attention to design, and the selection of the most suitable timber in the first instance. Experience has shown, for example, that for flooring blocks, subjected to an abrasive action, quarter-sawn (edge-grain) timber has a longer life than flat-sawn timber, and paving blocks usually last longer when laid on a resilient, rather than on a rigid, foundation. The corrosive action of sulphuric acid and hydrochloric acid fumes is sometimes regarded as a special form of mechanical wear. Timber brought into contact with hydrochloric acid gas fumes may be seriously weakened in a surprisingly short space of time, and the action of sulphuric acid on wooden cases of storage batteries is well known. Ordinary preservatives are in-effective against such forms of corrosion, but protection may be afforded by impermeable coatings of wax or other suitable materials. A few timbers, *e.g.*, abura and southern cypress, are relatively acid-resistant, and would probably prove more economical than the usual constructional timbers in positions subject to the slow action of corrosive substances. Where corrosive action is rapid, protection of timber by a mechanical barrier, such as wax or a suitable paint, is likely to prove more effective than dependence on the natural resistance of a few timbers to acid fumes.

Weathering

Another aspect of mechanical wear is the weathering of timber exposed to the elements. The constant wetting of the surface of wood by rain, followed by rapid drying when the sun comes out, results in checking and deterioration of the wood surface, which increases the hazard of infection by decaying agents, in the main wood-rotting fungi. Checking and slow disintegration of the surface constitute **weathering**; decay, which may follow, is a secondary condition. Hence, a mechanical barrier is essential to combat weathering, and for this a good oil paint is best. It is important to maintain such a surface, by regular renewal, before the paint film breaks down. The toxic properties of wood preservatives offer no protection against weathering, and such substances are, therefore, of no avail unless they also provide a semi-permeable or impermeable film as a mechanical barrier to the action of the elements. Weather-ing can occur, and persist for a long time, without decay developing: siding of buildings in Iowa that has never been painted has weathered, but is still quite serviceable after eighty years.

Fungi and Insects

Wood preservatives are used mainly to increase the resistance of timbers to insect attack and fungal infection; they are also used in the elimination, or control, of attack in progress, an aspect already referred to in Chapters 13 and 14. Many substances have been tried as wood preservatives; some have proved excellent and others worthless, but none can claim to be the best in all circumstances.

THE PROPERTIES OF PRESERVATIVES

It is important to keep in mind the special circumstances of each particular job when selecting a wood preservative. For example, a substance that is readily soluble in water may be excellent for indoor use but worthless for outside work, and, conversely, a substance with a pronounced odour may be quite satisfactory for outdoor work but totally unsuitable for indoor use. Extravagant claims on behalf of proprietary products should always be accepted with reservation; even the best preservative only prolongs the life of wood, it does not confer immunity from attack for ever.

The ideal wood preservative has yet to be found, but the properties desirable in such an agent may be enumerated, and are useful as a basis for comparison. It should be:[1]

1. Highly poisonous (toxic) to fungi and insects.
2. Readily penetrating into wood.
3. Chemically stable (*i.e.*, not readily volatilized or easily decomposed, and, for outdoor use, not easily leached out).
4. Easy to apply and not dangerous or harmful to those applying it or subsequently handling the treated timber.
5. Non-deleterious in effect on the timber treated.
6. Cheap and readily obtainable.
7. Non-corrosive to iron, steel, or other materials, according to circumstances.
8. Fire-resistant, or at least not liable to increase the inflammability of wood (this is of secondary importance in timber for fence posts, gates, and other similar purposes).

The following additional qualities are sometimes important, especially for indoor application:

9. Odourless.
10. Colourless and free from effect on subsequent painting or finishing processes.

The first three properties are essential qualities of any wood preservative, and the remainder are of diminishing importance, according to circumstances.

[1] Compiled from *The preservation of timber*, Trade Circular No. 27, Commonwealth of Australia, and 'The toxicity of preservatives against wood-destroying fungi', by K. St. G. Cartwright, *Forestry*, vol. v. p. 139, 1931.

CLASSES OF WOOD PRESERVATIVES

In general, existing preservatives may be divided into three classes: (a) the tar-oil group, *e.g.*, creosote, (b) water-soluble salts, *e.g.*, zinc chloride, silicofluorides, arsenic salts, copper salts, boron, and (c) solvent-type wood preservatives in which the toxic substances are dissolved in certain spirits or other volatile liquids.

Some preservatives consist of mixtures of two or more classes, and, in addition, there are volatile substances, the vapour of which is the toxic element.

(1) **The tar-oil group of wood preservatives.**   In this group is creosote, a complex substance derived from coal or wood distillation.   The toxic elements are frequently phenolic bodies, but it is apparent that several of the other 200-odd constituents may possess valuable toxic properties: some creosotes poor in phenolic bodies are quite good wood preservatives.   Not all 'creosotes' are equivalent in quality: some are the products of prolonged distillation that leaves only high-boiling tars of low toxic value.   Coal-tar distillates are likely to be superior to wood-tar derivatives.   To ensure a good grade creosote it is advisable to purchase on the basis of some accepted standard: in this connection the British Standards Institution's specification No. 144 of 1963, or the Australian draft specification No. K.55, ensure high-grade preservatives. The former specification defines three types of creosote, two being the products of distillation from vertical retorts (types A2 and B), and one from horizontal retorts (type A).   Service tests have failed to prove that any one type is superior to the others: type A is the heaviest. Creosote is sometimes mixed with Diesel oil where the latter is appreciably cheaper.   Provided adequate absorptions are secured, the mixture can effect worth-while economies in certain circumstances, *e.g.*, for railway sleepers limited in life by rail-cutting. Tests with mixtures in ratios of 75 per cent. creosote and 25 per cent. Diesel oil, 50/50, and 25/75, have established that reducing the percentage of creosote reduces the efficacy of the preservative. There is some evidence that the addition of fuel oil reduces subsequent splitting of the treated wood.

Although conforming to standard specifications, creosotes prepared from different coals vary in cleanliness, and in the amount of sludge produced when mixed with fuel oils.   One firm in the tropics imported creosote from Scotland because cleaning of the tankers in which the creosote was delivered was least costly with this particular creosote. Many creosotes are sold under proprietary names: some of these preservatives are excellent and others less so; greater cleanliness, and their being marketed in convenient one or five litre quantities, are the special merits of the best of these products, which tend to be very much more costly than non-proprietary creosote.

(2) **Water-soluble salts.** The common water-soluble wood preservatives include zinc chloride, sodium fluoride, and magnesium silicofluoride. Salts of copper, potassium, and arsenic are also extensively used, and boron. Some of these salts are essentially fungicides, and others are insecticides: copper salts are good fungicides, and arsenic preparations are good insecticides. Sodium pentachlorophenate, manufactured by Monsanto Chemicals Ltd, has given very good results as a fungicide when used as a 5 per cent. solution in water; it is marketed as 'Santobrite', or as 'Cuprinol for brickwork'. Its primary use should be for the sterilization of masonry and brickwork, and as a toxic barrier. Another similar product– sodium orthophenylphenate–manufactured by Coalite and Chemical Products Ltd, Bolsover, Derbyshire, and sold under a variety of trade names, has the advantage of being less corrosive to handle. Against moulds it is normally used at higher concentrations than sodium pentachlorophenate.

Several proprietary water-soluble preservatives are useful for small jobs, whereas, for larger programmes, it is obviously more economical to purchase the appropriate chemicals for dissolving in water on the site, in preference to buying ready-made solutions containing a high percentage of water on which distribution and freight charges are incurred.

Certain patented water-soluble wood preservatives employ two water-soluble toxic salts, and an oxidizing agent, with the object of depositing insoluble forms of these salts in the wood. The toxic salts are usually water-soluble copper and arsenic salts. Service tests of certain of these proprietary products have given encouraging results. As with any preservative, however, an adequate treatment is essential. Preservatives of this type are not ordinarily marketed to the public. Instead, the firms manufacturing the chemicals undertake the treating of timber under pressure in their own plants, or, alternatively, they supply the chemicals to firms owning approved plants, and the latter carry out the treatments, furnishing those who supply the chemicals with records of each treatment process. The Celcure and Tanalising processes are operated in this way in this country and overseas.

The special application of these and similar treatments is as an alternative to creosote, when freedom from odour, and a paint finish, are required. They are likely to be more economical than creosote when treatments are carried out at great distances from production centres because freight charges are incurred only on the toxic chemicals and not on the non-toxic solvent. It must not be overlooked that timber pressure treated with water-soluble salts has a very high moisture content immediately after treatment and must be dried before being put into service; for many purposes this may entail kiln-drying after treatment to obviate delays in the building or repairs programme.

A diffusion treatment, employing a highly soluble form of boron, was developed in the 1940s in New Zealand, primarily to combat insect attack, but the treatment has been found equally efficacious against fungi if timber is not subject to leaching subsequently. The treatment has to be applied immediately after conversion from the log, the timber being dipped for a period of two to ten minutes, depending on thickness, in a very strong solution of boron, and then covered up to permit of penetration of boron by diffusion, when excellent penetration, even of the heartwood, is secured. The timber is then air- or kiln-dried in the normal manner. The treated timber is marked under Borax Consolidated Limited's Registered Trade name of 'Timbor' (or Timborised timber). The process is now extensively used in Australia and the Scandinavian countries, and more recently in British Columbia and Africa.

(3) **Organic solvent wood preservatives.** These consist of toxic substances soluble in refined paraffins (and other petroleum oils) or white spirit; they are more expensive than water-soluble preservatives because of the higher cost of the solvent. Experiments indicate that permeability of this group of preservatives is usually good and most are reasonably permanent for both interior and exterior use; they are usually non-creeping once the solvent has evaporated and the treated surfaces can generally be painted when

dry. Most solvents have a strong odour and will taint foodstuffs without the food being in actual contact with the treated wood. They are inflammable, and some have dangerously low flash points, calling for especial care in use and in storage.

The spirit solvent preservatives are quick drying and the solvents usually have a less persistent and objectionable odour than the petroleum oil products. The latter especially may be detrimental to polished surfaces and, because of the fumes given off, masks and goggles should be worn when spraying. With the solvents at present in general use, it is possible that the spirit solvents have better penetrating qualities, but the oil industry is constantly conducting experimental work, which may provide 'oils' with better and better penetrating qualities. Oils are cheaper than the spirits as solvents.

The oil-solvent preservatives are of three distinct types: those whose toxic ingredients are wholly volatile, although complete volatilization may be prolonged over a period of two to three years, those whose ingredients are non-volatile, being retained more or less permanently in the wood, and those that include both volatile and non-volatile toxic ingredients. Quite apart from differences in merit of the many toxic ingredients used in oil-solvent preservatives, the class of product that should be used depends on the nature of the problem: eradication of active 'insect' infestation necessitates the use of a preservative containing volatile ingredients. If there is no risk of recurrence of infestation, and, assuming adequate, effective penetration, volatile substances alone suffice for dealing with an outbreak. For purely prophylactic purposes, that is, as insurance against possible future attack, whether by insects or fungi, and in combating fungal attack in all circumstances, the non-volatile toxic ingredients are the effective ones. Where active beetle attack is involved, a preservative containing both volatile and non-volatile ingredients is usually advisable: it is seldom possible to ensure complete penetration of the wood, and hence destruction of the pest in all stages, in one application, and the likelihood of re-infestation from any beetles that survive, or from an outside source at some later date, is usually appreciable.

Orthodichlorobenzene and paradichlorobenzene, and, more recently, the 'Gammexane' group of products (containing gamma BHC) are known to be highly effective volatile substances. D.D.T.

has been added to some commercial formulae. Pentachlorophenol
and copper naphthenate are examples of 'permanent' toxic ingredients.   In choosing a proprietary product, it is essential to
know to which of the three categories discussed above a particular
preservative belongs. Formulae of proprietary products tend to be
changed from time to time, which may invalidate the significance of
test results from Governmental laboratories – there is a constant
search for cheaper solvents, some of which are not capable of
retaining in solution the requisite proportions of the valuable toxic
ingredients.   A 5 per cent. solution of pentachlorophenol in an oil
solvent is the cheapest type of oil solvent preservative for use as a
fungicide or for porphylactic purposes, and 'Gammexane' emulsion
concentrate in 20 volumes of odourless kerosene is the cheapest form
for *in situ* use as an insecticide.   A 5 per cent. solution of pentachlorophenol in white spirit should be used when creeping of the oil or
staining must be avoided.

A more recent development has been the use of pentachlorophenol
as a bodied mayonnaise-type emulsion preservative, marketed under
the proprietary name of Woodtreat.   The emulsion consists of oil
containing a 5 per cent. solution of pentachlorophenol and 0·5 per
cent. of dieldrin dispersed in a much smaller volume of water.
With this formulation substantially better penetration is secured
than with similar formulations in normal spirit or oil solvents.
There is some risk that Woodtreat will creep, causing staining of
plaster in the vicinity, and if, for example, it is applied to the back
of timber 1 in. or more in thickness, it may be necessary to use
special primers before the opposite face of such timbers can be
successfully painted.

There is a fire hazard with organic solvent preservatives because of
the dangerously low flash points of some solvents. Containers of
preservative should not be stored where treatments are being carried
out; they should be stored in cool places so that there is no risk of heat
causing the solvents to volatilise in a partially full container, providing
an explosive vapour. Defective wiring for temporary lighting has
resulted in disastrous fires. It is advisable to cut off all electricity at
the mains, providing temporary lighting from cable complying with
B.S. No. 1003, fitted with flame-proof sockets and Buxton's certified
flame-proof lamps.   In addition, it is advisable to insist that two
water, gas-pressure Power-jet extinguishers and one 20 lb. dry powder

extinguisher are kept close at hand where treatment work is in progress. Smoking should be prohibited where the application of inflammable preservatives is in progress, and for, say, five days after the treatment has been completed. These somewhat drastic provisions are recommended because of the casualness of so many operatives today. If organic solvent preservatives are used on a considerable scale in unventilated roof voids, with say only one access hatch at one end of the void, the precautions outlined above may not suffice. The vapour of most of the carrier solvents in organic-solvent preservatives is heavier than air, and it will tend to accumulate at ceiling joist level, reaching dangerously high concentrations, and remaining a serious fire hazard for a considerable time. A naked light, *e.g.*, a blow lamp, taken into such a roof void, has been known to cause a series of explosions, resulting in the destruction of the greater part of the roof before the resultant fire could be brought under control. In such cases it is essential to provide additional temporary ventilation, by cutting holes in ceilings at several points. Until all traces of vapour have dispersed, no naked lights should be used anywhere on the premises. With the massive quantities of fluid used by some commercial firms the danger period may extend for much more than a week, even with substantial supplementary ventilation on the lines recommended. Provided adequate precautions have been observed there is no long-term fire risk from treated timber.

Finally, it is essential to ascertain what are the constituents in any proprietary formulation because certain chemicals can be deleterious to health. In particular, even quite small quantities of seekay wax may be injurious to those with liver complaints. No food should be either prepared or eaten in any room where a treatment has been carried out until there are no traces of odour – usually from the solvents – detectable.

(4) **Volatile substances, fumigants, and sterilization.** Other chemicals have an important place in prolonging the life of wood. These are substances that volatilize at ordinary temperatures, or that can easily be converted to a gaseous state. They are more strictly fumigants, rather than wood preservatives, since they are not retained in the wood, and, therefore, confer no protection against subsequent reinfestation. Moreover, their use is essentially restricted to control of insect, not fungal, activity. Against active

attack, they are effective, but their transitory character must be appreciated: they should not be regarded as 'wood preservatives'.

Kiln treatments must be considered as alternatives to fumigation, or to the use of volatile chemicals. Where kiln facilities exist, and the timber is such that it will not suffer from a kiln-sterilization treatment, the method is to be recommended because of its 100 per cent. efficacy when properly carried out. Moreover, kiln sterilization is effective both against fungal activity and insect infestation, whereas fumigation deals only with the latter. A temperature of 160° F. and 100 per cent. relative humidity, held for two hours, would be lethal to fungal infection and insects in timber up to 75 mm in thickness, but these conditions would be too severe for most manufactured wood-work. Temperatures as low as 115° F. and a relative humidity of 60 per cent., held for 36 hours, have been found effective for sterilizing *Lyctus*-infected timber; allowing for the lag period in heating up the wood to the temperature of the kiln, and a margin for safety, 46 hours for 25 mm material, up to 60 hours for 75 mm timber, are suggested as maximum periods. It seems probable that a temperature of 130° F. and 80 per cent. relative humidity, with a total exposure period of $2\frac{1}{2}$ hours for 25 mm and up to 7 hours for 75 mm material, will prove adequate against any form of insect infestation, but this has not yet been conclusively tested. Neither French polish nor turpentine varnish finishes are appreciably affected by a temperature of 130° F. and a relative humidity of 80 per cent., nor is plywood bonded with resin, casein, or blood-albumin glues for the periods necessary for sterilization. A table of alternative temperatures and relative humidities for the successful sterilization of *Lyctus*-infected timber is given in Princes Risborough Laboratory, Technical Note No. 43, November 1969.

Next to kiln sterilization, fumigation is the most effective means of eliminating active insect attack, whether in the egg, larval, or beetle stage. Portable articles are fumigated in special chambers; skilled operators are required to carry out the work because of the poisonous nature of the fumigants. The fumigants in general commercial use are hydrocyanic acid gas (hydrogen cyanide or HCN) and methyl bromide. Hydrogen cyanide is obtainable as a liquid, which has to be volatilized by suitable means, or the gas may be produced as a result of chemical reaction, inside the fumigating chamber. Fumigation is essentially a task for the expert, since so many factors are involved: in the gaseous state the substances may be harmless to polished surfaces and fabrics, whereas in the liquid state they may do considerable damage. Fumigation with hydrogen cyanide in the United Kingdom is governed by the Hydrogen Cyanide (Fumigation of Buildings) Regulations, 1938.

The application of toxic chemicals that volatize readily at ordinary atmospheric temperatures provide yet another method for combating active insect infestation. Chemicals such as orthodichlorobenzene belong to this group. The efficacy of these chemicals depends on the poisonous fumes reaching the insects in sufficient concentration to be lethal. It is usually necessary to time the application appropriately, *i.e.*, when the larvae are feeding or the adults are about to emerge. Further, it is advisable not to depend on a single treatment, but to repeat the application at the first signs of renewed activity. The difficulty of ensuring penetration of the fumes has to be overcome: a fountain-pen filler, small oil-can, or even a hypodermic syringe, to inject the liquid into flight holes or galleries, will ensure better penetration than surface applications. In addition, however, copious surface dressings should be applied to infected areas, and particularly in cracks, crevices, and joints. Many readily volatile substances are not suitable for use in inhabited rooms, because their fumes, if not actually poisonous to man, may induce sickness. Spirit-soluble preparations must, of course, be kept away from flames.

The Toxicity, Penetration, and Performance of Wood Preservatives

In BS 1282: 1945[1]: *Classification of wood preservatives and their methods of application*, it was stated that the following were the all-important factors determining the efficiency of wood preservatives:

1. Toxicity or killing power towards wood-destroying fungi or insects or both.
2. Penetrating power.
3. Performance, *i.e.*, resistance to leaching, evaporation, and chemical decomposition.

[1] Now BS 1282: 1959, with an amendment issued in 1961.

At the time this Standard was published the only British Standard relating to particular wood preservatives was BS 144: *Coal tar creosote for the preservation of timber*, revised in 1954 and amended in 1963. There are now BS 3051: 1958: *Coal tar oil types of wood preservatives (other than creosote to BS 144)*, with an amendment in 1963; BS 3452: 1962: *Copper/chrome water-borne wood preservatives and their application*, and BS 3453: 1962: *Fluoride/arsenate/chromate/ dinitrophenol water-borne wood preservatives and their application*. The British Wood Preserving Association has Standards for other preservatives, and, in accordance with past practice, it is to be expected that the Association's Standards will ultimately be issued as British Standards by the British Standards Institution.

Laboratory tests are available for measuring toxicity and four British Standards have been issued dealing with this aspect, BS 838: 1961: *Methods of test for the toxicity of wood preservatives to fungi;* BS 3651: 1963: *Method of test for toxicity of wood preservatives to the wood-boring insects* Anobium punctatum *and* Hyloptrupes bajulus *by larval transfer;* BS 3652: 1963: *Method of test for toxicity of wood preservatives to the wood-boring insect* Anobium punctatum *by egg-laying and larval survival*, and BS 3653: 1963: *Method of test for toxicity of surface-applied wood preservatives to the powder-post beetle* Lyctus brunneus.

Satisfactory means for measuring penetrating power and performance, are difficult to devise. Penetration is particularly difficult to determine with preservatives of the tar-oil group: even on a freshly-cut surface the oil spreads so rapidly that it is difficult to differentiate penetration from surface-spreading from excess preservative in the cell cavities on the periphery of the cut section. Exposure tests, using the same species of timber, and as nearly matched material as possible, for a series of tests, provide a yardstick for measuring the relative merits of different preservatives. Unfortunately, it is difficult to standardize the tests completely, and when all reasonable steps have been taken, Government laboratories may not be in a position to publish the results of tests on proprietary products. In the circumstances, strict comparison is impracticable, and selection of a preservative must depend on assessing each particular problem and deciding what are the most important conditions to be satisfied.

The Application of Preservatives

Having decided on the preservative, the problem of getting it into the wood remains to be settled. The four principal methods are (a) **brush coating**, (2) **dipping**, (3) **pressure treatments**, and (4) **diffusion**. But just getting some definite quantity of a preservative into a timber is no guarantee that the treatment will prove satisfactory; depth and uniformity of penetration are of the utmost importance. Uniformity of penetration is dependent largely on the anatomical structure of wood: sapwood absorbs preservatives more readily than heartwood, the early wood of softwoods better than the late wood, and timber of some species more readily than that of others – *e.g.*, Scots pine (European redwood) is much easier to impregnate than Douglas fir, apitong than lauan. The path of penetration of preservatives in hardwoods appears to be different from that in softwoods: in the latter there is appreciable movement through the pits in the cell walls, but in the former movement appears to be largely through the vessels and, in consequence, penetration is very much heavier in hardwoods along than across the grain. Depth of penetration is mainly important when checks are likely to develop after treatment, otherwise in theory the thinnest complete covering is all that is necessary. Wood below a treated skin is very liable to be exposed sooner or later and, when treated green, checks develop and expose wood below the depth of penetration to fungal and insect attack.

Certain general principles should be observed whatever the nature of the preservative or method of treatment. For example, because of its better absorptive powers, sapwood should be retained, and in larger-sized timbers those with a complete outer layer of sapwood should be selected for treatment in preference to those with one or more heartwood faces. Except when the diffusion method is used, better penetration is obtained with hot rather than with cold methods of application, and, therefore, except with chemicals that decompose on heating, preservatives should be applied hot whenever practicable. Except for timber to be treated by the diffusion method, when the treatment must be applied immediately after conversion from the log, it is desirable that timber should be at least air-dry before it is treated. This is because air-dry timber absorbs preservatives better than green wood. Moreover, if treated when air-dry, checking will have occurred before the timber is treated, and not afterwards, and the seasoning checks will facilitate impregnation. All trimming and boring of timber must be done before the wood is treated, or, when this is impossible to arrange, surfaces subsequently exposed should be given a heavy dressing of the preservatives; even with such precautions fungi and insects will usually launch their attacks through these weak places.

Brush coating. Brush coating is the simplest method of applying a preservative, but it is the least efficient. If, however, two or more coats are given, and the treatment is repeated every two or three years, some measure of protection is obtained, particularly with interior wood-work that is not exposed to mechanical wear. The method should, however, only be used when alternative treatments are impracticable, or temporary protection or a short service life alone is required. Wherever practicable, preservatives with good penetrative qualities should be selected.

Spraying and dipping. Spraying is usually not so effective as brush coating, but dipping is appreciably more effective, because the wood is in contact with an excess of preservative for the duration of the dipping process. If the length of the immersion period is sufficiently long, and the absorptive powers of the wood are good, a considerable quantity of preservative may be taken up in the dipping process, particularly if the preservative is one that can be heated and the bath or dipping tank is filled with hot liquid. When the preservative is applied by spraying there is a risk that the surfaces treated are not thoroughly wetted by the preservative, which may be held by surface tension on rough surfaces or particles of dirt or dust.

Open tank. A special form of dipping, in effect a modified pressure treatment, known as the **open-tank** method, is much more effective than brush coating or dipping. Except that it is more wasteful of the preservative than certain pressure treatments, the open-tank method gives results comparable with those obtained with pressure processes, provided the timber treated possesses good absorptive properties. The method requires a bath or tank with some form of heating apparatus, in which the timber to be treated can be completely immersed (Plate 61, fig. 1). The timber is submerged in the cold preservative, which is then gradually heated to a temperature of 160° to 200° F., which is held for ½ hour to 4 hours, after which the preservative is allowed to cool, with the timber still immersed. The peak temperature, and the period that the temperature is held, are varied according to the type of preservative and the absorptive powers of the timber. Heating causes the air in the cells of the wood to expand, some being expelled, and, as cooling begins, the air left in the cells contracts, and the external air pressure forces in the preservative to take the place of the air lost. Variation in the length of time the peak treatment-temperature is held, and the peak temperature itself, provide means of governing the degree of heaviness of the treatment and, consequently, the quantity of preservative absorbed by the wood.

The simplest form of open tank is a container that can be heated by external firing, but, because of the risk of the preservative catching fire, steam-heated coils inside the tank are much to be preferred. Various refinements, such as mechanical loading devices and separate cooling tanks, are sometimes introduced, but these are not essential. The method is to be recommended only where the quantity of timber to be treated annually is insufficient to justify the capital outlay on the more expensive, but more efficient, plant necessary for pressure processes.

Pressure processes. The most effective method of treating timber is the use of a pressure process for forcing the preservative into wood. With such processes penetration is obtained to a considerably greater depth than is possible by any other method. An open-tank treatment is in a sense a pressure process, but the effective pressure employed is of necessity less than one atmosphere. By using closed retorts, on the other hand, high temperatures, and pressures of several atmospheres, can be attained, the only practical limitations being the risk of mechanical damage to the timber. Temperatures in excess of 200° F., or pressures greater than 1·379 newtons per mm², are rarely used because of the adverse effect of such temperatures and pressures on woody tissue.

A pressure plant consists of a retort fitted with a door that can be hermetically sealed, a supply of steam for raising temperatures, and hydraulic pressure pumps for controlling the pressure inside the retort; in commercial-size plants loading is done on trucks outside the retorts (Plate 61, fig. 2).

The oldest pressure process, introduced in 1838, is the **Bethel** process, which is still in use today. In this process the retort is first loaded with timber and then hot preservative is run in and a pressure gradually built up; when it is estimated that an excess of preservative has been injected into the timber the pressure is released and the surplus liquid is drawn off. A common modification of the Bethel process is the application of a vacuum to the timber before running in the preservative. This, it is claimed, by drawing the moisture from the cell cavities, facilitates the subsequent entry of the preservative into the wood; technical opinion does not support this claim, however, holding that the vacuum period generally used in commercial practice is too short to affect the issue. The vacuum, however, serves two purposes: by increasing the effective pressure to which the timber is subjected, there is an increase in the pressure available for forcing the preservative into the wood, and, by establishing a vacuum in the retort, filling is facilitated, particularly if the storage tank is at a lower level than the treating cylinder.

The Bethel process and its modifications are known as **full-cell** processes; that is, if completely successful, not only is the preservative injected into the cell walls, but the cell cavities also are filled. Except with readily leachable substances, and timbers of low absorptive powers, an excess of preservative occupying the cell cavities is wasteful. To overcome this objection so-called **empty-cell** processes have been devised. One of these is the **Rueping** process, which involves the use of an initial air pressure, built up in the treating cylinder before the hot preservative is run in. This pressure compresses the air in the cell cavities of the timber. The hot preservative is then introduced and a still higher pressure is built up within the retort until an excess absorption of preservative is secured. The pressure is then reduced to atmospheric pressure, so that the air compressed in the cell cavities of the wood expands and ejects the surplus preservative. In this way a relatively deep impregnation is secured, with a smaller net absorption of preservative, compared with full-cell treatments. A modification of the Rueping process, known as the **Lowry** process, employs an initial, atmospheric-air pressure and a vacuum at the end of the run to extract surplus preservative; the amount extracted is comparatively small compared with that expelled at the end of the Rueping process.

In another pressure treatment, the **Boulton** process, an initial vacuum is applied to the timber while it is submerged in the heated preservative. This reduces the boiling point of the water contained in the wood, enabling it to evaporate at a temperature below 212° F. In this way considerable drying of the timber during the treatment can be effected at comparatively low temperatures, and without exposing the wood to the risk of serious degrade.

Incising. The relative ease with which different timbers absorb preservatives varies appreciably; beech and European redwood sapwood can be completely impregnated under pressure in 1 to 3 hours, but a similar treatment of Douglas fir, larch, or oak heart-wood, would not effect complete impregnation in several days. Moreover, a more severe treatment of recalcitrant timbers, using higher temperatures and pressures, does not overcome the difficulty, but merely secures increased absorption in patches. Various attempts have been made to solve the problem of treating such refractory timbers and, of these, an operation known as **incising** has proved the most successful. The process consists in making incisions parallel to the fibres to the depth of penetration required, and spaced sufficiently close together that the lateral spread of the preservative will ensure uniform and complete penetration to the depth of the incisions. By this means an increased absorption of 30 to 60 per cent. has been obtained with Douglas fir sleepers.

Other methods of treatment. The foregoing pages[1] have out-lined the more important methods of applying wood preservatives that are in general use, but one or two other methods are of interest. In the **Boucherie** process use is made of hydrostatic pressures: a water-soluble preservative is supplied from a raised tank, through a pipe, to the base of a log or baulk of timber fixed in a horizontal position; the hydrostatic head forces the liquid through the timber, the treatment being completed when the preservative appears at the top end of the log. The Boucherie process, with copper sulphate solution as the preservative, was used regularly in France for the treatment of telegraph poles and railway sleepers.

The impregnation of fence posts and telegraph poles *in situ* by diffusion has been tried in Germany. The preservative is injected, in paste form, at several points and spreads by diffusion through the

[1] See also page 294 for a brief description of the diffusion process.

wood: good penetration has been secured by this means with wet timber. The same principle is involved in the attempts to introduce preservatives into timber *before* the tree is felled, but the method is still in an experimental stage.

The impregnation of non-durable timbers with the alcohol extractives (infiltrates) of durable woods has not met with success, although it is known that these substances are responsible for rendering certain timbers naturally durable. The reason for this is, no doubt, that the infiltrates or extractives are more intimately associated with the cell-wall structure than can be reproduced by artificial impregnation.

Charring is a well-known method for protecting timber in contact with the ground. No experimental data have been collected to determine the efficacy of the method, but it is unlikely that charring would greatly prolong the life of timber. Charring destroys the accumulated food supplies stored in the parenchymatous tissue of the outer layers of a piece of wood, thereby rendering the charred timber less attractive to insects and fungi. Charring may possibly be justified for small round poles used as temporary fencing, when the cost of a proper preservative treatment cannot be entertained, but it cannot be ranked at all high among the various chemical methods of wood preservation.

ECONOMIC ASPECTS OF WOOD PRESERVATION

Much money is wasted annually through the indiscriminate use of wood preservatives, especially in tropical countries. This waste occurs chiefly through unnecessary application of brush coatings, but inadequate applications may also be wasteful. As a general rule, brush coatings are only justified in circumstances where repeated applications at regular intervals are practicable. Single coatings of inaccessible timbers are either unnecessary, or inadequate. This applies particularly to roof timbers: if conditions are so bad that such timbers require protection, a single brush coating will be insufficient. It could be argued that except for tile or shingle battens in really damp climates, it was not necessary to treat roof timbers. Now, however, that the quality of timber normally available for carcassing purposes is appreciably lower than formerly, and in particular it is likely to contain substantial quantities of sap-wood, there is a good case for using only *adequately* treated timber for roof members and battens. By adequately treated is meant either a pressure process, *e.g.*, Celcure treatments or Tanalising, or a diffusion treatment with boron, *i.e.*, Timborised timber. In those areas where the house longhorn borer is an active pest most Local Authorities lay down in their by-laws that the roof timbers shall be adequately treated, and in such areas pressure treated or Timborised timber should always be used for roof timbers and battens. Too frequent recourse to brush applications of wood pre-servatives in new work, etc. encourages the use of poor quality timber, and even the deliberate substitution of an inferior species to that specified, and for this reason wood preservation by chemical means should only be adopted in circumstances where it can be fully justified.

The economic aspects of wood preservation are more usually confined to weighing the advantages of using a lower-priced, non-durable timber, adequately treated with wood preservatives, against the more expensive timber possessing appreciable natural resistance to fungal and insect attack. In this sense the economic value of a preservative treatment becomes a matter of simple calculation: cost of maintenance and the annual charge on the material treated compared with the same figures for untreated material. The annual charge may be arrived at from the following formula:

$$A = P\frac{r(1+r)^n}{(1+r)^n - 1}.$$

Where A is the annual charge,

P is the cost of material plus erection,

r is the rate of interest, expressed in decimals, *e.g.*, 5 per cent. $=$ 0·05, and

n is the anticipated serviceable life.

Values for n should be based, wherever possible, on figures obtained from service tests.

Other factors that require consideration are the estimated life of a building before it may be presumed to become obsolete, or no longer suitable for the purpose for which it was erected, the salvage value of wood used in a building, and the saving that can be effected through using larger quantities of sapwood. There are undoubtedly many cases where only a short life is required, and in these circumstances it is wasteful to build a structure that will outlast, by many years, that required life.

The value of chemical treatments to enhance other properties than resistance to decay has already been discussed in connection with the hygroscopic properties of wood, *vide* pages 146 to 151 and 236. Similarly, improvement of mechanical properties by impregnation with synthetic resin would appear to have commercial possibilities in special circumstances, *vide* page 150.

The appreciable salvage secured from dismantling wooden buildings is one of the important advantages wood has over alternative building materials, and increasing the life of wood by the use of wood preservatives will enhance its value when the time for dismantling arrives. Increased use of sapwood, permissible as a result of adoption of adequate preservative treatments, may also prove economical compared with the higher cost of timber free from sapwood.

For timber used in salt water it is known that very heavy absorptions are required to secure even moderate extensions of serviceable life, and in such cases the use of chemical wood preservatives is rarely economical, compared with sheathing of wood with thin metal sheets. For piles in salt water, absorptions up to 250 kg per m³ have been used to secure a serviceable life of well under ten years. In such circumstances it would undoubtedly have been more economical to use comparatively light treatments, and to sheath the piles with Muntz metal. The object of the light treatment is to secure some resistance should the sheathing become damaged in the interval between regular inspections, which should be made three or four times each year. Evidence from Malaysia is to the effect that such sheathed piles have a very long life, although the timber has only moderate resistance to marine borers: sheathed merbau piles were still perfectly sound after 27 years' service. Unprotected greenheart would certainly not have lasted nearly so long in the particular circumstances.

Another aspect of the economics of wood preservation, outside the control of the individual user, is of the utmost importance to him. If the use of wood preservatives permits of the utilization of many species that would otherwise be valueless, exploitation of the forests as a whole will be less costly. Further, the availability of adequate timber supplies will tend to keep prices of the most favoured species at a steadier level. Both these factors apply particularly in the tropics, where the number of species in any unit area of forest is large, and the number commercially exploitable without the use of wood preservatives is very small.

CHAPTER 16

The Eradication of Fungal
and Insect Attack

TIMBER used as railway sleepers, fence posts, power-line poles, and the like is inevitably exposed to conditions favouring attack throughout its service life. By contrast, certain conditions of service ensure that timber never becomes attacked: piling timbers in deep water or timber in the abnormally dry atmosphere of the Egyptian tombs will remain free from attack indefinitely. Much more timber is used in circumstances between these extremes of certain attack and complete immunity, giving rise to the need for prophylactic measures or problems of eradication.

Timbers predisposed to attack by sap-stain fungi, 'pin-hole' borers, or 'powder-post' beetles are very liable to suffer deterioration unless appropriate precautions or prophylactic measures are taken. The sapwood of those timbers susceptible to 'mould' fungi are liable to become stained unless logs are converted immediately after felling, and the converted material is then dried so rapidly that the fungal 'spores' are not given the opportunity to germinate. It is often not possible to arrange for sufficiently rapid drying, in which case the use of chemical dips provides a practicable solution. The sodium salts of certain chlorinated phenols are particularly effective against both the blue-stain fungi and other moulds. Examples of highly successful proprietary products using this group of chemicals are Dowicide P,[1] and Santobrite.[2] Certain organic mercurial salts have been found very effective in controlling blue-stain, but less so against green moulds: Lignasan[3] contains ethyl mercuric chloride as the toxic agent. Two per cent. solutions of

[1] Manufactured by Chemical Treatments Co., 1604 Per Marquette, New Orleans, Louisiana.

[2] Manufactured by Monsanto Chemicals Ltd., Victoria Street, London, S.W.1.

[3] Marketed by E. I. du Pont de Nemours & Co., Wilmington, Delaware.

borax have also given good protection against blue-stain fungi. These chemicals, used in a dipping bath or trough through which the timber passes as it comes off the saw, provide a practicable solution of the blue-stain problem. The period of immersion is brief: 20 seconds is usually quite sufficient. After dipping, the timber must be given a chance to drain before it is piled in a stack. With some of the tropical timbers, where blue-stain infection is not confined to the sapwood, prevention of stain in planks 75 mm in thickness and upwards has often not been secured by chemical dips alone. Great care has to be taken to drain the planks before piling, often necessitating exposure to the sun, with frequent turning to obviate surface checking. Besides the handling cost involved, the yard space is rapidly taxed to capacity with planks draining and drying preparatory to piling. The most satisfactory procedure with such timbers is kiln-drying, but these facilities are often not available, when the alternative is to cut as thin timber as possible: 25 mm or 37 mm boards can usually be air-dried, after dipping, without staining.

'Pin-hole' borers present a somewhat similar problem to that of sap-stain fungi, in that attack is dependent on the existence of sufficient moisture in the wood to support the growth of the ambrosia fungus on which the pin-hole borer larvae feed. Some species attack standing trees, so that the damage is done before the trees are felled, and no methods have as yet been devised for dealing with infestation in this stage. Much infestation, however, undoubtedly occurs after felling, while logs are lying in the forest or at the mill awaiting conversion. Extraction immediately after felling, followed by immediate conversion at the mill, and proper piling to ensure rapid drying of the converted timber, would unquestionably minimize the widespread damage done by pin-hole borers. The rapidity with which these borers attack the logs of some species, coupled with the difficulties of organizing extraction and conversion, makes dependence on rapid extraction and conversion to avoid infestation uncertain, and chemical sprays have been used with some success. Disappointing results were obtained with chemicals that have proved effective against sap-stain fungi, and creosote and certain proprietary tar-oil preservatives proved a positive attraction to some ambrosia beetles. The gamma isomer of benzine hexachloride, available as a powder miscible with water or as a dispersion in oil, on the other hand, has given encouraging results. The water-miscible powder appears to give surprisingly good protection at considerably lower gamma isomer concentrations than the oil dispersions, but it is easily washed off by rain;[1] it has the advantage of being much less expensive than the oil dispersions. Although chemical prophylactic measures are still very much in the experimental stage, there is little doubt that effective treatments against much pin-hole borer infestation can be devised. It is, however, doubtful whether the general use of chemicals will prove practicable: where it is essential to apply the treatments in the forest, the difficulties of organizing supplies of the preservative, and the supervision of their application, may prove unsurmountable. Where a short delay between felling and the prophylactic treatment is unimportant, it should be possible to organize the application of an effective short-term treatment at log-collecting depots – these are matters that remain to be investigated thoroughly. From somewhat tentative investigations, it would seem that some timbers are attacked by some species of pin-hole borers within a matter of a few hours of felling, whereas, with other timbers and other borers, a few days may elapse before extensive attack occurs.

The countering of 'powder-post' beetle attack presents a problem intermediate in complexity between elimination of sap-stain fungal infection and 'pin-hole' borer attack on the one hand, and eradication of fungal or insect attack in timber in service. Attack does not normally occur in the log, but *Bostrychid* beetles will attack timber very soon after conversion, and *Lyctus* as the timber becomes drier. Good yard hygiene, on the lines discussed on page 274, can play a very important part in minimizing the depredations of this pest, but, until good practice is general throughout the country, resort to chemical dips or spraying is often advisable. Successful 10-second dips are 5 per cent. borax at a temperature of 180° F., and 1 or 2 per cent. aqueous solutions of 'solfocide' (sodium pentasulphide in liquid form) at 190° F. Solvent-type preservatives have also been tried on seasoned timber: a 3 per cent. solution of pentachlorophenol, in a light fuel oil of the kerosene range, and an immersion period of 3 minutes, has given good results. In this country, the available evidence points to the need for purely transitory protection against *Lyctus* (apart from eradication of outbreaks when these

[1] Browne, F. G., *The Malayan Forester*, No. 4, 1949.

occur): were joinery, furniture, etc., delivered free from infestation, the likelihood of attack occurring in service would be remote. This, of course, presupposes that manufactured articles will not be stored in already contaminated premises. In effect, precautions are especially necessary while timber is in stick for drying, or awaiting manufacture. For these circumstances, it has been found that a 2 per cent. emulsion of D.D.T. in water, applied by spraying to piled timber, is effective; complete coverage of the sapwood must be secured. The spray is prepared from miscible oil concentrates in solvent naphtha or xylene containing some specified amount of D.D.T. diluted with water to the D.D.T. concentration required. No penetration of the timber is aimed at, or secured, with this spray, so that when timber is worked or re-sawn it will require spraying anew if it is likely to be exposed to infestation before being made up.

The question is often posed as to whether decay and 'wood-worm' infestation in buildings is more prevalent today than formerly, and the answer is undoubtedly in the affirmative. The increased incidence of 'decay' is attributable to rather different causes from those responsible for more widespread depredations of insect pests. Formerly, many outbreaks of fungal infection could be directly related to constructional defects, although many buildings that fell short of good standards of construction managed to escape attack. The war years and economic factors have materially changed the position. Delays in making good war damage, and the large quantities of water used in putting out fires, have caused well-constructed buildings to become the victims of widespread fungal attack; in buildings that fall short of good construction, the consequences were naturally the more serious and inevitable. Further, normal maintenance was in abeyance for six or seven years, and properties were often uninhabited for long periods, so that any small defects that developed tended to assume considerable importance. Fuel rationing, the shortage of domestic staff, and financial stringencies – factors likely to persist – have operated similarly, especially in larger houses: with only some accommodation in regular use, the woodwork of unheated and unventilated rooms will pick up moisture wherever defects in construction exist, ultimately creating conditions favourable to fungal growth. That the whole of such accommodation was formerly used, well heated and adequately ventilated, enabled border-line cases of bad construction to escape penalties.

Availability of moisture is, in practice, the factor that governs liability to fungal infection, although different fungi vary in regard to their 'total' moisture requirements. With insect pests, favourable factors are more varied: the death-watch beetle, the wood-boring weevil, and the wharf borer thrive only on timber that has first been attacked by fungi, although the death-watch beetle will spread to sound wood. The common furniture beetle is less exacting, but shows a preference for sapwood. The *Lyctus*, or powder-post beetle, is the most exacting of all, confining itself to the sapwood of certain hardwoods, and then only if adequate supplies of starch are present, if the timber is sufficiently seasoned, and if the pores or vessels are large enough to permit egg laying.

ERADICATION OF FUNGAL DECAY

Recommendations for dealing with 'dry rot' are to be found in the Bible (Leviticus xiv. 34-48). The fungal origin of the 'plague' or 'leprosy' was not known until centuries later, but the passage is of interest in that it brings out two important points: the need for establishing that attack is still active, and the need for drastic remedial measures – rather too drastic in the light of modern knowledge. The fungus would not have been *Merulius lacrymans* as the temperatures in Palestine are too high for this species. The Old Testament writer overlooked an all-important point: namely, the importance of tracing the source of moisture that gave rise to fungal infection in the first place. By comparison, it is less important to identify the species of fungus, although it is, of course, essential to determine whether *Merulius lacrymans* is involved, since this fungus calls for altogether more drastic remedial measures than any other.

The three most common faults, when dealing with outbreaks of fungal attack, are to do too little 'site' investigation, to do too extensive replacements, and to ignore the fundamental cause, namely a supply of moisture. It is essential to determine the full extent of infection, but knowing what to look for, and where to look for it, will minimize the amount of opening up to be done. Time spent on the careful examination of the exterior of a building is usually well repaid. The more obvious points in regard to damp-proof courses, and levels of flower-beds and paths, are generally understood. The importance of an adequate number of air bricks is also usually appreciated, although it is sometimes overlooked that their number may be adequate and yet the air bricks are ineffective: they may be obscured by a plate or joists on the inside. Many air bricks have, at best, only 50 per cent. of voids, and a floor joist or plate can reduce this 50 per cent. to almost nothing. Water from above – from defective gutters, rainwater heads, or down pipes – can cause as widespread devastation as indifferent ventilation below the ground floor. Hence, the condition of the rainwater disposal arrangements, and their efficacy, warrant close examination: staining of external walls, the growth of algae, and the condition of pointing – or evidence that such matters have recently received attention – will often indicate where the search for fungal infection should be directed inside the building. Parapet gutters, 'internal' valley gutters, and lead or asphalt 'flats' are other fruitful sources of trouble, and any defects in these features, or signs of past patching, call for particular attention being given to the condition of timber beneath such weak spots. Past history can be most relevant: flooding from burst pipes, when the property is unoccupied, or water used in putting out fires, can cause the most extensive dry-rot. Areas of new slates or tiles should arouse suspicion. Armed with clues on the lines discussed above, tracing of infection is simplified. Corrugations in skirtings and panelling are obvious defects to look for, but even bowing of such timbers, if on the opposite side of a wall where 'defects' exist outside, should not be given the benefit of any doubt: removal of such timber often reveals surprisingly extensive, 'unsuspected' decay.

Having located infection, and its extent, it is essential to determine where the wood obtained a supply of moisture sufficient to render it fit to support fungal growth. Very often an outbreak has more than one 'focal' point, which means more than one source of moisture, and it is all-important that these should be detected. Moisture will not travel upwards nearly so far as it will travel downwards: infection in basement or ground-floor rooms will be sustained by a different source of moisture from infection in the same premises but on the first or higher floors.

Until the source of moisture is traced, the appropriate remedial measures cannot be laid down. For example, moisture arising from constructional defects can only be eliminated if these defects are corrected, or, when this is impracticable or too costly, by recognizing that no timber should be used in repairs that will not be isolated from the source of moisture by a water-impervious barrier.

Badly ventilated basement floors, with no site concrete and no damp-proof courses, present just the problem envisaged. However thoroughly the wood is stripped out and replaced by new, fungal infection will reappear *unless* moisture can be excluded from the under-floor space, and good ventilation be provided; heat sterilization of the surrounding surfaces is purely transitory, and the liberal use of wood preservatives will, at best, defer the date of reinfection. In practice, it will generally prove too costly to remedy major construc-tional defects, and the solution will lie in using inorganic materials in repairs. Affected basement floors should be replaced with solid floors, incorporating an impervious membrane, wood blocks, laid in a bituminous mastic, can be used as the floor finish. Thermoplastic tiles would be suitable and cheaper than wood blocks.

When the floor area is large, as in a gymnasium or concert hall, decay has been known to occur, in spite of the existence of damp-proof courses in all walls, and provision of the normal number of air bricks. For example, 9·67 cm² of air brick per 0·3 metre run of wall does not ensure a constant area of air bricks per unit volume of underfloor air space – the ratio falls as the area enclosed by the external walls increases.[1] In a building 9 metres by 4·5 metres, with 9·7 cm² of air brick per 0·3 metre run of wall, and 0·3 metre between site concrete and the under surface of the floor boards, 870 cm³ of air bricks are dealing with 12·15 m³ of air, whereas for a building 45 metres by 9 metres, and the same ratio of air bricks per unit length of wall, 3481 cm² of air bricks would have to deal with 127·44 m³ of air. In effect, it is sometimes impossible to ventilate the underfloor space of large areas adequately, and moisture must be excluded by providing a waterproof barrier in the site concrete.

Decay resulting from what can be regarded as temporary sources of moisture presents an entirely different problem from that arising from what may be called chronic dampness, and the appropriate remedial measures should be dependent on the identity of the fungus. Defective rainwater disposal arrangements, the raising of flower-beds above damp-proof courses, the blocking of air bricks, and the temporary flooding of normally dry sites (as in combating a fire)

[1] I am indebted to Dr W. P. K. Findlay for drawing my attention to this point.

are the chain of circumstances most likely to establish conditions favourable to 'dry rot' infection, and, in particular, to *Merulius lacrymans*. Persistent plumbing leaks, defective flashings, and recurring condensation, on the other hand, usually makes timber too wet for *Merulius* attack, and one of the so-called 'wet rots' is more likely to develop. Cure of the cause of dampness in the examples enumerated presents no great difficulty, and such action alone will, in many cases, arrest further decay. It is, of course, necessary to cut back affected wood to sound material, and to use well-seasoned timber in repairs; if there is a risk of dampness recurring from neglect of maintenance in the future, the use of pressure-treated timber in repairs should be considered. Timber that has been exposed to attack, and is only very slightly decayed, can be rendered safe by a brief kiln-sterilizing treatment; application of preservatives to such wood is unlikely to kill all traces of fungus. When panelling or valuable flooring has become sufficiently damp to support fungal growth, it is usually advisable to dismantle such timber and to dry it, otherwise splits may develop as the timber dries *in situ*, or, with floors, compression set may have been induced (see page 135). If the back of panelling is only very slightly decayed, and the quantity of panelling does not warrant a kiln-sterilization treatment, the application of Woodtreat should be considered, in spite of the possibility of Woodtreat working through to the face of the panelling, when short oil type metallic grey primers have to be applied before repainting.

Where *Merulius lacrymans*, the common 'dry rot' fungus, is the causal agent, more drastic measures than those outlined above are likely to be essential. This fungus penetrates masonry and brickwork, where it can remain dormant for long periods if its minimum moisture needs can be satisfied. In practice, walls that have become saturated are likely to retain sufficient moisture to sustain *Merulius* for some years after the original source of water has been cut off, particularly if there are even quite small pieces of timber hidden in the wall to provide the essential food material: fixing blocks for down pipes on the outside of a wall may suffice, and bond timbers will, of course, ensure vigorous continuing attack. The time it takes for thick walls to dry out explains the recurring attacks so frequently experienced after an outbreak of *Merulius lacrymans* infection: as soon as the new timber has had time to absorb sufficient moisture from the wall to raise its moisture content to about 20 per cent.,· conditions are ripe for active fungal hyphae to attack the new timber provided. The common precautionary measure of heat sterilization of walls with a blow-lamp is totally inadequate: raising the temperature of one face of a $4\frac{1}{2}$ in. brick wall to 900° C., and holding that temperature for four hours, will only raise the temperature of the opposite face to about 50° C. The surface application of chemicals is unlikely to be any more effective, and elaborate irrigating of walls is costly in labour, and in the large quantities of chemicals absorbed, besides being somewhat uncertain. Moreover, the boring of holes to permit of irrigation treatments may well weaken old brickwork, necessitating much rebuilding. The mycologists of the Princes Risborough Laboratory, in collaboration with research officers at the Building Research Station, Watford, have evolved a toxic plaster and paint that promise to provide a measure of security at reasonable cost.[1] No attempt is made to

[1] The formulae for the plaster, paint, and solution, as recommended by the Government chemists, are set out below. Important points are the use of an appropriate grade of zinc oxide and dry sand, otherwise the proportions are not 'critical' in the sense that small errors in weighing, such as may occur in ordinary practice, are unimportant. Rather more solution than is required for gauging the plaster or paint is necessary, as surfaces to be treated should be wetted with the solution before the plaster or paint is applied. In gauging the plaster, care should be exercised in adding the solution as the correct consistency for applying the plaster is approached, because a small amount of fluid at that stage has a marked effect on the consistency of the mix. It is important to cut out all timber in walls to be plastered or painted. Experience has shown that it is essential to mix the solids and the solution off the site because otherwise there are never enough clean containers to hand when required; gauging is, of course, done on the site just prior to use. Preservation Centre for Wood Ltd, of 24 Ossory Road, London, S.E.1 maintains a stock of ready mixed ingredients.

Gauging Solution for Paint and Plaster	*Parts by Weight*
Fused zinc chloride (technical)	8
Boric acid	1
Ammonium chloride	1
Water	20

Solid Ingredients

Paint		*Plaster*	
Zinc oxide (BS 254 Type 1)	2	Zinc oxide (BS 254 Type 1)	1
Talc	4	25/50 mesh sand	5
Whiting	6	52/100 mesh sand	2
		Whiting	1

Approximately 9 parts by weight of mix to 2 parts by weight of solution is suggested.

destroy deep-seated fungus in the affected wall; instead, a non-crazing barrier is erected, which the fungus will not cross. The basis of this method is zinc oxide, gauged with a solution of zinc chloride. The other ingredients of the plaster are sand and whiting, and of the 'paint' talc and whiting. Boric acid and ammonium chloride are added to the zinc-chloride solution to act as retarders. The solid ingredients of the plaster are gauged with the solution, and applied to the wall in the same manner as any other plaster; any gypsum plaster must first be removed. The surface to be treated may require to be hacked to provide a key. Surfaces so rendered can be decorated when dry in the normal way, or, if an adequate key is provided, a setting coat of a calcium sulphate plaster can be applied. The ingredients are relatively expensive, but only a thin coat is required. It is, however, important that the toxic barrier should be continuous where it is necessary to introduce new timber close to a contaminated wall. Where floor joists and skirtings are involved the barrier should extend from 15 cm above the skirting to 15 cm below the bottom of the floor joists, and if there is a cornice with timber grounds in the room under, then the barrier should be extended to 15 cm below the cornice. A similar barrier should be applied all round window and door openings, including reveals, soffits, and sills, and for 15 cm on face all round these openings, and timber fixings for the frames should be replaced by inorganic fixings, e.g., breeze concrete blocks. Where it is necessary to build timbers into affected walls, the wall holds should be rendered with the plaster, but it is always preferable not to take timbers into contaminated walls. Instead they should be supported by hangers built into the wall, the holds for which should be included in the continuous toxic barrier of zinc oxichloride plaster. The paint is made up in the same way as any cement paint; that is, the solid ingredients are first mixed, and gauged immediately before use with the solution, the ratio being 12 parts by weight of solids to 9 parts by weight of the solution; two coats are recommended, with an interval of twenty-four hours between each coat. The use of zinc oxichloride paint or plaster does not do away with the need for eliminating the source of moisture in future. Further, it is essential to ensure that no bond timbers are left in a contaminated wall. In older buildings it is advisable to hack off a strip of plaster from floor to ceiling to ensure that no bond timbers are overlooked.

To sum up: control and eradication of fungal infection in buildings is first and foremost dependent on tracing and eliminating the source of moisture. Subsequent steps depend on whether the fungus is *Merulius lacrymans* or one or other of the less virulent fungi. Toxic chemicals have a part to play, but in a secondary rôle. The work of eradication is likely to necessitate the assistance of carpenters, bricklayers, and plasterers, often the plumber and tiler as well, and is, therefore, essentially a job for the building contractor, fully aware of the importance of each step. Thoroughness is the keystone of success.

ERADICATION OF INSECT INFESTATION

In temperate climates insect infestation of timber on land means beetle attack; in salt or brackish water, marine borers are the destructive agent. Control and eradication of beetle infestation are dependent on an understanding of the life cycle of different species and their food requirements. The life cycle of any species is liable to be prolonged if its preferred food supply is deficient. In dealing with insect attack it is all-important to determine which pest is at work before attempting eradication: drawing-pin holes from the blackout days have been mistaken for beetle infestation. The use of wood preservatives, applied *in situ*, will often prove the only practicable method for dealing with insect infestation, when brushing the timber free from dirt is an essential preliminary, to secure thorough 'wetting' of the wood.

If the damage is the work of the 'pin-hole' borer (ambrosia beetle), no remedial measures are necessary: such 'wormy' timber is perfectly safe to use, since the damage will not get worse, and cannot infect other wood. These pests are mainly tropical, hence infestation and the full extent of the damage done occur before such timbers are exported. Control of pin-hole borer infestation rests with those exploiting tropical forests, and the appropriate measures are discussed on pages 311 and 312. 'Pin-worm' infestation can usually be differentiated from other forms of insect attack by the galleries running at right angles to the grain; the galleries of different species vary from 0·05 to 3 mm in diameter.

Longhorn borers are also mainly forest pests, although one species, the house longhorn (*Hylotrupes bajulus*), attacks converted softwoods. With the exception of the last-mentioned species, attack is best dealt with in the forest: logs must not be left lying on the ground, and if extraction is likely to be delayed, logs should be barked immediately after felling. The house longhorn is a serious pest in parts of Sweden, Denmark, and Germany, and has caused damage in parts of Surrey and Essex. It attacks the sapwood of seasoned softwoods (the heartwood is not completely immune). Extensive damage is likely to have occurred before the first flight holes are detected, because the life cycle is relatively long (3 to 11 years), and indications of attack, other than flight holes, are often wanting, or are easily overlooked (*e.g.*, blister-like swellings on the surface of infested wood). Attack generally originates in the roof timbers and attics, from which it may spread to other timbers throughout the building. Remedial measures are likely to involve replacing appreciable quantities of timber, coupled with application *in situ* of wood preservatives. Heat sterilization and fumigation are used on the continent in combating outbreaks, but facilities for heat sterilization are not available in this country, and fumigation is impracticable in most houses, particularly in built-up areas. The use of wood preservatives calls for very thorough applications to ensure that all timber surfaces are adequately treated: surface applications secure only very shallow penetration of wood by the fluid used. Inspection, for signs of renewed infestation in ensuing years, is advisable. Oil-solvent wood preservatives are appropriate: a 5 per cent. solution of pentachlorophenol and 0·35 per cent. of gamma B.H.C. in a suitable oil or spirit solvent is an economical preservative that can be made up in quantity at much less cost than most proprietary products. Where the pest is prevalent it is wise to use only pressure-treated timber in repairs and for all new work; brush-treatments give only short-term protection.

The death-watch beetle is almost always associated with decay, although, once established, the beetle may extend its attack to sound wood. Hence, in dealing with death-watch beetle infestation, it is imperative to deal with the decay too, since the cause of this will have been responsible for the subsequent, and 'secondary', beetle infestation. It is also important to decide whether the beetle infestation is still active; 'It is a common feature of damage by the death watch beetle for attack to cease before all the available timber has been destroyed, and this is no doubt due to absence of the conditions of moisture and fungal decay now known to be suitable for attack and which must once have been present in the building'.[1] It is not always easy to determine whether attack is continuing, although clear-cut rims to the flight holes, from which bright-coloured frass is spilling out, indicate the recent emergence of adult beetles. A search of the ground beneath attacked timber during the emergence period (April to June) is helpful: live beetles will usually be found if attack is still active. The presence of large numbers of live, steely blue beetles (*Corycorinetes coeruleus*), predatory on the death-watch beetle, also provides evidence of continuing attack. If investigation shows that attack has ceased, the lavish use of wood preservatives is obviously unnecessary. In many cases of continuing attack, too, cutting off supplies of moisture, which will bring decay to an end, may be as effective as attempting to eliminate the beetle infestation. In dealing with serious devastation the first step is to conduct a thorough check for any signs of decay, the cause of which must be eliminated; next, replacement of all structurally weakened timber; and, finally, treatment *in situ*, with an oil-solvent preservative, of timber that has been exposed to infestation, to destroy any remaining infestation; surface applications should be supplemented by injecting flight holes with a pressure spray. Eradication is likely to necessitate 'repeat' treatments in succeeding years, but these can be confined to areas of continuing active infestation, provided thorough inspections will be made in April or May each year.

Two other beetles infest decayed building timbers: the common wood-boring weevil, often found in basement floors,[2] and the much larger wharf borer. Both these pests are secondary to fungal decay, and their control by chemical means (wood preservatives) should not be attempted. Exclusion of the source of moisture, coupled with the cutting out of decayed and infested wood, may well suffice. Damage done by the wood-boring weevil is sometimes mistaken for furniture-beetle attack; the galleries run longitudinally, and are of about the same diameter as those of the furniture beetle, and the frass excreted by the feeding larvae is gritty but finer than that of the furniture beetle. If the galleries are searched, or heavily attacked timber

[1] Leaflet No. 4: *The death watch beetle*, issued by the Department of the Environment, Princes Risborough Laboratory.

[2] Exceptionally the wood-boring weevil may attack decayed timber at other than basement level, *vide* the grounds of window-boards of a 2nd floor window in Smith Square, S.W.1 where the defective external stone sill had induced an outbreak of 'dry rot'.

is broken up, it is usually possible to find, if not whole beetles, at least a snout, which puts identification as weevil damage beyond doubt.

The common furniture beetle is a pest of sound, seasoned timber, but, even with softwoods, infestation is largely confined to the sap-wood unless decay is also present. It was formerly thought that softwoods were unlikely to be attacked until they had been in service for about 15 years, and hardwoods not within the first 20 to 30 years. As a result of more detailed studies of the life cycle of the common furniture beetle, entomologists have revised their earlier views, concluding that initial infestation may occur as soon as the timber is seasoned, although it may not be discovered until attack has been in being for some years (see also page 276). The larvae excrete a gritty frass, which contains elongate pellets. In dealing with outbreaks, it is important to determine whether the infestation is still active, because only then are remedial measures necessary: inspections should be made in the early months of the 'flight' season, *i.e.*, May and June. With experience, fresh infestation can be detected by the bright colour of the frass, which appears to be spilling out of the flight holes, and the margins of the holes are clear-cut. With age the frass becomes discoloured, and the rims of the flight holes burred over. Two distinct tasks are involved when deal-ing with active infestation: eradication of existing infestation and protective measures against reinfestation. Wherever practicable, heat sterilization is the most effective method to employ (see page 298 for recommended schedules). Obviously, only movable timber can be so treated, *e.g.*, panelling, flooring, furniture, and small wooden articles. Sterilization calls for only low temperatures, at moderately high humidities, for relatively short periods, which should not damage glue joints or polish finishes. The treatment is done in an ordinary timber-drying kiln, which must, however, be in charge of a skilled operative. Were it possible to ensure that all infested timber in a building was sterilized, there would be no need for subsequent protective measures. This, however, is rarely prac-ticable, and sterilization needs to be supplemented by treating all surfaces with the pentachlorophenol gamma/B.H.C. preservative previously mentioned or some other oil-solvent preservative. For application to other than structural timbers, the solvent should be white spirit or odourless paraffin. A small-scale test should be made be-fore applying the preservative to polished surfaces. After sterilization, there is little point in treating polished surfaces, unless these are riddled with flight holes, because eggs are not laid on such surfaces. Before refixing panelling or skirtings, the grounds should be examined: they are almost certain to be attacked, and replacement with pressure-treated timber (Tanalised or Celcure treatments are appropriate) or Timborised timber is sound practice. Next to heat sterilization, fumigation is likely to prove the most effective control method. This work must be done by specialist firms; methyl bromide is an appropriate fumigant.

In practice, many outbreaks of furniture-beetle attack will have to be dealt with by *in situ* treatments. Timber should be brushed down prior to treatment, or a powerful vacuum cleaner can be used; any heavily infested edges or surfaces should be cut back to sound wood. The preservative should be applied with a flat brush in two coats with at least twenty-four hours between each coat. Inaccessible timber should be sprayed, using a small pressure spray. Where infestation is heavy, one or two flight holes in each group should be injected. The preservative should not be allowed to run down onto plaster surfaces. In theory, all timber that has been exposed to infestation should be treated, but this is rarely economically practicable. Provided thorough inspections are made regularly each year, treatment can be confined to actually attacked timber, and immediately adjacent members; it is advisable to take up flooring to permit of treating joists and plates beneath. In most buildings there is likely to be much hidden timber requiring treatment, *e.g.*, joists and plates, roof timbers behind sloping plaster ceilings. Some opening up is essential to determine the severity of attack in such timbers. If these are not structurally weakened, the use of smoke generators [1] provides an economical method of eradicating attack in confined spaces, *e.g.*, roof voids. Holes require to be cut in plaster ceilings, or occasional floor boards taken up, to permit of inserting ignited generators, but this involves much less making good than would be involved were it necessary to take up floors or hack down ceilings. The 'smoke' given off by these generators is only effective against emerging adults; no penetration of the wood is secured. The generators should be used about once each month during the 'flight' season of the pest.

[1] Messrs. Waeco Limited, High Post, Salisbury, Wilts., and Imperial Chemical Industries manufacture smoke generators.

Unless the amount of sapwood in carcassing timbers is excessive, the likelihood of the timbers being structurally weakened by furniture beetle attack is remote. There is, however, no justification for ignoring active infestation, but careful consideration should be given to the necessity for carrying out general insecticidal treatments. If the treatment is really thorough the cost of 'making good' after a complete treatment may be appreciably greater than any real damage the pest could possibly do in the next half century or more. Further, it is quite common to find ample evidence of old attack, but no continuing active infestation, although by no means all sapwood in adjacent timbers has become affected, indicating that even widespread infestation may die out completely while there are still ample food supplies available to support active attack of the pest. Structural damage is more likely to be found in old period cottages, where the ceiling joists and rafters were little more than half-round poles. With the passage of time all the sapwood may have been destroyed, when there is insufficient sound wood left for structural purposes. In such circumstances an insecticidal treatment is pointless, and the attacked timbers have to be renewed, although this may necessitate stripping and recovering roofs, and taking down and renewing ceilings.

There are certain ancillary precautions that are worth observing, even by residents of modern blocks of flats containing little or no structural timber. On purchasing or inheriting valuable furniture it is a wise precaution to treat each piece prophylactically, applying the preservative thoroughly to all unpolished surfaces, paying particular attention to joints of drawers and the like. Second-hand furniture should be carefully inspected for any traces of flight holes before it is brought into the home. Wicker-work articles, e.g., linen baskets and waste-paper baskets, old plywood packing cases, and discarded articles stored in lofts, should always be suspect because these are ideal breeding grounds for the common furniture beetle. A surprising number of outbreaks can be traced to such sources, and hence the frequency with which attacks are found to be concentrated around access hatches to roof voids, and in cupboards under stairs. The articles mentioned should be carefully inspected at the end of May or early in June each year, and any infected article is best burned, as this is a far more certain method of eradicating attack than any chemical treatment.

The *Lyctus* beetle is the common powder-post beetle, the larvae of which feed on the starch stored in the sapwood of some timbers. Only hardwoods are attacked, and then only species of timber with pores large enough for egg-laying (about 0·1 mm in diameter): seasoned, or nearly fully seasoned, timber is selected by the adult for egg-laying.   In effect, only some timbers are attacked, and then only the sapwood of such timbers, provided it contains sufficient stored starch.   The frass is a very fine powder, which feels like flour when rubbed between the fingers.

Eradication of powder-post beetle infestation is rather more difficult than dealing with furniture-beetle attack, because the damage is usually more deep-seated and extensive by the time it is detected, and the attack soon ceases to be confined to a gallery system along which the volatile toxic ingredients of insecticides can travel in concentrated quantities.   Where practicable, heat-sterilization is the most effective method of eradication, and, provided no timber is excluded from sterilization, further precautions are likely to be unnecessary.   Heat-sterilization is only practicable for 'portable' articles, and in many cases the infestation has to be dealt with *in situ*. Where there is no real risk of re-infestation, the feasibility of fumigation should be considered; fumigation must be carried out by specialists.   If the site conditions preclude fumigation, and attack is widespread, it may be necessary not only to replace all attacked timber, but all pieces of susceptible timbers containing sapwood.

To sum up: identification of the particular pest at work is important, but it is of no less importance to establish whether or not there is continuing, *active* infestation.   Thousands of pounds are being expended annually in applying wood preservatives indiscriminately to the timbers of our churches and other buildings, often when infestation is no longer active, and even may have been 'dead' for upwards of a century or more.   With some insects, it is often sufficient to concentrate on eradicating decay, which will dispose of the beetle pests too.   Heat-sterilization is the most certain method of killing all stages of infestation, and is recommended in all appropriate circumstances.   The *in situ* use of wood preservatives has an important place in dealing with areas of continuing, active infestation, and where it is essential to prevent re-infestation.   It must not be overlooked that preservatives do not restore the strength properties of attacked timbers, and a single treatment is unlikely to secure 100 per cent. success. The proper cleaning of wood prior to treatment, and the thoroughness with which the preservative is applied, are all-important. The initial cost of a preservative is not a good yardstick for assessing its efficacy; a disclosed formula, whose proper cost can be accurately determined, is to be preferred.

CHAPTER 17

Grading of Timber

GENERAL PRINCIPLES

THE inherent variability of natural products presents many difficulties in their marketing, particularly since competition and mass production have brought a high degree of uniformity in rival, manufactured materials. Timber producers have long found it advantageous to study the variability of wood, and some have evolved sets of rules and grading marks that have come to be regarded as a guarantee of high quality. Unfortunately in some countries there is little standardization in the quality of timber from different mills, and some producers are found to be inconsistent in their grading over a period of years. In the interests of all parties, attempts have been made on the part of various Governments to standardize the grading of many natural products, but timber has generally escaped such beneficial action, and in this country there has been no concerted effort on the part of the industry as a whole to introduce grading rules.

A set of rules applicable to organic materials must of necessity be to some extent arbitrary, and the rules will invariably be subject to the personal factor in interpretation. This accounts in part for the delay in the universal acceptance of grading rules for timber. The first set of rules was issued as long ago as 1764, when Sven Alversdon of Stockholm defined four grades of Swedish pine, *i.e.*, 'best', 'good', 'common', and 'culls', and there have been the Hernosand rules since 1880, but timber has not been obtainable graded in accordance with these written rules. In 1941 the British Standards Institution published grading rules for stress-graded timber; Part I, revised in 1944 as BS 940, Part I: 1944, covered European redwood and European whitewood, and BS 940, Part II: 1942, applied to North American timbers. These Standards, and BS 1175: 1944 :

Sizes of stress-graded softwood timber, were subsequently replaced by BS 1860: 1952: *Structural softwood, measurement of characteristics affecting strength,* which should be read in conjunction with British Standard Code of Practice CP 112 (1967): *The structural use of timber in buildings.* BS 1860 was revised in 1959 as BS 1860, Part I: 1959: *Structural timber, measurement of characteristics affecting strength, Part I, Softwood.* BS 1860 is at present under revision and will shortly be issued to include stress-grading rules on the lines of those set out in BS 1175: 1944. In 1964 grading rules for sawn home grown timbers were issued: BS 3819: 1964: *Grading rules for sawn home grown softwood,* and similar rules for home grown hardwoods are in course of preparation. Other British Standards dealing with quality of timber include BS 1297: 1952: *Grading and sizing of softwood flooring*; BS 1186: Part I: 1952: *Quality of timber and workmanship in joinery, Part I: Quality of timber;* BS 2482: 1954: *Timber scaffold boards,* and BS 1129: 1966: *Timber ladders, steps, trestles and lightweight stagings.* There are also several BS Codes of Practice covering timber utilization problems, *e.g.,* CP 201 (1951): *Timber flooring.* Timber is not as yet ordinarily sold under these grading rules in the United Kingdom, but for large contracts it should be obtainable if so specified. If it is decided to specify that timber shall comply with a particular British Standard it is advisable not to attempt to enlarge on the provisions of the Standard unless there are very good reasons for so doing because this can lead to the inclusion of contradictory clauses in an architect's Specification or in a Bill of Quantities, giving rise to disputes with contractors when items are delivered on site that conform only in part with the conditions laid down in the contract documents.

In America and Canada timber has been graded for many years in accordance with special, written grading rules, agreed on by the different sections of the industry as a whole. These rules have the great advantage of having been drawn up by those thoroughly acquainted with the quality of timber ordinarily produced in the mills of each particular region, and the rules are revised and modified to meet changing market needs. In consequence, the rules meet consumers' requirements, but remain essentially workable. This is an important point, and one that is noticeably absent in rules drawn up by consumers, say, in London, for use in Africa or Asia. However desirable it may be that timbers should be of some particular quality, only a small proportion of the outturn of a region will be found to conform with a set of arbitrary rules that have not taken into account the range in quality ordinarily encountered in the mill. With data on the last-mentioned point, the problem of drawing up a set of rules becomes practicable; standards can be devised that fit the outturn, while meeting the essential requirements of the consumer. The alternative, of drawing up rules to meet consumers' demands, is usually unsatisfactory: much timber just fails to come up to the standard of each particular grade and has to be degraded to the next lower grade; slight modifications in the rules, shifting the emphasis on certain grading criteria, would often make such degrading unnecessary, and yet would ensure retention of emphasis on qualities of most importance to consumers. Rules drawn up by producers must, however, take sufficient regard of market requirements, otherwise purchasers will have no confidence in the products offered.

North American published rules are lengthy documents that cover every aspect of sorting the mixed outturn of sawn timber obtained by breaking down logs in a saw-mill. For the home market the rules still use Imperial Units, but exports to the United Kingdom are quoted in straight conversions to metric units. Personal judgement still comes in grading, although it is specifically laid down that such judgement must not supersede stated conditions in the rules themselves. It has been found that whereas two experienced graders may differ somewhat as to the grade of individual pieces of timber, when whole consignments are judged, such individual differences are evened out. The aim is to achieve standardization in quality, and this will more readily result from intelligent interpretation of the rules than rigid adherence to the letter – hence the importance of skilled judgement in graders.

Grading is done from the worst face of every piece of timber, unless otherwise specified. It is usual to allow a small margin, say, 5 per cent., of below-grade material to cover the human factor in grading. Lengths, widths, and thicknesses are standardized, with maximum allowances laid down for variations from these standards. **Standard lengths** are even numbers of 8 ft and up in 2 ft increments for softwoods, but both odd and even feet from 4 to 16 ft apply to hardwoods, with not over 50 per cent. of odd lengths. **Standard thicknesses** and **standard widths** are similarly prescribed in the rules, but in these two dimensions the sizes are **nominal**; that is, the actual widths and thicknesses are less, within specified limits, than the named dimensions.

Many grades of hardwoods are determined on a defect basis; that is, the different defects are given a numerical value, and, according to the grade, different maximum scores are allowed, depending on the size of the piece. In the National Hardwood Lumber Association's rules a knot $1\frac{1}{4}$ in. in diameter constitutes a **standard defect**, four pin-worm holes or their equivalent also constitute one standard defect; knots $2\frac{1}{2}$ in. in diameter count as two standard defects, and so on. Alternatively, the grade is fixed by the percentage of clear timber obtained by cross-cutting or ripping, or both, to exclude defects; the percentage of clear timber so obtained, the number of separate **cuttings** to obtain such timber, and the minimum sizes of pieces to count as cuttings, vary in the different grades. There are many grades, *e.g.*, Firsts & Seconds, No. 1 Common, No. 2 clear & better, Merchantable; firsts and seconds are frequently combined as one grade with a minimum percentage of firsts to be included.

To ensure uniformity of grading as far as possible, several of the larger lumber associations in North America employ corps of men who grade the produce of all the members. In this way grading is made entirely independent of the millers, and the public is assured of a minimum standard, wherever the purchase is made. This has been tested by observers from the Forest Products Laboratory, Madison, by surveys carried out in many States. There is still adequate scope for individuality, by attention to detail not specifically covered by grading rules.

THE BALTIC TRADE

Before discussing the standardized rules in use it will be as well to consider the position of those countries that have not adopted standard rules. Chief of these, as far as the British market is concerned, are the Baltic countries, including Russia, which supplied the bulk of the carcassing timber used in the United Kingdom up to the time of the Ottawa Conference in 1933, and these countries still supply the greater proportion of U.K. needs. Grading in these countries is in essentially the same state as it was seventy or more years ago, with the important difference that there has been a marked falling-off in quality of timber exported. The only possible exception to this statement is Finland since 1937: whereas the Hernosand rules applied to one district of Sweden, the Forest Products Association of Finland sponsored a committee to draw up grading rules for that country as a whole. A set of rules was published in 1936–7, but how much timber graded in accordance with the rules reached the United Kingdom is not known. It is, therfore, generally true to assert that every manufacturer follows his own rules, the actual grading, or **bracking** as it is called, being done by men with a lifelong association with timber. The grades used are 1sts, 2nds, 3rds, 4ths, 5ths, and 6ths, and 'unsorted', the last being a mixture of grades better than 5ths. The bulk of the timber from the Scandinavian countries is graded as 'unsorted' or 5ths, and that from Russia as 1sts, 2nds, or 3rds. In the Scandinavian countries heavy cutting in the past has necessitated the opening-up of new areas of forest, with the result that the nature of the raw material coming to the mills has changed. Shippers who have been in the habit of obtaining supplies from one locality year after year are obliged to go to different localities each year, and the brackers, with only empirical experience to guide them, are dependent on their own judgement for maintaining continuity of quality. Moreover, when it has been customary for the total production to yield certain percentages of each quality for many years, there is a tendency to secure the same percentages, irrespective of any fall in quality in the mill intake. Nowadays the produce of successive years, shipped under the same marks, may vary appreciably and, in consequence, the purchaser can no longer depend on particular shipping marks to secure the type of timber required.

The practice of architects, surveyors, and engineers, of specifying their requirements in considerable detail, does little to alleviate the position. Until quite recently, the specifications frequently imposed ridiculous limitations. For example, the contractor was frequently called on to supply timber 'straight in the grain, free of sapwood, knots, and all other defects', and clauses were sometimes added to exclude 'dead' wood, and 'blue-stain''. Such specifications are, and always have been, impossible of fulfilment. Timber is a natural product and is never absolutely free from defects or minor blemishes, many of which do not impair its usefulness. Some latitude, for example, in straightness of grain is permissible for most purposes, the exclusion of sapwood is rarely essential and is frequently impossible, and knots cannot be avoided except in timber from the outside of really large trees. For many purposes 'blue-stain' is unimportant, and it has been established that sound wood from dead trees is in no way inferior to that from living trees, and as it is drier it may be actually superior. As dead trees are more liable to attack by fungi and insects than living trees, their timber should be inspected for any signs of infection. If the expense is justified such stock should be kiln-seasoned before use.

In effect, such stereotyped specifications as are so often used are quite valueless, failing to perform the essential function of a specification, which is to define the quality of materials required. The first essential is that the specification shall be capable of fulfilment, and, secondly, that the conditions laid down are appropriate to the particular case, when it becomes the duty of those responsible to ensure that the specification is, in point of fact, implemented.

THE AMERICAN AND CANADIAN TRADE

The practice in the U.S.A. and Canada is for the producers of different classes of timber to form themselves into Associations, and for the Associations to issue grading rules for their products. Thus in the U.S.A. nearly all hardwood timber is graded in accordance with the rules of the National Hardwood Lumber Association. Softwoods, being more widely distributed, are handled by a larger number of Associations, each with its own rules. For example, there are the Southern Pine Association, the Northern Pine Manufacturers' Association, the Southern Cypress Manufacturers' Association, the Californian Redwood Association, and the West Coast Lumbermen's Association; each is situated in a different geographical region, and handles different timbers. In Canada the Associations are fewer, but the principles remain the same. Douglas fir, western hemlock, western spruce, and western red cedar, for example, are graded in accordance with the grading rules adopted by the British Columbia Lumber and Shingle Manufacturers, Limited. Canadian hardwoods are graded in accordance with the American Association's rules, and in 1942 the Maritime Lumber Bureau of Nova Scotia issued grading rules for Eastern Canadian spruce.

The Commonwealth Trade

Attempts to introduce many new timbers from tropical countries to the United Kingdom market were handicapped by the absence of recognized standards of quality. This led to official action, and a sub-committee of the Imperial Institute Advisory Committee on Timbers was appointed to examine the position. As a result, grading rules and standard sizes for tropical hardwoods were evolved on American principles. These rules were published by the Imperial Institute; they were divided into three sections:

- A. Hardwoods from countries other than Canada and New Zealand.
- B. Canadian hardwoods.
- C. New Zealand hardwoods.

The rules were concerned in detail only with timbers in section 'A', the rules of the National Hardwood Lumber Association being accepted for grading timbers in sections 'B' and 'C'. Three divisions were made in section 'A': (1) *Standard Grades* (two qualities), (2) *Wormy Guides* (three qualities), and (3) *Grades for Shorts, Squares, Strips, Quarter-sawn stock* (seven qualities).

The rules have been in existence sufficiently long to permit of their practical assessment, and for their shortcomings to become apparent. It was unfortunate that section A did not follow the American rules more closely, and that they were not evolved from actual study of mill outturn. As a result, the rules provided for an ideal standard, but with many timbers only a small percentage of the outturn was of this highest standard, and the balance graded not to the intermediate grades, but to the lowest grade provided for. This was because the grades were not well-balanced from the mill output standpoint. Modifications were required to ensure that the outturn could be graded into a series of balanced, progressively lower grades, from prime downwards. The American principle of combining two consecutive grades, stipulating a minimum percentage of the higher grade, would have assisted in ensuring workable rules. No timber is reaching the United Kingdom today graded under the old Empire Rules. Instead, certain regions, *e.g.*, Malaysia and Sabah, have developed their own printed rules, based on the American clear-cutting principles, and others—at least in theory—apply the National Hardwood Lumber Association's Rules.

It remains to be seen how any fixed grading rules will work out in practice in the tropical countries, except where the principles of grading were well established under European guidance before these overseas territories became independent. Expansion of production has necessitated expansion of grading facilities, and personnel of adequate expertise have obviously not always been available. The introduction of rules is, however, a step in the right direction, but the overseas territories in Africa and Asia have a long way to go before grading there can be expected to be in as satisfactory a position as it is on the North American continent.

STRESS-GRADED TIMBER

Co-operation between producers is necessary to ensure the working of grading rules, but, as has already been indicated, certain other conditions must be fulfilled. Firstly, the grades must be so balanced that they fit the range in quality of the timber to which the rules apply; secondly, the rules must fill the requirements of the market; and thirdly, there must be a proper appreciation of the significance of defects. Modern grading rules differ in details, but all are based on the evaluation of defects or blemishes.

The significance of defects depends on the purpose to which timber is put, and on the kind of timber. For example, blemishes that mar the appearance of wood, but do not reduce its mechanical strength, are unimportant in structural timbers, and, conversely, defects that reduce strength properties without appreciably spoiling the appearance of a piece of timber are serious. With decorative timbers the reverse is true, and a defect that reduces strength but does not disfigure the timber, is less important than one that has little effect on mechanical properties but is relatively conspicuous. These facts have long been recognized, and are allowed for in the rules as they stand today. A new departure is the development of **stress-grading rules**, aimed at ensuring that the timber in each grade shall have certain minimum strength qualities. Again, the evaluation of visible defects is the principle behind the stress-grading rules; the theory has been discussed earlier, *vide* pages 183 to 188.

British stress-grading rules were first published as BS 940: Part I:

1941: *Grading rules for stress-graded timber*, recognizing three grades: 1200 lb. f., 1000 lb. f., and 800 lb. f., for European larch, European redwood, and European whitewood. This Standard was subsequently replaced by BS 1860: Part 1: 1959[1]: *Structural timber, measurement of characteristics affecting strength*, and BS Code of Practice CP 112 (1952): *The structural use of timber in buildings*. CP 112 (1952) was substantially revised, and re-issued under the same title, but as CP 112 (1967). CP 112 (1967) has now become CP 112 (1967) Part 1, and a metric version has been published as Part II, 1971. A Part III on Truss Rafters is in course of being published, and a Part IV on Fire Resistance of Timber Structures is in preparation. CP 112 (1952) was, however, the edition referred to in The Building Regulations 1965, and should, therefore be read in conjunction with these Regulations. A revised edition of the Building Regulations was published in 1972. BS 1860 was revised for publication in 1967, but it was withdrawn with a view to its re-issue in metric units. BS 1860, as revised in 1967 was, however, incorporated as Appendix A in CP 112 (1967). Softwood species have been grouped in three species groups: S1 – Imported Douglas fir and pitch pine and home-grown Douglas fir and larch; S2 – imported unmixed and commercial western hemlock, parana pine, redwood, whitewood, and Canadian spruce, and home-grown Scots pine: and S3 – home-grown European and sitka spruce and imported western red cedar.

For each group there are five timber grades: basic and 75, 65, 50, and 40 grades, and for laminated timber three grades: LA, LB, and LC. Tables of 'green' (*i.e.*, exceeding 18 per cent. moisture content) and 'dry' (*i.e.*, under 18 per cent, moisture content) stresses for the three grouped softwoods for each grade are provided, and similar tables for individual softwoods and hardwoods.

Appendix A in CP 112 (1967) sets out the maximum permissible variations in size, dependent on the thickness of the piece. Nominal sizes of sawn timber and the finished sizes of surfaced timber are to be determined as at 20 per cent. moisture content, with appropriate allowances for moisture contents above and below 20 per cent. Visual factors to be taken into account in determining grade are:

---

[1] B.S. 1860 is expected to be reissued in the near future. In the revised form there will be two grades for visual stress grading: SS, special structural grade and GS, general structural grade, and two machine-graded grades, MSS and MGS.

(i) *Rate of growth:* varying from a minimum of 8 rings per in. of radius in the 75 grade to a minimum of 4 rings per in. in the 50 and 40 grades, and no limits in the laminated grades or in hardwoods.

(ii) *Fissures* (*i.e.*, checks, shakes, and splits): are to be measured as a factor of thickness, as are resin pockets.

(iii) *Slope of grain:* permissible limits range from not steeper than 1 in 14 in the 75 grade, down to 1 in 6 in the 40 grade, and from 1 in 18 down to 1 in 8 in the laminated grades.

(iv) *Wane:* limits for wane are not exceeding one-eighth of any surface in the 75 and 65 grades and one-quarter in the 50 and 40 grades, with no wane permitted in the laminated grades.

(v) *Wormholes:* which by definition in BS 565: 1949 is a general term for a hole, irrespective of size, caused by any wood-boring insect, are permissible in all grades provided there is no active infestation.

(vi) *Sapwood:* whether bright or blue-stained, is not regarded as a structural defect and is therefore acceptable.

(vii) *Other defects:* all pieces showing fungal decay, brittle-heart or other abnormal defects affecting strength are excluded from all grades.

(viii) *Knots:* five kinds of knots are recognized – splay, arris, edge, margin, and face knots, and tables are provided showing the sizes of knots, in relation to the width of the piece, permitted in the four grades. A separate table sets out the maximum widths for permissible knots in laminated timber.

Provision has been made for a Composite Grade, being a mixture of grades containing at least 75 per cent. of 50 grade and better, the balance being 40 grade. A study of the grading rules suggests that they are decidedly complex, and grading is likely to be an arduous undertaking. In fact these stress-grading rules are no more difficult to apply in practice than other grading rules once the grader has acquired sufficient knowledge and experience. It is estimated that an experienced man should be able to grade between ten and twelve standards of 165 cu. ft per standard in a day. A future development in stress-grading is the use of mechanical and electrical methods which have passed beyond the purely experimental stage.

The possibility of developing a machine for automatically stress-grading timber was investigated at the Princes Risborough Laboratory. It was established that there is a correlation between the deflection of timber under a known load and its ultimate strength. The relationship was studied on commercial samples of Baltic redwood from three different sources, and the following conclusions were arrived at:[1]

(1) A very good relationship, with correlation coefficients of better than 0·70, between deflection and ultimate strength in bending.

(2) That the relationships are equally good for deflections recorded under loads applied to the face of a piece as they are when the load is applied on the edge.

(3) That deflection gives a better index of the strength of a piece of timber containing knots than does a measure of the knot size or area.

A machine was devised to measure deflection of timber subjected to a small fixed load when fed through the machine. Although the strength of timber in depth is usually most important – as in timber used as beams or joists – there were practical reasons for preferring determination of deflection in timber when laid flat. Hence the, relationship between the ultimate strength on edge and 'stiffness' measured both on edge and when laid flat were investigated. Using test specimens $1\frac{3}{4} \times 3\frac{3}{4}$ in. in section, supported over a span of 52·5 in., and under loads equivalent to bending stresses of 300 and 600 lb. per sq. in., the following correlation coefficients between stiffness and ultimate load were obtained:[2]

'(a) stiffness in depth correlated with load in depth, 0·836;

(b) stiffness on the flat correlated with load in depth, 0·784;

(c) sum of the reversed stiffnesses on the flat correlated with load in depth, 0·794; and

(d) smallest of the two stiffnesses on the flat correlated with load in depth, 0·787.'

Machines have been developed in Australia and America for commercial use, and these have been investigated at the Princes

---

[1] Sunley, J. G., and Curry, W. T., 'A machine to stress-grade timber', *The Timber Trades Journal*, 26 May 1962.

[2] Sunley, J. G., and Hudson, W. M., 'Machine-grading of lumber in Britain', *Forest Products Journal*, April 1964.

Risborough Laboratory. In his paper—*Mechanical stress grading of timber*—presented to the Symposium of Non-destructive testing of concrete and timber, organized by the Institute of Civil Engineering, 11–12 June, 1969, Mr W. T. Curry, Dip. Eng., refers to four commercial grading machines then available, two manufactured in America and two in Australia. On grounds of cost, and because the American machines test only planed timber, Mr Curry concluded that the Australian machines are the most suitable for European conditions . . . 'The most suitable machine for European conditions is the Australian Computermatic machine, which is a development from their original Microstress machine. The timber is now tested on edge, as for joists instead of, as previously, on the flat. The timber is passed through the machine being tested between supports at 0·9 metre, or 3 feet. The Computermatic machine is an advance on the Microstress, in that timber has to be passed only once through the machine, not twice as formerly.'

Mr Curry points out that the work that has been done on machine grading at laboratories throughout the world is not always comparable because of differences in testing techniques, 'but the following general conclusions can be arrived at:

(a) Changes in moisture content have no significant influence on the stiffness (E1) of timber. As the moisture content decreases, modulus of elasticity increases, but this is offset by a reduction in the second moment of area of the section caused by shrinkage.

(b) The same regression line for strength on modulus of elasticity may be used for a number of different species of timber.

(c) There are acceptable correlations for compression and tension strength and modulus of elasticity as a board, but not for shear strength.

(d) Size of section influences strength—the larger the section the lower the apparent ultimate stress, but with no corresponding reduction in modulus of elasticity. The basic relations used for control setting of a machine should therefore be based on a standard size of section and appropriate modifications made to the corresponding grade stress values to correct for the effect of size.

(e) Different regression equations apply for forecasts of bending strength as a board and as a joist.'

Later in his paper Mr Curry pointed out that the advent of machine grading offers an opportunity to examine the current specification of grades for timber: 'The need for four grades must be questioned, and the emphasis which, in visual grading, is placed on bending strength is hardly appropriate to a material which, in use, is more often than not limited by deflection considerations. With machine grading, higher modulus of elasticity values are associated with the better grades, whereas in the present situation the same values of modulus of elasticity are assigned to a species irrespective of grade. The classification of timber on the basis of modulus of elasticity could, in fact, lead to more efficient utilization, and a further development of grading machines to enable them to determine an average modulus of elasticity for each piece of timber would certainly enhance their value.'

Stress grading was also discussed at the Timber Trades Club on 5 October, 1972 (*vide* Timber Trades Journal, 14 October 1972, pp. 31, 33, and 36). Speakers agreed that grading to the requirements of BS Code of Practice 112, 1967, was not really a commercial proposition, whereas machine grading was. Mr Jan Korff, Assistant Divisional Engineer, Structural Division Department of Architecture and Civic Design of the Greater London Council echoed the views from Mr Curry's paper quoted above to the effect that the advent of machine grading offers an opportunity for re-examining the need for four grades. Mr Korff pointed out that a commercial parcel of Swedish fifths, when machine graded, 'contained on an average 80 per cent. of pieces of outstanding quality, 15 per cent. of good structural quality and only 5 per cent. of rejects'. This is certainly a very different conclusion, compared with the impression gained from visual inspection of a typical parcel of carcassing fifths.

The Sontrin Timber Selector and Stiffness Testing Machine is more a selection rather than a testing machine. It tests the piece only at mid-span, not throughout its length over short spans of 0·9 m, but for certain end uses this may be sufficient. The manufacturer's literature states that: 'The machine subjects the member to a predetermined central load and senses the deflection. Permissible deflections have been calculated and are tabulated on the charts and graphs supplied with the machine. If the deflection figure is not reached or exceeded the machine will accept the member and mark the timber adjacent to the point of application of load. If the deflection is exceeded, the machine will reject the member. This will be indicated visually by the illumination of a red warning light. No marking of the rejected member takes place.'

The whole subject of non-destructive testing of timber was reviewed by Mr J. G. Sunley, M.Sc., at the Symposium held on 11–12 June, 1969—Mr Sunley's paper was not confined to machine stress grading but covered the different types of moisture meters, and certain other equipment only suitable for laboratory use. Mr Sunley concluded: 'Visual methods of measurement and grading of timber and non-destructive methods of estimating moisture content are commercially available and widely used. Mechanical grading systems are just coming into commercial operation. Sophisticated non-destructive density estimators have been widely used in laboratories. Work is at present taking place in various laboratories on the use of sonic and ultrasonic techniques in assessing timber quality, and this may well form the base for the next generation of grading machines.'

It will be appreciated that the modern Computermatic stress grading machine has substantial practical advantages over visual grading. It will not only make for economy in the use of timber, but will justify the claims that have been made that timber is an engineering material.

SPECIFICATIONS

The existence of standardized rules does not obviate the necessity for specifications to define the purchaser's precise requirements, but in the absence of such rules the rôle of specifications is even more important. They serve the dual purpose of defining a client's reasonable demands and the contractor's legitimate liabilities. The simplest course is to specify a particular grade of timber and, where necessary, to add clauses to meet the special requirements of a particular case. This is possible only when recognized grades of the required timber exist – a condition seldom obtaining at present in Britain. Alternatively, the grade may be more vague and the conditions more explicit. Up to the second world war it was sufficient to specify 'unsorted, Finnish prime from the Kotka or Uras districts' to ensure a reasonable standard, suitable for carcassing work, although there was nothing like the same assurance of quality, compared with conditions existing thirty years earlier. Similarly, 2nd-quality Archangel was likely to be entirely satisfactory for joinery purposes, but with the same reservation as applied to carcassing material. It is possible that eventually a few ports will revert to prewar standards and be recognized for the high quality of their shipments, but it is most unlikely that this state of affairs will be at all widespread. Moreover, any general tendency to restrict specifications to material from only a few ports will inevitably put enhanced values on such timber.

In the absence of grading rules it is essential to restrict the number of defects that will be accepted, and to define these in precise terms in the specification. In this connection the stress-grading rules are an invaluable guide, even in cases where the rules as a whole are not adopted. With a progressive lowering in quality of ungraded timber, it becomes imperative to check the suitability of selected sizes by means of the simple formulae available. If the scantling sizes call for a value for f. above $5 \cdot 516$ f per mm^2, then nothing lower than the minimum quality provided for in the stress-grading rules is suitable. Alternatively, if a value of only $3 \cdot 447$ for $4 \cdot 137$ f is required it is almost certain that timber is being used uneconomically.

Besides placing some restrictions on the grade of timber to be supplied, certain general conditions should always be laid down in a specification, together with clauses governing subsequent constructional stages. No standard form can cover all requirements, but instructions under each of the heads given below should be considered in every timber specification.

General. – All timber is to be correct as to species and quality specified. State whether sizes are full, bare, or scant, with limitations in regard to what will be tolerated, unless these points are covered by a British Standard and it is specifically stated that the timber is to comply with a particular Standard.

Time of delivery. – Timber to be delivered on the site before any building operations, other than clearing of the site, are commenced, and to be properly stacked under cover in a suitable position.

(Note: if kiln-dried timber is specified. the storage period should be reduced to a minimum, or delivery should be as and when the material is required.)

Storage of timber on the site. – Proper stacking to include the provision of a roof over the stacks of timber, and protection at the sides from driving rain and direct sun; foundations of stacks to be described. The timber in the foundations, and for stickers, to be thoroughly sound, and if of a species susceptible to powder-post beetle attack, absolutely free from sapwood. Size and spacing of stickers to be defined. Stacks of 25 mm material and under to be adequately weighted, and all projecting lengths of timber to be supported beneath and covered above.

Framing of timber. – Timber joints to be well and accurately cut, and all members properly framed together in accordance with drawings.

(Note: full-size, detailed drawings should be supplied for each type of timber joint required, and where members are to be spiked the size and number of nails to be used should be specified.)

Finishing, including wood preservatives. – Provision for inspection before wood is painted or brush-coated with preservatives should be provided for.

In the absence of appropriate British Standards it is necessary to specify requirements in considerable detail. The following paragraphs, appropriate in countries where suitable standards are not available, cover reasonable requirements for (a) a naturally durable, heavy constructional timber, (b) a moderately durable, medium-heavy constructional timber, (c) a general utility timber, *e.g.*, Douglas fir, European redwood, red meranti, (d) shingles, and (e) plywood; it is assumed that the plans will incorporate sound principles in regard to the provision of adequate ventilation, and in countries where white ants occur, termite-proof measures. The author ventures to put forward a list of standard defects at the end of the timber specification; local conditions may suggest modifications of this list. Timbers to be used should be specified by their standard trade names; defects known to be common in the selected timbers should be specially referred to, and the extent to which such defects will be admitted must be defined.

Naturally durable, heavy constructional timbers. All timber in contact with the ground is to be of [a named timber], or other approved hardwood, and all scantlings are to be thoroughly sound, full to thickness, well cut and with parallel edges. The timber is to be planed all round, and is to be absolutely free of defects except in regard to the following, which shall be allowed:

(a) Bends not exceeding 25 mm in 3·65 m run.

(b) Sapwood not exceeding ⅛th of the width of any face and restricted to not more than two faces; the sapwood is to be adzed off before fixing.

(c) 'Pin-worm' holes, *i.e.*, galleries not exceeding 0·78 mm in diameter, shall not count as a defect provided they are old and inactive and that the surrounding timber is thoroughly sound.

(d) One standard defect in lengths of 2·43 to 3·65 m, two standard defects in pieces 3·65 to 4·87 m long, and three standard defects in pieces over 4·87 m long. For the purpose of this specification sapwood and 'pin-holes' are excluded from the list of standard defects, being limited by clauses (b) and (c) respectively.

Moderately durable, medium-heavy constructional timbers. (i) *Shingle or tiling battens*: [a named timber], or other approved timber, is to be used for shingle battens; it is to be thoroughly sound, absolutely free from sapwood, 50 mm by 25 mm nominal sizes, well cut, and with parallel edges; the following defects will be permitted:

(a) Bends not exceeding 25 mm in 3·65 m run.

(b) Not more than twelve 'pin-worm' holes, *i.e.*, galleries not exceeding 0·78 mm in diameter, will be allowed in any foot run on the worst face.

(c) One standard defect in lengths of 2·43 to 3·65 m, two standard defects in pieces 3·65 to 4·87 m long, and three standard defects in pieces over 4·78 m long. For the purpose of this specification sapwood and 'pin-holes' are excluded from the list of standard defects, being limited by clause (b) and the preamble.

(ii) *Framing and other carcassing timber*: [a named timber], or other approved timber. All timber other than posts in contact with the foundations, shingle battens, flooring and joinery, is to be of [a named timber], or other approved timber selected by the architect. It is to be thoroughly sound, well cut, and with parallel edges, framing to be planed all round; sizes are nominal.

Every piece is to be absolutely free of defects except in regard to the following, which will be allowed:

(a) Bends not exceeding 25 mm in 3·65 m run.

(b) Sapwood not exceeding ⅛th of the widest face and 12 mm of any one other face of the same piece.

(c) One standard defect in lengths under 2·43 m, two standard defects in pieces 2·43 to 3·65 m long, three standard defects in pieces 3·65 to 4·87 m long, and four standard defects in pieces over 4·87 m long. For the purpose of this specification, sapwood is excluded from the list of standard defects, being limited by clause (b).

General utility timbers. Flooring and joinery is to be of [a named timber], or other approved timber selected by the architect. It is to be thoroughly sound, well cut, and have parallel edges; flooring is to be planed on one face and both edges, and all other timber in this section is to be planed all round; sizes are nominal. Every piece is to be absolutely free of defects except in regard to the following, which will be allowed:

(a) Bends not exceeding 25 mm in 3·65 m run.

(b) Sapwood not exceeding 12 mm on any face, but allowing a total of 25 mm plus the thickness of the piece if confined to one edge.

(c) Not more than four 'shot-holes', *i.e.*, galleries up to 3 mm in diameter, in any 0·3048 m run, and an average of not more than one per 0·3048 m run of total length in 50 per cent. of the material, the remainder to allow an average of not more than two 'shot-holes' per 0·3048 m run of total length.

(d) Except in flooring timber, slight 'spongy heart' will be allowed provided no visible compression failures are present; flooring timber is to be free from all visible traces of 'spongy heart'.[1]

(e) One standard defect in lengths under 2·43 m, two standard defects in lengths of 2·43 to 3·65 m, three standard defects in lengths of 3·65 to 4·87 m, and for standard defects in lengths of over 4·87 m. For the purpose of this specification sapwood and 'shot-holes' are excluded from the list of standard defects, being limited by clauses (b) and (c) respectively.

Shingles. The shingles are to be best quality, split (or quarter-sawn), seasoned [a named timber] shingles, purchased from a firm selected by the architect; shingles are to be pre-bored before fixing (this clause applies to billian or other hardwood shingles and not to western red cedar). They are to be laid to a lap equal to one-third of their length (a greater lap may be allowed if double coursing is specified), with a clear space of 1·6 mm to 3 mm between each shingle; no vertical joint in any three successive rows shall coincide but shall be staggered not less than one-third of the width of a shingle.[2]

Plywood. Inner cores of plywood shall be of Douglas fir or other approved softwood, or of a timber immune to powder-post beetle attack; the surface veneers shall be free from sapwood unless of a species immune from powder-post beetle attack.[3] The cementing matrix shall be a casein glue or other approved adhesive. All external plywood to be resin-bonded plywood of exterior grade quality.

[1] If the flooring timber selected is a species susceptible to powder-post beetle attack, complete freedom from sapwood must be specified; this precaution also applies when such timbers are to be used for joinery, panelling, and fittings.

[2] If stained or treated shingles are required, such treatments are to be applied before fixing. If the species is one susceptible to powder-post beetle attack, freedom from sapwood must be specified unless the shingles are to be pressure-treated.

[3] If the plywood available does not conform to this specification, the edges of the sheets must be adequately treated with a suitable wood preservative.

SUGGESTED STANDARD DEFECTS

The following is a list of standard defects:

Item	Equivalent number of standard defects
Knots	
One sound knot 16 mm to 31 mm in diameter or equivalent	1
One sound knot over 31 mm to 62·5 mm in diameter or equivalent	2
One sound knot over 62·5 mm to 87·5 mm in diameter or equivalent	3
Two knots, each under 16 mm in diameter or equivalent	1
Three knots, all under 16 mm in diameter or equivalent	2
Worm holes	
One or more 'pin-worm' holes, *i.e.*, galleries up to 0·78 mm in diameter, in a group not exceeding 31 mm in diameter	1
One or more 'shot-holes', *i.e.*, galleries up to 3 mm in diameter in a group not exceeding 31 mm in diameter	1
One bore-hole (or gallery) other than a 'pin-worm' hole or 'shot-hole' not exceeding 6 mm in diameter	1
One bore-hole (or gallery) other than a 'pin-worm' hole or 'shot-hole' but exceeding 6 mm in diameter	2
Splits	
One end split, or splits at each end, not exceeding one twenty-fourth of the total length of the piece; each split opening out not more than 12 mm.	1
One end split, or splits at each end, not exceeding in total length in millimetres the lineal measure of the piece in 300 mm; each split opening out not more than one twenty-fourth of the length of the piece	2
Sapwood	
Sap on species not exceeding 137 mm in thickness:	
(i) If 150 mm or over in width, not exceeding 19 mm on any face, but allowing a total of 37 mm plus the thickness of the piece if confined to one edge	1
An additional 6 mm of sapwood as above shall be considered as equal to a total of two defects.	
(ii) If under 150 mm in width, 12 mm on any face, but allowing a total of 25 mm plus the thickness of the piece if confined to one edge	1
An additional 6 mm of sapwood as above shall be considered as equal to a total of two defects.	
Sap on pieces exceeding 37 mm in thickness:	
(i) Sapwood on not more than three faces and in width on any face not exceeding ⅛th of the width of that face	1
(ii) Sapwood on not more than three faces and in width on any face not exceeding ¼ of the width of that face	2

(Note: it will be apparent that the grade of a piece of wood can be improved if defects are cut out by trimming the piece; trimming is regularly practised in American and other up-to-date mills.)

Other conditions than quality of the timber are of practical signifi-cance. Chief among these are the moisture content of the timber and the condition of the building at the time the timber is to be installed. Requirements as to moisture content are specified on pages 143 and 145, but equally important is the state of dryness of the building. It is sometimes economical to install temporary heating before fixing second fixings or, preferably, de-humidifying units, *vide* page 145. Special instructions must be given regarding the drying of timber to the correct moisture content if air-conditioning is to be in-stalled: only two ways of drying the timber adequately are available, one being in a kiln and the other in an air-conditioned chamber. Furniture suitable for ordinary atmospheric conditions will suffer serious degrade in air-conditioned rooms if the moisture-equilibrium conditions of the latter are lower than normal, which they are most likely to be.

The recommendations made above provide for considerable stiffening of specifications for timber acceptable for building work; more latitude than is allowed at present may be granted in a few directions, *e.g.*, the use of sap-stained timber where sapwood is not objectionable, the use of pin-worm material. Where appropriate British Standards exist the timber should be required to conform to the appropriate Standard, the number and date of which should be cited. If this procedure is adopted it should suffice to specify the Standard and the name of the timber required, indicating, where they exist, whether any permissive clauses in the Standard shall apply. Where timber absolutely free from sapwood is required, *e.g.*, joinery and flooring in *Lyctus*-susceptible hardwoods, it should be so specified, and a rider added that the condition will be rigidly enforced. In general, such meaningless phrases as 'well-seasoned timber' should be omitted and the moisture content range required specified.

CHAPTER 18

Wood as an Engineering Material

VARIABILITY in wood undoubtedly imposes limitations on its uses, but an understanding of the variation discussed in Chapter 8 is probably of more importance to the grower of timber, that is, the forester, than it is to the consumer, be he architect or engineer. Fortunately, with the knowledge that variability exists, and some knowledge of its extent, it has been possible to overcome the limitations imposed. The data for strength properties, and the development of stress-grading rules from such data, coupled with the use of timber connectors and modern adhesives, have rendered strength variations of secondary importance in modern timber design. Moreover, these developments place timber today in the field of engineering materials, more modern even than steel or reinforced concrete. In fact, it is the newest material available, with many advantages on economic grounds over its competitors. This has attracted the interest of a few structural engineers who see an almost limitless new field for wood. The writer has borrowed freely from the writings of one of these, Mr P. O. Reece,[1] in the following paragraphs.

Timber has been used as a structural material from the earliest times: examples can be traced to the neolithic era, dating back to about 8000 B.C., but it was not until 1678, when Robert Hooke established the fundamental relation between stress and strain, that any adequate theory was available to guide designers. For another two hundred and fifty years little scientific knowledge accumulated to aid timber utilization; it remained a material for the craftsman, and the engineering formulae developed for it in this period were

---

[1] 'Recent Experience in the Design of Timber Structures', a paper read at a meeting arranged by the Architectural Science Board of the Royal Institute of British Architects and published in the *Journ. R.I.B.A.*, vol. 51, No. 5, 1944.

based on little more than empirical practice. Between the two world wars the whole position changed: an immense volume of data relating to the strength properties of timbers was collected, but these data have, as yet, been very sparingly applied.

Regarded as an engineering material, the approach to wood must be quite different from past practice: there are many yard-sticks to

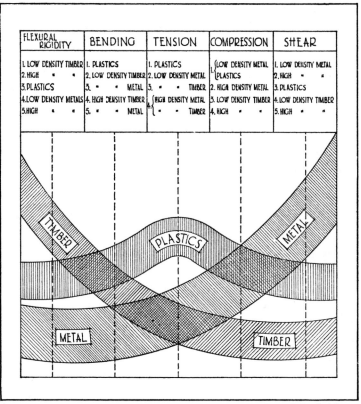

FLEXURAL RIGIDITY	BENDING	TENSION	COMPRESSION	SHEAR
1. LOW DENSITY TIMBER	1. PLASTICS	1. PLASTICS	1. {LOW DENSITY METAL, PLASTICS	1. LOW DENSITY METAL
2. HIGH " "	2. LOW DENSITY TIMBER	2. LOW DENSITY METAL	2. HIGH DENSITY METAL	2. HIGH " "
3. PLASTICS	3. " " METAL	3. " " TIMBER	3. LOW DENSITY TIMBER	3. PLASTICS
4. LOW DENSITY METALS	4. HIGH DENSITY TIMBER	4. {HIGH DENSITY METAL, " " TIMBER	4. HIGH " "	4. LOW DENSITY TIMBER
5. HIGH " "	5. " " METAL			5. HIGH " "

FIG. 45. The order of merit of structural materials: 'The broad shaded bands lie one above the other in order of merit, and varying in their relative positions, depending on the nature of the stress. Thus, for flexural rigidity the order is timber, plastics, and metals; in shear, metals, plastics, and timber; and so on . . . in each band the upper limit represents low density forms of the particular material, and the lower limit the high density forms' (*loc. cit.* p. 163)

By courtesy of P. O. Reece, Esq.

apply, but one of the first described by Reece is **specific strength**; that is, the ratio of strength to unit weight. A second criterion must be that of cost; the comparison here is the ratio of weight to cost.

Specific strength must, of course, be separately calculated for strength in compression, in tension, and in shear. Moreover, as few structural members are used in pure compression, tension, or shear, flexural considerations that cannot be divorced from size and geometrical properties of these members must be studied. It has been established that, for geometrically similar sections of equal weight, the cross-sectional area varies inversely as the density, and the section modulus varies inversely as the density raised to the power of 1·5. The specific strength in bending is, therefore, the stress divided by the density raised to the power of 1·5. When deflection is a governing condition, strength is a function of the flexural rigidity factor: modulus of elasticity multiplied by the moment of inertia of the section, or simply EI. The moment of inertia varies inversely as the square of the density. Hence, the specific strength of members governed by deflection limitations is the modulus of elasticity divided by the square of the density.

Working from the foregoing premises, Reece examined the comparative efficiency of five different structural materials subjected to different kinds of stresses. From this analysis he constructed the pictorial order of merit chart illustrated in Fig. 45. From the chart the superiority of timber in flexural rigidity is immediately apparent, rendering it pre-eminent for all structural components that can fail from elastic instability, e.g., slender, slightly-loaded columns; long, lightly-loaded beams; stressed-skin construction, where the covering or panel-filling material is utilized to stiffen the framework; and, in general, all components that are lightly loaded in relation to their size.

Next, Reece has provided a comparison between the loads that solid, circular-section, timber struts of different lengths will carry, and those that rolled steel joists of equal weight and similar lengths would support, vide Fig. 46; the advantage is with steel for very short lengths, but becomes reversed as the slenderness ratio comes into play.

Another interesting aspect of the peculiar properties of wood, which has come to light as a result of scientific testing of timber, is the effect of duration of loading. If a value of 100 is taken for the ultimate stresses for loads sustained over periods of 40 to 50 years, for a period of one month the figure would be 130, for an hour over 150, for five seconds nearly 200, and for impact loading 250.

These results, compared with similar data for other materials, show timber to be particularly suitable under short-time loading conditions, as in roof construction, towers, pylons, high single-storey buildings, of the hangar type, where the main problem is the accommodation of high wind loads for very short periods. Similarly,

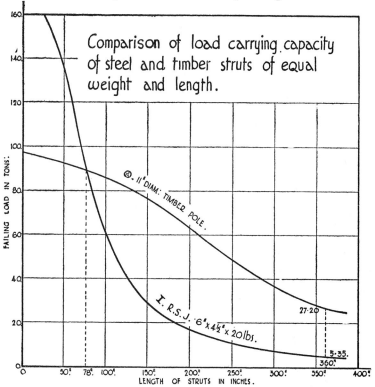

FIG. 46. Strength properties of wood and steel compared

By courtesy of P. O. Reece, Esq.

most floors, other than warehouse floors, are loaded to capacity only for brief periods, and this makes timber a particularly suitable material for floor construction.

Analysis has confirmed the correct usage of timber by craftsmen, except for excessive generosity in sectional area of members. On the other hand, when steel became available its introduction into timber trusses for tension members, but not for the short members subjected to compression, was apparently at variance with correct usage, which suggests the retention of timber for tension members

and their replacement by steel for short, thick compression members.
The reasons for the foregoing practice are bound up with the serious
effect on tensile strength of knots, and the loss in strength that
results from cutting away of timber to form joints.   More recent
developments have provided a solution to both objections, namely
stress grading, timber connectors, and modern adhesives.

TIMBER CONNECTORS

The ease with which timber can be fabricated has contributed
enormously to its usefulness, but joints and fastenings have always
been the weak link in timber construction.   Many carpentry joints
necessitate reducing the cross section of a member, thereby reducing
the strength appreciably compared with the full sectional area.
Elaborate carpentry joints that reduced loss in strength of jointed
members to a minimum were developed by our ancestors, but for
most purposes these became altogether too expensive when labour
costs began their upward trend after the Industrial Revolution.
Instead, bolts and sometimes only a few nails, replaced good practice, frequently with loss in strength, because there was no suitable
method available to replace the older empiricism.   Simple bolt
fastenings, although preferable to random nailing with an unspeci-
fied number of nails of indeterminate quality, reduce the strength
properties of each piece of wood by the amount of timber cut away
to take the bolts.   Moreover, the strength of the joint is less than
that of the bolt, as failure is generally induced either by shear
through the timber at the bolt-hole, or through crushing of the
timber bearing on the bolt itself.

The high stresses around the bolt-hole were early recognized to be
a weak point of bolted joints: solutions were sought by means of
bushings around the bolt, aimed at increasing the bearing area.
Efforts made in this direction have given rise to modern rings,
toothed plates, and variously shaped discs.

An American patent was granted as early as 1889 for a toothed
plate for joining timber, and it is recorded in the U.S. Department
of Commerce bulletin, *Modern connectors for timber construction* (1933),
that even earlier than this cast-iron rings were used in American
bridge construction.   The years 1916 to 1922 produced urgent
constructional problems that resulted in real progress being made in

the evolution of suitable mechanical devices for improving timber joints, which are now generally referred to as **timber connectors,** although the majority are actually pieces of metal. Originating in Europe, modern timber connectors reached the U.S.A. in 1930, since when rapid advances have been made, and the scope for timber construction has been greatly widened.

More than sixty different types of connectors have been patented in Europe, and in several cases U.S. patents have also been obtained. All embody the same principle, which is to increase the area over which the load at the joint is transmitted, thereby minimizing

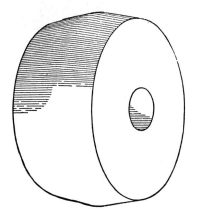

FIG. 47. The Kübler dowel   FIG. 48. A bulldog timber connector

crushing of the timber by the bolt, and the tendency for shear failure at the bolt-hole. Connectors are, however, conveniently divided into two general types: connectors stressed chiefly in shear, *i.e.*, short dowels and auxiliary connectors other than dowels, and those that have to sustain a certain amount of bending in addition to shear, *i.e.*, long bolts and dowels.

The first timber connector was the **Kübler dowel** (Fig. 47), evolved in Germany; appropriately enough, it was made of wood. The dowel itself, which was of hardwood, was doubly conical, being widest in the centre, and therefore at the point of contact of the two pieces of wood to be jointed; it was bored longitudinally to receive a bolt. Both timbers had to be recessed to take the dowel, which necessitated much cutting to house the conical ends of the dowel satisfactorily, and to secure a flush bearing surface between the two members.   The dowel, however, increased the bearing surface appreciably, compared with a single bolt, thereby reducing crushing of the timbers by the metal bolt.

The Kübler dowel was later followed by a steel ring, working on the same principle, but weakening the timber less by reason of the small amount of cutting necessary for recessing a thin ring, compared with a thick, conical dowel.   Moreover, cutting of the recess was greatly simplified and much less costly.   About the same time O. Theodorsen, of Oslo, patented the **bulldog timber connector** (Fig. 48), which, with slight modifications, is still widely used.   The bulldog connector consists of a round or square steel plate, punched at the centre to take a bolt, and with the perimeter turned alternately upwards and downwards to provide triangular teeth.   The timbers to be jointed are bored to take a bolt, the connector is inserted between the two members, and the bolt slipped through.   As the nut is tightened the teeth of the connector bite into the wood.   It is a comparatively simple matter to tighten the bolt when the connector is used with softwoods, but with hardwoods it is usually necessary to hammer the teeth of the connector into one of the timbers, and to use a bolt with a shank of high tensile steel that will permit of the nut being tightened while the teeth are pressed into the opposite member.   Once drawn together, with the teeth of the connector ring embedded, the special bolt can be replaced by an ordinary one.

Bolt fastenings of connectors are used in conjunction with large metal washers or plates to prevent the bolt head or nut from pulling through the bolt-hole.   Power-operated tools have been designed for cutting the special shaped recesses required for different connectors, and for boring the bolt-holes, which greatly facilitate the fitting of connectors.   Power-operated assembly tools are available for further accelerating assembly of modern timber connectors.

Connectors stressed chiefly in shear come into one or other of the following categories:

(1) **Split ring connectors** (Plate 62) fit into pre-cut grooves of opposing members, which in turn are drawn together by a centre bolt which is completely independent of the ring.

(2) **Toothed-ring connectors** (Plate 63), the teeth of which are forced into the timber as the two members are drawn together by the securing bolt, which (as with the split ring connector) is completely independent of the toothed ring.

(3) **Claw-plate connectors** (Plate 64, fig. 1), a development of the bulldog connector, which fit into pre-cut recesses; the claw plate has protruding teeth that are pressed further into the wood as the bolt is tightened. Claw plates are used singly in timber to metal joints, and in matched pairs – male and female claw plates – in wood-to-wood joints. In the Teco claw plates[1] for timber-to-timber joints the outside hub (on the face opposite the teeth) of the male plate consists of a central boss that slips into the recess of the hub of

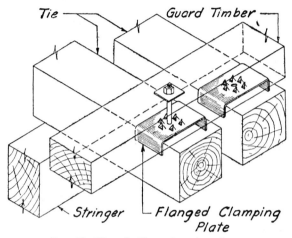

Fig. 49. Use of a Teco clamping plate
By courtesy of the Timber Engineering Co.

the female plate. A large bolt fits the hub snugly, the connector being flush with the adjacent surfaces of the members joined.

(4) **Shear plates** (Plate 64, fig. 2) are also flush-fitting connectors for wood-to-wood or wood-to-metal joints. A circular recess, with a sunken margin, is cut in the timber (in both timbers for a wood-to-wood joint) to receive the shear plate, which consists of a circular steel disc, bored centrally to take a bolt and with a raised rim on one or both sides. The plate is tapped into place with a hammer, and the bolt dropped into position and tightened.

In the split ring and toothed-ring connectors the load is transmitted by shear, more or less independently of the bolt, whereas the claw-plate and shear-plate connectors are dependent on the bolt for transmitting load by shear from member to member.

[1] Manufactured by the Timber Engineering Company, 1319 Eighteenth Street, N.W., Washington, D.C.

Other modifications of the bulldog connector on the market are the Teco spike grids (Plate 65), and the Teco clamping plates (Fig. 49) designed for special uses; the illustrations are self-explanatory.

Connectors, other than nails and bolts, that have to sustain a certain amount of bending as well as shear are of two types. In the Meltzer steel tack, developed in 1910, several steel pins of small diameter replace a few large-diameter bolts. Holes of the exact diameter of the pins are drilled through the members to be joined, and the pins inserted; the pins are without heads or nuts, friction being counted upon to hold members together. Meltzer joints of strength comparable to bolt joints effect appreciable savings in weight of metal used, the more numerous steel pins weighing half that of the requisite number of bolts. The second type is the Cabröl method employing a metal pipe. The joint is bored to take a hollow pipe, the ends of which are covered, after fixing, to exclude moisture. Metal bearing plates are used between the pipe and the wood-filler blocks that transmit the stresses to other pins.

Scholten[1] has listed the principal advantages of connector joints, which may be summarized as follows:

1. Relative high efficiency of joint compared with carpentry joint.
2. Relatively simple and practical application.
3. A minimum number of units or pieces to handle in the erection stages. (Compared with single bolts there may be more pieces in the assembly stage when connectors are used.)
4. Adaptability to prefabrication for subsequent field assembly.
5. Connectors give a better performance when used under adverse conditions than bolts or nails (water-proof glues may be superior).
6. Improved appearance of connector joint over exposed metal strapping.
7. Greater fire resistance of connectors over strapping because embedment of connectors in wood reduces amount of metal exposed to fire temperatures.

On the debit side, the special tools required, preferably power-operated, for cutting recesses for the split ring, claw-plate, shear-plate, and bulldog connectors, are a disadvantage in small jobs because of their cost.

[1] John A. Scholten, *Timber connector joints, their strength and design.* U.S. Dept. Agric. Tech. Bull. No. 865, 1944.

In evaluating the merits of connector joints it is desirable to compare the strength of such joints with the various alternative methods of metal fastenings. Reece[1] has done this, comparing the bearing strength in a simple double lap joint of Douglas fir, 50 mm minimum thickness, provided by various types of fastenings. He found that equal bearing strength is provided by –

> Two 62 mm bolted split ring connectors
> Ten 12 mm diameter bolts
> Eight 50 mm bolted hardwood dowels
> Eighteen 9 mm diameter wood screws
> Twenty-six 3 mm diameter wire nails

Reduction in the number of fastenings required effects considerable savings in assembly costs, but, compared with nails, the advantage of connectors lies in the greater reliability with which the strength of the joints can be determined.

Working loads of connectors elude precise mathematical calculations, and have had to be built up from test data. The process has required accurate and comprehensive strength tests of actual joints, which have provided figures suitable, when modified by appropriate factors of safety, for use as safe working loads in constructional design.

The evolution of a strong and reliable joint was responsible in itself for widening the scope of timber construction, by making possible designs that were quite impracticable so long as much weight was wasted in framed members that could develop at the joints only a fraction of their overall strength properties. Moreover, connectors permitted the use of many small-sized members, instead of a few large-dimension ones, greatly widening the field from which the raw material could be drawn, and, incidentally, the cost of the necessary timber. These facts, coupled with the era of scientific investigation of the strength properties of wood that resulted in the evolution of stress grading, have transformed wood into a precision material that an engineer-designer can use as an alternative to steel or reinforced concrete. Often timber will provide a better solution from the engineering standpoint, and invariably at less cost, in timber-producing countries.

[1] P. O. Reece, *loc. cit.*

ADHESIVES

The most recent aid to the utilization of wood as an engineering material has come from developments in adhesives since 1930. Gluing of wood to wood has been practised since very early times, but modern glues are very different from the joiner's glue-pot that has persisted, often in more senses than one, for hundreds of years. Even primitive glues, however, had a considerable influence on the utilization of wood: the Egyptians practised marquetry with veneers of contrastingly-coloured woods around 1500 B.C. The art of veneering was revived in the 17th and 18th centuries as a result of the invention of wood-cutting machinery: the first mechanically-operated saw came into use about 1650, a circular saw was patented in 1775, and a bandsaw in 1808. In 1793, Sir Samuel Bentham patented a series of wood-working machines, one of which was designed to produce veneers intended for gluing together, and therefore heralding the product we now know as PLYWOOD (see pages 363 to 366).

Research into the properties of adhesives has been responsible for extending the scope for timber and products made from timber. Scientific investigation has established that adhesion between two solid bodies may be of two kinds: (1) **natural or specific adhesion** produced by molecular forces of the same kind as those holding together the molecules of any solid body, and (2) **mechanical adhesion** by the setting of an adhesive that has obtained a key by filling crevices in two adjacent glued faces. The natural adhesion between two super-finished surfaces is of a very high order, *e.g.*, the force required to seperate two planes of glass, when in contact, is in the region of 1416 kgm per mm².

Certain glues function in both ways, and with most modern glues the adhesive has a shearing strength greater than that of wood. There are additional factors that influence the strength of glued joints: firstly, the smoother prepared surfaces are made prior to gluing, the better will be the results; and secondly, with two surfaces offering the minimum rugosity, maximum joint strength is secured with the thinnest possible glue line. These findings are in keeping with the theory of natural or specific adhesion. It has, however, been established that in gluing birch veneers with phenolic glues penetration of the veneer by the glue, and hence mechanical interlocking, does undoubtedly occur. The time-lag between preparing the surfaces for gluing and completing the operation is also important, although the chemical and physical reasons for this are obscure.

In the evaluation of joint strength it is particularly important to take into account the glue-line thickness. With special-purpose adhesives, satisfactory joints have been made between two surfaces up to 1·25 mm apart, but the conditions prevailing in such **gap joints**, as they are called, are very variable and often quite impossible to predict. Crazing and discrete points of contact are defects to be guarded against: selection of a suitable adhesive and attention to processing technique are important in such circumstances.

So long as the glue undergoes no change the strength of a glued joint tends to be governed today by the dimensions and physical properties of the wood used, and by the shape of the joint, rather than by the qualities of the particular glue selected, although, as has just been indicated, gluing technique is important. Glues with the requisite initial strength have now been evolved that are also proof against damp, the ravages of micro-organisms, and the lapse of time; they are as stable and durable as the timbers they join. Not all modern glues possess these desirable characteristics and, of the most durable, which are phenol formaldehydes or derivatives or homologues of phenol reacted with suitable aldehydes, some have the disadvantage of necessitating high temperatures and pressures to secure setting; these requirements are impracticable in many gluing operations.

Glues may be classified in the following categories:

Animal glues: skin or hide glues, bone glue, extracted bone glue, and rendered glue. – The adhesive in all cases is an organic substance called collagen. The merit of these glues is simplicity in use. Being subject to attack by micro-organisms, they lack durability, and lose their adhesive properties if exposed to damp.

Casein glues are derived from skimmed milk, the adhesive being a protein product. To improve the water resistance of these glues it is now usual to add various compounds such as sodium and calcium hydroxides, sodium silicate., and certain copper salts More recently, formaldehyde and urea have been added to give 'improved' casein glues. The group generally has the merit of simplicity in use, good strength, and lasting qualities. The resistance and durability of properly mixed casein glues, properly applied to adequately prepared surfaces, are such as to justify the continued use of these glues for interior work. Exposure to damp is liable to cause chemical breakdown and bacterial attack, with consequent loss in strength of the glued joint. Heavy laminate structures, made up with casein glues, have, however, been used for exterior purposes with success over quite long periods.

Extracted soya bean flour is another protein glue of modern origin with similar properties and limitations to casein glue.

Starch and soda silicate glues. – The basic ingredients are dry cassava flour, caustic soda, and water, to which may be added various chemicals to improve the low water resistance of this type of glue. In the absence of damp conditions the adhesive qualities are good, but the glues are not suitable for brush spreading.

Synthetic resin glues. – Phenolic and urea formaldehydes are at present leaders in the field of synthetic resin glues; they usually require high temperatures and pressures to secure good bonding. Many modified products have been evolved, aimed at reducing the cost of these glues, which is relatively high, and overcoming the need for high temperatures and pressures. Most of the modifications so far available are at the expense of extreme durability, resistance to damp, and bacterial action, and a wide range of temperatures: for the present the only absolutely reliable water-impervious glues are the derivatives of phenol. Already, however, several melamine glues have met the very stringent tests of BS 1203 and 1204 in regard to water resistance.

There are other types of synthetic resin glues than the phenolic and urea formaldehydes that may prove to be of high importance, but data from exposure tests of sufficient duration are not available to permit of a final assessment of their true worth. For example, formvar, a derivative of polyvinyl alcohol, gives a joint strong enough to pass the A.I.D. Test for propeller manufacture, and polyvinyl acetate, and modifications of this, also give a range of useful adhesives. More recently, vinyl derivatives, used in association with synthetic and natural rubber lattices, have come to the fore.

Modern glues have greatly increased the scope for wood, and

laminate construction (gluing of relatively thin layers of wood together) has entered many fields previously restricted to solid wood construction or alternative materials, proving themselves superior and less costly (see also page 366 for discussion of blockboards and some other forms of laminate construction). The next stage is **built-up construction** or **glued laminated construction** – combining laminations, adhesives, and connectors, which gives freedom to the designer, besides scope for using material that would otherwise be too small for structural purposes. Large spans can be bridged with built-up arches and girders much more satisfactorily than with solid timber, and at considerable savings in cost, compared with steel or re-inforced concrete. All these developments have hinged on pro-gressive improvement of glues. In Reece's words: 'Gluing does for timber what welding does for steel; it enables joints to be made without cutting any material out of the members joined and makes it possible for the designer to take advantage of all the economies associated with monolithic construction'.[1]

Summarizing the position: fundamental research into the strength properties of wood has pointed the way to rapid visual assessment of the strength properties of individual pieces of timber by means of stress-grading rules. This makes possible the allocation of safe work-ing loads within narrow limits to each member of a composite timber structure, thereby securing the utmost economy in use of wood. In other words, timber becomes an alternative to steel and concrete in engineering design: the special advantages of each for any particular problem can be accurately weighed. The development of timber connectors and modern adhesives overcome one of the major difficul-ties formerly inherent in timber construction, namely the weakness of the joints or fastenings, and the modern methods of jointing give joints of calculable efficiency.

The combination of the two factors referred to above has already resulted in great strides being made in the extended constructional use of wood. The future would appear to hinge on the problem of world timber supplies.[2] If a permanent shortage of wood can be avoided the possibilities are immense.

---

[1] *Loc. cit.*

[2] Shortage of timber may arise from shortage of labour: the whole of the annual increment of Pacific coast forests, for example, is not being exploited for this reason.

CHAPTER 19

Manufactured Wood Products

This book is primarily concerned with timber as such, whereas a substantial proportion of the total annual output of 'round wood' from the world's forests does not reach the sawmills. Unfortunately, it is not easy to arrive at the true statistical position because total output is recorded in different units for different products. The best available data are those published by the Food and Agricultural Organisation of the United Nations. Over 43 per cent of the total output of 'round timber' is used as firewood, mostly in an extremely wasteful manner so that only a relatively small proportion of the heat value of the wood burned is utilized.

Next to sawn timber and fuel wood comes wood pulp, both mechanical and chemical, which supplies two different industries, namely paper and some forms of man-made fabrics, *e.g.*, rayon. Chemical processes, whether for paper or man-made fibres, seek to extract the cellulose from the pulp, leaving the lignin, infiltrates, and resins as waste products. Whether the pulp is mechanical pulp, *i.e.*, ground up logs, or chemical pulp, the end products do not resemble timber.

After sawn timber, firewood, and wood pulp, plywood is the next most important outlet for round logs. The F.A.O. statistics bring plywood, particle board, and fibre boards together, and here the astonishing change is in the output of particle board, *i.e.*, chipboard, between 1960 and 1970. Output of plywood has virtually doubled in that decade, but particle board has increased more than fivefold. The figures in 1000 m³ for 1970 are plywood 32,607, 1000's m³, particle board 18,127, 1000's m³ and fibreboard 7,778, 1000's m³.

PLYWOOD

Plywood may either be sliced for decorative end uses, or peeled, which is the method generally used when appearance of the finished product is not the primary consideration. Initially the cutting of veneers was an attempt to widen the decorative scope of natural wood;

it is not a resort adopted to cover up inferior materials or workmanship. Glue was an adjunct to this decorative use of wood, and the then existing glues were adequate for the purpose. The manufacture of plywood for constructional ends was a much later development, dependent on the improvements in adhesives that occurred.

'Modern' plywood was the outcome of improvements in machinery, but even in the first American patent taken out in 1868 specific mention is made of the improved strength properties of the resultant product, compared with solid wood:

'The invention consists in cementing or otherwise fastening together a number of these 'scales'[1] or sheets, with the grain of the successive pieces, or some of them, running crosswise or diversely from that of others . . . The crossing or diversification of the direction of the grain is of great importance to impart strength and tenacity to the material, protect against splitting, and at the same time preserve it from liability to expansion or contraction.'[2]

The use of plywood as a trade name for the material may be traced to the war years 1914–18. The earliest uses were for furniture, and later, joinery and packing-case material; its use for constructional purposes was dependent on the evolution of water-proof glues – synthetic resin products – that made their appearance in 1930. Previous to this, existing glues often secured, at least for a time, a bond between two timber surfaces equal to the cohesion of the cells in a piece of wood; many glued pieces of inadequately seasoned wood, e.g., wide shelves, window boards, have split instead of separating along the glue line. The qualification 'for a time' is, however, all important; it restricted the use of glued material constructionally.

The types of glue available are described on pages 358 to 361; not all these are suitable for use in the manufacture of plywood. Four types of glue are recognized in B.S.1455:1956: *British-made plywood for general purposes* and B.S.1455:1963: *Specification for plywood manufactured from tropical hardwoods*:

"*Type W.B.P.: Weather and boil-proof.* Adhesives of the type* which by systematic tests and by their records in service over many years

[1] Scales was the name given to the sheets of veneer.

[2] Quoted from *Modern Plywood* by T. D. Berry, Pitman Publishing Corporation, New York and Chicago, 1942, p. 26.

* At present only certain phenolic adhesives have been shown to meet this requirement.

have been proved to make joints highly resistant to weather, micro-organisms, cold and boiling water, steam and dry heat.

Type BR: Boil resistant. Joints made with these adhesives have good resistance to weather and to the boiling water tests, but fail under the very prolonged exposure to weather that Type WBP adhesives will survive. The joints will withstand cold water for many years and are highly resistant to attack by micro-organisms.

Type MR: Moisture-resistant and moderately weather-resistant. Joints made with these adhesives will survive full exposure to weather for only a few years. They will withstand cold water for a long period and hot water for a limited time, but fail under the boiling water test. They are resistant to attack by micro-organisms.

Type INT: Interior. Joints made with these adhesives are resistant to cold water but are not required to withstand attack by micro-organisms.''

Type WBP: Weather and boil-proof. Weather and boil-proof glues are phenol-formaldehyde products, suitable for withstanding the most severe exposure conditions without deterioration. It is important to ensure that the veneers themselves are resistant to decay. The mistake is often made of specifying exterior grade plywood, bonded with WBP adhesives, without specifying the species of timber that are acceptable. If the face veneers are birch, or other similar perishable timbers, the plywood is certainly not suitable for exterior use, notwithstanding the fact that the adhesive is weather and boil proof. This is the distinction secured with marine grade plywood, *vide* B.S. 1088 & 4079: 1966 *Specifications for plywood for marine craft. Metric units.* This Standard lists eleven timbers, the heartwood of which is sufficiently durable in its natural state to comply with the requirements of Clause 3 of B.S. 1088; among the eleven timbers are African mahogany, idigbo, makoré, omu, light and dark red meranti, sapele and ultile.

Type BR: Boil-resistant adhesives. These are urea-melamine formaldehyde adhesives. The addition of melamine to urea-formaldehyde resins increases resistance to immersion in boiling water; they are immune to attack by micro-organisms.

Type MR: Moisture-resistant and moderately weather-resistant. These adhesives are urea-formaldehyde resins, providing a high bonding strength in the dry state, and capable of withstanding prolonged soaking in water at normal temperatures, and for a few minutes in boiling water. They are immune to attack by micro-organisms. In the manufacture of plywood, the urea-formaldehyde resins may be extended by the addition of cereal flour or blood albumen at the expense of the strength of the bond and resistance to micro-organisms. *Type INT: Interior.* These adhesives are typically casein glues consisting of a mixture of the curds of milk, hydrated lime and certain other chemicals. Extracted soya bean flour is another protein glue, comparable to casein glue. Although heavy laminate structures made up with casein glues have been used for exterior pupo.ses with success over quite long periods, this must not be interpreted as justifying the use of casein-glued plywood for exterior sheathing under conditions of full exposure to the weather; for such conditions even plywood bonded with the so-called water-proof glues is inadequate, and only genuine exterior grades should be employed in circumstances of full exposure.

There is an extensive literature on plywood, apart from leaflets and bulletins, and the British Standards. British Standards that deal with plywood that have not yet been cited are: B.S. 3493: 1962: *Information about plywood;* B.S. 3583: 1963: *Information about blockboards and laminboard;* B.S. 3444: 1972: *Blockboard and laminboard;* B.S. 1203: 1963: *Synthetic resin adhesives (phenolic and aminoplastic) for plywood;* B.S. 3842: 1965: *Specification for treatment of plywood with preservatives;* B.S. 4512: 1969: *Methods of tests for clear plywood.* Those requiring a comprehensive introduction to the subject will find the Timber Research and Development Association publication *Plywood: Its manufacture and uses* most helpful. This publication covers many aspects of the subject, including properties, manufacture, adhesives, types of plywood, and the structural use of plywood.

The valuable properties of plywood stem from its method of manufacture, namely with the grain of alternate veneers at right angles to one another. This reduces movement with changes in temperature and relative humidity; in round figures, the movement of plywood is only about 1/10th of the movement of solid timber over any given range in temperatures and relative humidities. Taking the extreme range in equilibrium moisture contents of timber in heated buildings as being from 8 to 15 per cent as between winter and summer – this range can be expected to produce a dimensional change in timber of 2.2 per cent; with plywood exposed to a similar range in equilibrium moisture contents, the dimensional changes would be only about 0·22 per cent. Another important property of plywood is its resistance to splitting, which permits of nailing and screwing relatively close to the edges of the sheets. Plywood is stiffer than many other materials, including mild steel sheet, when compared on a weight-for-weight basis. Generally plywood has a high strength/weight ratio. Because of the absence of joints over large areas, plywood improves the insulation value of the construction.

Another obvious advantage of plywood for certain purposes is the large size of sheets available; a typical size is 2440 mm × 1220 mm (8 feet by 4 feet), but larger sheets are obtainable. Thicknesses range from 3 mm to 25 mm. Plywood much over 6 mm in thickness will be made up of more than three veneers; 25 mm plywood may contain as many as nine veneers.

Apart from three, five, and multiple plywood, blockboard, laminboard, and battenboard may be classified as types of plywood. 'Blockboard consists of a core of wood made from strips up to 25 mm wide placed together, with or without glue between each strip, to form a slab which is sandwiched between outer veneers of 2–3·5 mm with their grain direction at right angles to the grain of the core. . . . When the length exceeds the width, the blockboard should be 5-ply construction: a core of the same construction as that mentioned above, a veneer on each side of about 2 mm thickness running with grain at right angles to the direction of the grain in the core, and a thinner outer ply with grain running parallel to that in the core.'[1]

Laminboard. This is similar to blockboard but the core is built up of strips of wood 1·5 to 7·00 mm thickness, glued face to face to form the core, with facing veneers as with blockboard. Because of the amount of glue used in the manufacture, laminboard is both heavier and more expensive than blockboard.

Battenboard. This is a variation in which the core is built up of strips, usually not exceeding 75 mm in width.

IMPREGNATED AND DENSIFIED WOOD

Impregnated and compressed wood. Much work has been done in recent years on the chemical treatment of wood aimed primarily

[1] *Plywood: its manufacture and uses. Metric* published by the Timber Research & Development Association, May, 1972, page 13.

at securing dimensional stability, although some of the treatments secure other improvements, *e.g.*, in strength properties, including hardness and resistance to abrasive wear. This is not the place to go into this subject in any detail, as, apart from numerous papers, bulletins, and a substantial amount of trade literature, there are several well-documented text books, *e.g.*, R. H. Farmer: *Chemistry in the Utilization of Wood*, Pergamon Press, 1967. The development of synthetic resin-forming chemicals opened up this new field for the chemical treatment of wood. The principal use with wood products was to increase the stability of wood against shrinkage and swelling, while securing a significant improvement in certain mechanical properties. For example, impregnation of maple to the extent of 20 per cent of the dry weight of wood, besides reducing shrinkage and swelling permanently, increased side-hardness by 40 per cent over untreated wood. Such chemical treatments have a limited application: their effectiveness depends on the resins being deposited in the fine cell-wall structure; they are, therefore, only suitable if applied under pressure.

Several kinds of compressed wood have been developed in recent years: **staypak** is wood compressed at sufficiently high temperatures and moisture contents to cause the lignin to flow, thereby relieving internal stresses. It is stated to have advantages over densified wood (**improved**) and resin-treated compressed wood (**compreg**) in that, although it will swell appreciably under conditions that cause swelling in wood, it will return to practically the original compressed thickness on drying to the original moisture content. Improved wood is wood compressed under conditions that do not cause flow of the lignin cementing material in the cell wall structure; it may be made from solid wood or veneers preferably assembled with a synthetic resin glue. Compreg is resin-treated wood that is stabilized in the compressed form by the resin; the process has been applied to veneers. Staypak, improved wood, and compreg were American products that found specialist uses during World War II. Products known as Improved wood were made in this country, *e.g.*, by F. Hills & Sons Limited, of Stockton-on-Tees. Messrs Hills & Sons' product was an 'improved wood,' used in the manufacture of Jablo airscrew blades. The product is described in *Aircraft Production*, April, 1942. The veneers used were remarkably thin – 0·6 mm thick – which were bonded together in batches of 50 or more (approximately 1 in. thick) with a synthetic resin glue, under considerable pressure and at a high temperature. The timber used was Canadian birch, and the finished product had a specific gravity ranging from 0·9 to 1·3. Other producers of improved or compressed wood in the war years were Airscrew Company Limited, of Weybridge, Surrey and Saro Laminated Wood Products Limited of London, the factory being at Folly Works, Wippingham, Isle of Wight. These firms' products are described in *Wood* May, 1939. The Airscrew Companys' product was known as **Jicwood.** Although hope was expressed by the Companies at the time that new uses would be found after the War for these 'improved wood' products, demand was not forthcoming and only Permali Limited, incorporating Harden Richmond Limited, have continued in production.

Permali Limited of Gloucester produce four products – Permali (in three grades), Permawood, Hy-du-lignum, and Jabroc. All are manufactured from veneers, and all except Permawood are impregnated with synthetic resins; they are densified under pressure. The densest form of Hy-du-lignum has a specific gravity of 1·31. Permali Data Sheet No. DS2021 describes Permali as a high-voltage insulating material offering unusual scope to the designer of electrical equipment. Its maximum mechanical strength can be developed in any required direction 'by varying the arrangement of the wood veneers and thereby the collective disposition of their grain structure.' Varying the arrangement of the veneers with reference to the direction of the grain is a feature of several Permali products. Hy-du-lignum found a use in World War II for variable-pitch propellers for service aircraft and today is being used for helicopter main and tail rotor blades. It is also replacing metal in press tools and jigs. Other uses enumerated by the manufacturers are 'insulating fishplates, fans, lapping discs, bossing and tinmans' mallets, rounder blocks, connecting rods, guillotine cutting sticks, cutting formes and bench tops.' Jabroc is described as a 'highly densified wood veneer laminate . . . used extensively in the manufacture of press tools . . . it is only one-fifth of the weight of steel yet more than half as strong . . . [it] possesses great toughness – withstanding a high degree of wear and tear with excellent resistance to attack by moisture or the effects of climate conditions.'

Staypak, improved wood, compreg, and the Permali products can hardly be considered as 'wood'; they are manufactured products, the raw material for which is wood, usually in veneer form; in their finished state, their origin is still apparent. In addition to improved stability, the strength properties are not unnaturally very considerably increased, compared with untreated wood, rendering the finished product suitable for a variety of new purposes, besides being a superior product for certain old uses. Inevitably they are relatively costly.

CURIFAX WOOD

The various forms of densified laminated wood are not by any means new processes, having been developed rather before the outbreak of World War II, although it was not until the War years that they became commercial products. By comparison, the Applied Radiation Chemistry Group at Harwell has developed Curifax – timber pressure-impregnated 'with a carefully formulated mixture of plastic-forming monomers. The treated timber is then exposed to the action of intense radiation, which brings about the controlled and uniform polymerization of the monomers throughout the wood structure. No catalyst or solvent is used. In the finished material, the cured resin, free from any active residual materials, is intimately bound to the wood fibres and fills virtually all the pores and voids of the original timber' (Leaflet issued by Applied Radiation Chemistry Group, AERE, Harwell, Didcot, Berks).

Unlike the densified laminated woods previously described, Curifax is produced as dimension timber, up to 3·66 m (12 ft.) in length. This is because of the penetrating qualities of the monomers used, coupled with the penetrating power of radiation. The Harwell leaflet already quoted gives the following strength and other data for Curifax wood, relative to untreated wood:

Compression strength	70	to	140	per cent
Static bending strength	40	to	70	per cent
Elastic modulus	12	to	33	per cent
Radial shear strength	25	to	95	per cent
Tangential shear strength	30	to	110	per cent
Radial and tangential hardness	6	to	12	times
Longitudinal hardness	5	to	7	times
Abrasion resistance	2	to	9	times
Density			0·85 to 1·15, depending on timber and monomer loading	

Water uptake: typically from 100 per cent uptake on dry weight of untreated wood to 10 per cent uptake on Curifax wood.

The Harwell leaflet gives as some of the uses for Curifax wood the following: technical woodwork, form and pattern making, templates, jigs and press tools, machinery parts, shuttles, bobbins and spindles, industrial and decorative flooring, speciality wood-ware, turnery, sports goods, musical instruments, brush handles, table and industrial cutlery handles. A typical example of Curifax wood block flooring is to be seen in the entrance hall of the Princes Risborough Laboratory—some of the blocks were manufactured at Harwell, and some at the Princes Risborough Laboratory.

PARTICLE BOARD

Wood particle board, or chipboard, is the newest of all the man-made sheet materials. According to Laidlaw & Hudson,[1] 'Commercial chipboard production began in Switzerland and Germany as early as 1941, and in the U.K. in 1946–47.' The growth of the industry in Europe has been phenomonal, from 12,000 metric tonnes in 1950 to an an estimated 5,060,000 metric tonnes in 1969. Chipboard consumption is now highest in West Germany, which has recently overtaken Switzerland, with Canada, U.S.A., and the U.K. very far behind the two leaders.

The bulk of the raw material for chipboard manufacture is wood, but flax or hemp shives are also used. In addition to the wood particles, synthetic resin adhesives, typically to the extent of 8 to 10 per cent of the weight of wood particles, water, and a small proportion of wax – 0.25 to 1.0 per cent – are used in the manufacture of chipboard.

The Timber Research & Development Association has issued a very comprehensive booklet entitled *Particleboard in Building, A guide to its manufacture and use*, which I have drawn on freely in the account that follows. Five types are recognized: General purpose, Interior – structural, Interior – non-structural, Exterior – structural, and Exterior – non-structural. The last two types are not yet available.

There are four types of platen-pressed boards: (i) single layer,

[1] *Chipboard developments in Europe* by R. A. Laidlaw & R. W. Hudson, Timberlab Papers, No. 20, 1969.

(ii) three layer, (iii) multi-layer, (iv) graded density. The density of single layer boards depends on the nature of the particles used – with wood particles the density is of the order 600 kg/m³ and more, whereas boards made from flax shives are of the order of 350 kg/m³ and upwards. Three layer boards consist of thin, relatively high density layers top and bottom, with a core of larger particles of lower density. Three layer boards are inferior in strength properties to single layer boards, and their screw-holding qualities are lower. The merit of the three layer boards, lies in the high density of the fine surfaces top and bottom, which are suitable for painting or veneering, and the saving in weight may be an advantage. Multi-layer boards have basically similar properties to three layer boards, but the surface finish is carried to finer limits. In the Timber Research & Development Association's booklet, it is stated that multi-layer boards would 'in some instances meet the requirements of "interior structural" classification, and would offer special advantages in "general purpose" or "interior non-structural" applications.' *Graded density*. 'These are boards with graduated change in structure, and are midway between single and three layer boards with some of the attributes of both. The particle elements vary in size, and are distributed through the board in such a manner that they decrease in size from the centre thickness of the board to the outside.' Apart from the four types of platen-pressed boards described above, some particleboards are 'made by extrusion through a die. The particles lie with their larger dimension mainly at right angles to the direction of extrusion.'

The sources of raw material for chipboard manufacture given in the Timber Research & Development Association's booklet are:

(i) Veneer peeler cores and forest thinnings.
(ii) Planer shavings, and turnery and other chippings from woodwork production.

(iii) Edge rippings and offcuts, slabs, etc., from sawmills.

The third category are usually thicker particles, more suitable as core material. It is stated that 'coniferous woods with relatively high natural resin contents have some advantage over broad-leaved hardwoods in that there is generally less tendency to absorb adhesive, but generally speaking, softwoods and low density hardwoods are more frequently used, the primary consideration being local availability and cost'.

The bulk of the chipboard produced to date has been intended for use in dry situations, *i.e.*, interior uses, furniture, flooring, and partitions, and for these purposes urea-formaldehyde (UF) resin adhesives are extensively used. The Princes Risborough Laboratory has recently been engaged in the development of an exterior grade of wood chipboard[1] for which the urea-formaldehyde boards are quite unsuitable. It is stated in the paper cited that 'The objectives of Timberlabs work on exterior chipboards are two-fold: 1. To determine the factors affecting the behaviour of chipboards when exposed to severe environments; and to develop rapid methods of assessment; and 2. To develop a board which is resistant to such environments, and to study its behaviour in accelerated laboratory tests and when exposed to the weather.' This paper pointed out that the bulk of the U.K. production then (1971) 'is bonded with urea-formaldehyde (UF) resin and is only suitable for interior applications. However, the addition of melamine-formaldehyde (MF) to UF adhesives leads to greatly improved durability of the glue, and phenol-formaldehyde (PF) resins, of the type used in WBP-bonded plywood, are completely unaffected by moisture.'

When interior grades of chipboard are used in atmospheres with high relative humidities, the chipboard deteriorates rapidly – swelling occurs, which is not reversible on return to lower relative humidities, and ultimately the chipboard delaminates, losing all its strength properties. This makes the UF grades unsuitable for decking under flat roofs, even if the risk of the weathering material splitting is remote.

Accelerated ageing tests were designed at the Princes Risborough Laboratory, and for this purpose commercially produced boards were used. 'The material tested included two brands manufactured with PF adhesives, one with MF/UF resin, and one made using the waste product from some types of sulphite pulping mills as the adhesive (SL). For comparative purposes, one UF bonded board was also included.' The boards bonded with sulphite waste are dark olive green in colour; although weaker in strength properties than boards of the same thickness bonded with conventional synthetic resin adhesives, the bond in sulphate waste bonded boards is extremely stable.

A second report on tests on particleboard under adverse conditions

[1] R. W. Hudson: *Timber & Plywood Particle Board Supplement*, November 24, 1971.

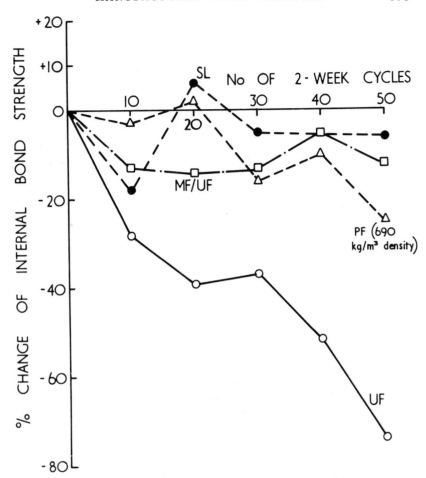

FIG. 50. Effect of cycling samples of commercial chipboard between 30% and 90% RH: internal bond strength.

was published in November, 1972.[1] Fig. 50 illustrates the results obtained from exposure to fifty 2-week cycles in conditioning chambers maintained at 30 and 90 per cent relative humidity. On presenting these data, Mr Beech drew attention to the fact that, as used in buildings, the edges and surfaces of chipboard panels 'are usually

[1] J. C. Beech, B.Sc.(Eng.): *Performance of wood particleboard under adverse conditions.* Timber & Plywood Particleboard Supplement, November 15, 1972.

covered with decorative laminates or paint films, which greatly retard the effect of humidity changes in the atmosphere.'

There is an extensive literature on chipboard manufacture, and several British and German Standards, which are cited in the *Selected Bibliography*, under Chapter 19, pages 414 to 415. In the Timber Research & Development Association's booklet, it is stated that particleboard is ideal for ground floor constructions 'provided the requirements of the Building Regulations (1965) are complied with'. It is essential that the membrane below suspended chipboard ground floor floors shall be really effective, and the flooring must qualify with the requirements laid down in BS. 2604 Amendment 3, relating to impact tests. Boards must meet:

'(a)   Static tests in which point loads are gradually applied, and
 (b)   Impact tests in which point loads are suddenly applied.'

Two Bulletins issued by the Greater London Council deal with chipboard: one part of *Bulletin 50* deals with strength properties of chipboard, and one part of *Bulletin 51* deals with moisture resistance of chipboard.

There is no doubt that chipboard or particleboard has great possibilities but it is all-important that users should appreciate its limitations. Interior grades are particularly prone to rapid deterioration if exposed to high relative humidities or persistent condensation, but it would appear that the limitations of the material can be overcome using appropriate adhesives for adverse site conditions.

FIBRE BUILDING BOARDS

Finally, mention must be made of fibre building boards, which have been in production longer than chipboard or the various forms of impregnated and densified wood. There is a separate Trade Association—FIDOR—concerned with all types of fibre building board. This organisation has issued numerous leaflets and test data sheets on the various types of boards.

The British Standard that deals with these products has recently been revised and reissued in three parts: BS 1142: Part 1 (1971): *Methods of test*; Part 2 (1971): *Medium board and hardboard*; and Part 3 (1972): *Insulating board softboard*. The British Standards classify fibre building boards primarily on a density basis. There are two broad distinctions, namely **softboards** or **insulation boards,** and the heavier hardboards, which are sub-divided into **medium, standard,** and **tempered hardboards.** Most softboards and all hardboards have this in common: they are manufactured from pulped wood. It was necessary to make the distinction that, with softboards, not all boards are wood fibre boards: some, *e.g.*, celotex, are made from sugar cane stems.

Fibre building boards differ from other forms of sheet material discussed in this chapter in that, except for tempered hardboards, no chemicals or adhesives are used in their manufacture. There are various processes which, in the main, felt together the pulped fibres under heat and pressure. In FIDOR's *Fibre building board information— Board types: manufacture*, it is stated that 'Fibre building board is the only wood-based sheet material which is *reconstituted* as opposed to solid timber reassembled, *e.g.*, cross veneers of plywood or bonded particles in particle board'.

Fibre building boards may be made from softwoods or hardwoods, depending on the availability of supplies locally. The raw material is derived from sawdust or forest thinnings, and undersized sawmill stock, up to about 200 mm in diameter. The first stage in the manufacture of all types of hardboard and softboard is to reduce the raw material to chips, when steam under pressure is introduced, commencing the softening of the chips before breaking them down into fibres. One of two methods is used for producing fibres: (i) the 'explosion' process, using high pressure steam, or (ii) the mechanical defibrator. The two processes are illustrated in Fig. 50. The descriptions of the manufacturing processes that follow have been culled from FIDOR's *Fibre building board information—Board Types: manufacture.*

After separation into fibres, the fibres are added to water to produce a 95 to 97 per cent solution or 'slurry' of fibres in water. This slurry has to be constantly agitated to prevent the fibres from settling out. Additives may be added at this stage, *e.g.*, a water repellent size, a solution of alum, certain fungicides or insecticides, and occasionally a phenol formaldehyde resin. Felting is the next stage, and is described as follows: 'the slurry is laid out on to the forming machine which consists of a steadily moving belt of perforated metal, normally 1200 or 1500 mm in width . . . [where] the felting or interlocking of the fibres takes place.' As the pulp (or 'wet lap' as it is now called) progresses, water is removed through the mesh screen by (1) drainage under gravity, (2) the squeezing action of thicknessing rollers, (3)

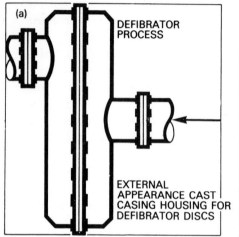

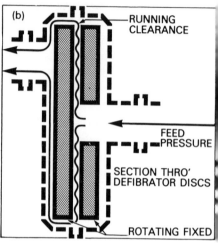

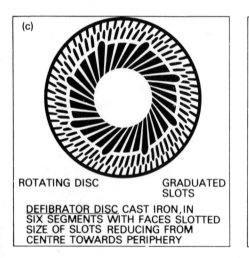

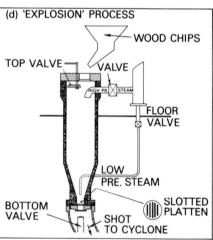

FIG. 51. Methods for producing fibres in fibre board manufacture. (a), (b), and (c) The mechanical defibrator process, (d) The explosion process.

(a), (b), and (c) *by courtesy of FIDOR*
(d) *by courtesy of Masonite Ltd*

suction pumps. The partially dry 'wet lap' is then cut to approximate sheet lengths and transferred to (a) 'Drying ovens for the production of non compressed insulating board', or to (b) 'Heated multi-platen press(es) for the production of hardboard and medium board.' Pressing is done at a temperature of about 200°C, often at a heavy initial pressing of 3 N/mm^2, followed by a reduced pressure. 'A typical pressure cycle time for 3·2 mm hardboard is 7 to 8 minutes.' After pressing, the boards have to be subjected to a conditioning process to increase the moisture content to 7 to 9 per cent. Oil tempering and Heat Treatment may be applied after removal of the sheets from the presses and before subjecting them to Conditioning. 'These processes are designed to increase the strength and water/moisture resisting properties of hardboards and medium boards.'

Insulation boards are typically thicker than hardboards, the range being from 8 to 30 mm in thickness, compared with 2 to 6 mm for most hardboards. Insulation boards may be impregnated with bitumen, when they are particularly suitable for providing the thermal insulation between a vapour barrier of bituminous felt and a 3-ply felt roof. Hardboard is extensively used for 'facing' the core material of flush doors, and for providing a level floor in an old building.

Savory[1] has investigated the resistance to fungal attack of four types of standard hardboard: Brand A—a board made by the 'explosion' process; Brand B—a board made mainly from hardwoods, *Eucalyptus* species; Brand C—a typical Swedish board made from the softwoods pine and spruce; and Brand D—a board of British manufacture. Savory summarises his findings as follows:

> The species of timber utilised and the method of manufacture of the board do not appear to influence resistance to attack by wood-rotting Basidiomycetes [*Merulius lacrymans*, *Poria vaillantii*, *Coniphora cerebella* and *Lenzites trabea*]. The extent of attack of all four boards was about half that obtained in Scots pine sapwood, thus their resistance to attack was broadly comparable with that of a 'moderately durable' timber. In contrast, boards made from softwood were more resistant to attack by wood-rotting microfungi than the board made from hardwood though even this was somewhat more resistant than Scots pine sapwood. Similarly, there was variation amongst the softwood boards, that made by the 'explosion' process being the most resistant. None of the boards supported the growth of common superficial moulds at all readily.

[1] Savory, J. G., *The fungus resistance of standard hardboard*, Timberlab Papers, No. 15 (1969).

It will be seen that resistance to fungal attack is surprisingly good, compared with durability of the raw materials from which the boards are made. It is to be presumed that the processes employed to reduce the raw materials to fibres wash away some of the food material in solid wood attractive to fungi. The hygroscopicity inherent in wood, however, is not lost in its conversion to softboards or hardboards.

PLATE 62

A split ring connector, showing the portable pneumatic drill for cutting the recess and boring the bolt hole. Manufactured by the Timber Engineering Co., Washington, D.C.

By courtesy of J. MacAndrews & Forbes Ltd.

PLATE 63

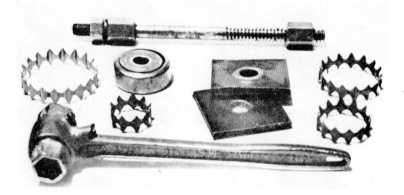

Fig. 1. Toothed-rings, washers, and assembly apparatus

Fig. 2. Tightening a toothed-ring connector joint

By courtesy of the Timber Engineering Co., Washington, D.C.

PLATE 64

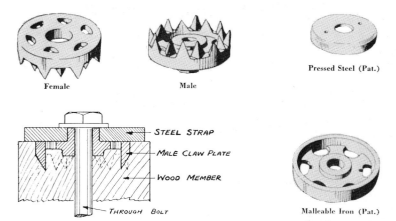

Female   Male

Pressed Steel (Pat.)

STEEL STRAP

MALE CLAW PLATE

WOOD MEMBER

THROUGH BOLT

Malleable Iron (Pat.)

FIG. 1. Top left: male and female claw plate and below, the use of a male claw plate. Top right: two types of shear plates

FIG. 2. Shear plates in position

By courtesy of the Timber Engineering Co., Washington, D.C.

PLATE 65

The Teco spike grid. A horizontal member secured to a
round pole by means of a spike grid

By courtesy of J. MacAndrews & Forbes Ltd.

List of Botanical Equivalents of Common or Trade Names used in the Text

In many cases it is not possible to give a single botanical name because the latter is often applied to the timbers of more than one species or even genus ; such cases are given below as 'spp.', no attempt being made to list all the species that may provide commercial supplies. It will be seen that in a few cases the botanical name cited below differs from that used in the text. This is because such names have recently been revised, but the former name is the better known, and more likely to be found in the literature.

abura = *Mitragyna ciliata* Aubrev. and Pellegr (syn. *M. stipulosa* Kuntze)

afzelia = *Afzelia* spp.

agba = *Gossweilerodendron balsamiferum* Harms

afrormosia = *Pericopsis elata* van Meeuwen (syn. *Afrormosia elata* Harms)

alder = *Alnus glutinosa* Gaertn.

American whitewood, *see* whitewood, American

apitong = *Dipterocarpus* spp. (Philippines)

ash = *Fraxinus* spp.

ash, American = *Fraxinus americana* L., *F. pennsylvanica* var. *lanceolata* Sarg., *F. nigra* Marsh.

ash, European = *Fraxinus excelsior* L.

ash, mountain, *see* oak, Tasmanian

balsa = *Ochroma lagopus* Sw.

beech = *Fagus sylvatica* L.

birch = *Betula* spp.

blackwood, African = *Dalbergia melanoxylon* Guill. et Perr.

blackwood, Australian = *Acacia melanoxylon* R. Br.

box, Cape = *Buxus macowani* Oliv.

box, European = *Buxus sempervirens* L.

boxwood, Ceylon = *Canthium dicoccum* Merr.

brush box = *Tristania conferta* R. Br.

camphorwood, Borneo, *see* kapur
camphorwood, East African = *Ocotea usambarensis* Engl.
camphorwood, Formosan = *Cinnamomum camphora* Nees et Eberm.
canary whitewood, *see* whitewood, American
cedar = *Cedrus* spp.
cedar, Central American = *Cedrela mexicana* Roem.
cedar, cigar-box, *see* cedar, Central American
cedar, South American = *Cedrela* spp. including *C. odorata* L.
cedar, western red = *Thuja plicata* D. Don
chengal = *Balanocarpus Heimii* King
cherry = *Prunus avium* L.
chestnut, American = *Castanea dentata* Borkh.
chestnut, sweet = *Castanea sativa* Mill.
coachwood = *Ceratopetalum apetalum* D. Don
cornel, Turkish = *Cornus* spp.

dahoma = *Piptadeniastrum africanum* (Hook. f.) Brenan (syn. *Piptadenia africana* Hook. f.)
deal, Baltic, *see* redwood, European
deal, red, *see* redwood, European
Douglas fir = *Pseudotsuga* menziesii (Mirb.) Franco (syn. *P. taxifolia* Brit., *P. douglasie Carr.*

ebony = *Diospyros* spp., *Maba* spp.
ekki = *Lophira alata* Banks ex Gaertn.
elm = *Ulmus* spp.
elm, common = *Ulmus procera* Salisb.
elm, Dutch = *Ulmus hollandica* var. *major* Rehd.
elm, European = *Ulmus procera* Salisb.
elm, wych = *Ulmus glabra* Hudson (non Miller)

fir = *Abies* spp.
fir, Douglas, *see* Douglas fir
fir, noble = *Abies nobilis* Lindl.
fir, silver = *Abies alba* Mill.

gaboon = *Aucoumea klaineana* Pierre
gedu nohor = *Entandrophragma angolense* C. DC.
greenheart = *Ocotea radiaei* Mez
guarea = *Guarea cedrata* Pellegr., *G. Thompsonii* Sprague et Hutch.
guarea, scented = *Guarea cedrata* Pellegr.
gum, American red = *Liquidambar styraciflua* L.
gum, spotted = *Eucalyptus maculata* Hook. and *E. citriodora* Hook.
gurjun = *Dipterocarpus* spp. (Andamans, Burma)

hazel = *Corylus avellana* L.
hemlock = *Tsuga* spp.
hemlock, eastern = *Tsuga canadensis* Carr.
hemlock, western = *Tsuga heterophylla* Sarg.
hickory = *Carya* spp. (syn. *Hicoria* spp.)
holly, American = *Ilex opaca* Ait.
holly, European = *Ilex aquifolium* L.
hornbeam = *Carpinus betulus* L.

idigbo = *Terminalia ivorensis* A. Chev.
iroko = *Chlorophora excelsa* Benth. et Hook. f.
ironbark = *Eucalyptus crebra* F. v. M., *E. paniculata* Sm., *E. siderophloia*
    Benth., *E. sideroxylon* A. Cunn., *E. fergusoni* R. T. Bak.
ironbark, grey, *see* ironbark

jarrah = *Eucalyptus marginata* Sm.
jelutong = *Dyera costulata* Hook. f.

kapur = *Dryobalanops aromatica* Gaertn. f. (Malaysia), *Dryobalanops*
    spp. (Sabah, Sarawak)
keledang = *Artocarpus lanceifolius* Roxb.
kempas = *Koompassia malaccensis* Maing. ex Benth.
keranji = *Dialium* spp.
keruing = *Dipterocarpus* spp. (Malaysia, Sabah)
kokrodua, *see* afrormosia

larch = *Larix* spp.
larch, European = *Larix decidua* Mill.
lauan = *Shorea* spp., *Pentacme* spp., *Parashorea* spp. (Philippines)
lauan, dark red, *see* lauan
lauan, light red, *see* lauan
lignum vitae = *Guaiacum officinale* L., *G. sanctum* L.
lime = *Tilia vulgaris* Hayne
logwood = *Haematoxylon campechianum* L.

mahogany = *Swietenia* spp.
mahogany, African = *Khaya ivorensis* A. Chev., *K. grandifoliola* C.
    DC., *K. anthotheca* C. DC.
mahogany, Brazilian = *Swietenia ? macrophylla* King
mahogany, Central American = *Swietenia macrophylla* King
mahogany, cherry, *see* makoré
mahogany, Cuban = *Swietenia mahagoni* Jacq.
mahogany, Gabon, *see* gabon
mahogany, Honduras, *see* mahogany, Central American
mahogany, Philippine, *see* lauan

mahogany, sapele, *see* sapele
mahogany, Spanish, *see* mahogany, Cuban
makoré = *Tieghemella heckelii* Pierre ex A. Chev. (syn. *Mimusops heckelii* (A. Chev.) Hutch. et Dalz.)
mansonia = *Mansonia altissima* A. Chev.
maple = *Acer* spp.
maple, Pacific = *Acer macrophyllum* Pursh
maple, Queensland = *Flindersia brayleyana* F. v. M., *F. pimenteliana* F. v. M.
maple, rock = *Acer saccharum* Marsh (principally)
melawis = *Gonystylus warburgianus* Gilg.
meranti = *Shorea* spp. (Malaysia, Sarawak)
meranti, red = *Shorea* spp. (Malaysia, Sarawak)
meranti, white = *Shorea* spp. section *Anthoshorea* Brandis (Malaysia)
meranti, yellow = *Shorea* spp. section *Richetia*
merbau = *Intsia palembanica* Baker
mersawa = *Anisoptera* spp. (Malaysia)
mountain ash, *see* oak, Tasmanian
mujua = *Alstonia congensis* Engl.
muninga = *Pterocarpus angolensis* DC.

oak = *Quercus* spp.
oak, American red = *Quercus rubra* var. *pagodaefolia* Ashe, *Q. borealis* Michx. f., *Q. borealis* var. *maxima* Sarg., *Q. falcata* Michx., *Q. shumardii* Buckl.
oak, American white = *Quercus alba* L., *Q. montana* Willd., *Q. lyrata* Walt., *Q. prinus* L.
oak, Australian, *see* oak, Tasmanian
oak, Australian silky = *Cardwellia sublimis* F. v. M., *Grevillea robusta* A. Cunn.
oak, Austrian, *see* oak, European
oak, English, *see* oak, European
oak, European = *Quercus robur* L., *Q. petraea* Liebl.
oak, Tasmanian = *Eucalyptus gigantea* Hook. f., *E. obliqua* L'Hérit., *E. regnans* F. v. M.
oak, Turkey = *Quercus cerris* L.
obeche = *Triplochiton scleroxylon* K. Schum.
olive, East African = *Olea hochstetteri* Bak.
opepe = *Nauclea diderrichii* (De Wild et Th. Dur.) Merrill (syn. *Sarcocephalus diderichii* De Wild.)

pear = *Pyrus communis* L.
peroba = *Aspidosperma* spp.
persimmon = *Diospyros virginiana* L.

pine, Columbian, *see* Douglas fir

pine, Corsican = *Pinus nigra* var. *calabrica* Schneid.

pine, long-leaf pitch = *Pinus echinata* Mill., *P. palustris* Mill., and *P. taeda* L.

pine, maritime = *Pinus pinaster* Ait.

pine, Oregon, *see* Douglas fir

pine, pitch, *see* pine, long-leaf pitch

pine, Scots = *Pinus sylvestris* L.

pine, white, *see* pine, yellow

pine, yellow = *Pinus strobus* L.

poplar = *Populus* spp.

poplar, black = *Populus nigra* L., *P. candensis* Moench var. *serotina* Rehd., and *P. robusta* Schneid.

poplar, black Italian = *Populus serotina* (Hybrid)

poplar, Canadian = *Populus* balsamifera L. (syn. *P. tacamahaca* Mill.), *P. grandidentata* Michx.

poplar, grey = *Populus canescens* Sm.

poplar, white = *Populus alba* L.

punah = *Tetramerista glabra* Miq.

purpleheart = *Peltogyne* spp.

redwood, *see* redwood, European *and* sequoia

redwood, Baltic, *see* redwood, European

redwood, Californian, *see* sequoia

redwood, European = *Pinus sylvestris* L.

redwood, Kara, *see* redwood, European

rengas = *Melanorrhoea* spp.

resak = *Vatica*, spp. *Cotylelobium* spp. (Malaysia)

robinia = *Robinia pseudoacacia* L.

rosewood, Indian = *Dalbergia latifolia* Roxb.

sandalwood = *Santalum album* L.

sapele = *Entandrophragma cylindricum* Sprague and *Entandrophragma* spp.

satin walnut, *see* gum, American red

satinwood, East Indian = *Chloroxylon swietenia* DC.

satinwood, West Indian = *Fagara flava* Krug.

sepul = *Parishia* spp.

sequoia = *Sequoia sempervirens* Endl.

seraya = *Shorea* spp., *Parashorea* spp. (Sabah)

seraya, Borneo white = *Parashorea* spp.

seraya, white = *Shorea* spp., *Parashorea* spp. (Sabah)

sneezewood = *Ptaeroxylon obliquum* (Thunb.) Radlk.

spotted gum, *see* gum, spotted

spruce = *Picea* spp.
spruce, Canadian = *Picea glauca* Voss. (principally)
spruce, European = *Picea abies* Karst.
spruce, Norway, *see* spruce, European
spruce, Sitka = *Picea sitchensis* Carr.
swamp gum, *see* oak, Tasmanian
sycamore = *Acer pseudoplatanus* L.

tali = *Erythrophleum guineense* G. Don. and *E. ivorense* A. Chev.
tallowwood = *Eucalyptus microcorys* F. v. M.
Tasmanian oak, *see* oak, Tasmanian
teak = *Tectona grandis* L. f.
tembusu = *Fagraea gigantea* Ridl.
terentang = *Camnosperma* spp.
tulip poplar, *see* whitewood, American
tulip tree, *see* whitewood, American
turpentine = *Syncarpia laurifolia* Ten.

walnut = *Juglans* spp.
walnut, African = *Lovoa trichilioides* Harms (syn. *L. klaineana* Pierre
    ex Sprague)
walnut, Australian, *see* walnut, Queensland
walnut, Nigerian, *see* walnut, African
walnut, Queensland = *Endiandra palmerstonii* C. T. White
whitewood = *Picea abies* Karst. and *Abies alba* Mill.
whitewood, American = *Liriodendron tulipifera* L.
whitewood, canary, *see* whitewood, American
willow = *Salix* spp.
willow, crack = *Salix fragilis* L.
willow, cricket bat = *Salix alba* var. *coerulea* Sm.
willow, white = *Salix alba* L., *S. viridis* Fr.

yang = *Dipterocarpus* spp. (Thailand)
yellow poplar, *see* whitewood, American

List of the Commoner Hardwood Tree Genera with the Families to which they belong

Where possible the families given are for the most part those recognized by Hutchinson

Acacia – Leguminosae
Acalypha – Euphorbiaceae
Acanthopanax – Araliaceae
Acer – Aceraceae
Achras – Sapotaceae
Acioa – Rosaceae
Ackama – Cunoniaceae
Acradenia – Rutaceae
Acrocarpus – Leguminosae
Acrodiclidium – Lauraceae
Acronychia – Rutaceae
Actinodaphne – Lauraceae
Adenanthera – Leguminosae
Adenodolichos – Leguminosae
Adina – Rubiaceae
Adinandra – Theaceae
Adinobotrys – Millettia
Adiscanthus – Rutaceae
Aegiceras – Myrsinaceae
Aegle – Rutaceae
Aesculus – Sapindaceae or
   Hippocastanaceae
Aextoxicum – Euphorbiaceae
Afraegle – Rutaceae
Afrolicania – Rosaceae
Afrormosia, see Pericopsis
Afzelia – Leguminosae

Agauria – Ericaceae
Agelaea – Connaraceae
Aglaia – Meliaceae
Agonandra – Opiliaceae
Agrostistachys – Euphorbiaceae
Ailanthus – Simaroubaceae
Alangium – Alangiaceae
Albizzia – Leguminosae
Alchornea – Euphorbiaceae
Alchorneopsis – Euphorbiaceae
Aleurites – Euphorbiaceae
Alfaroa – Juglandaceae
Allantoma – Lecythidaceae
Allophylus – Sapindaceae
Alnus – Betulaceae
Alphitonia – Rhamnaceae
Alphonsea – Annonaceae
Alseodaphne – Lauraceae
Alstonia – Apocynaceae
Altingia – Hamamelidaceae
Amanoa – Euphorbiaceae
Amblygonocarpus – Leguminosae
Amoora – Meliaceae
Ampelozizyphus – Rhamnaceae
Amphimas – Leguminosae
Amyris – Burseraceae
Anacardium – Anacardiaceae

Anacolosa – Olacaceae
Anaphalis. – Compositae
Andira – Leguminosae
Aneulophus – Erythroxylaceae
Angelesia – Rosaceae
Angophora – Myrtaceae
Angylocalyx – Leguminosae
Aniba – Lauraceae
Anisophyllea – Rhizophoraceae
Anisoptera – Dipterocarpaceae
Anneslea – Theaceae
Annona – Annonaceae
Anodopetalum – Cunoniaceae
Anogeissus – Combretaceae
Anona,[1] see Annona
Anonidium – Annonaceae
Anopyxis – Rhizophoraceae
Anthocephalus – Rubiaceae
Anthocleista – Loganiaceae
Anthostema – Euphorbiaceae
Antiaris – Moraceae
Antidesma – Euphorbiaceae
Antrocaryon – Anacardiaceae
Apeiba – Tiliaceae
Aphanamixis – Meliaceae
Aphananthe – Ulmaceae
Aphania – Sapindaceae
Apodytes – Icacinaceae
Aporosa – Euphorbiaceae
Aporosella – Euphorbiaceae
Aporrhiza – Sapindaceae
Apuleia – Leguminosae
Aquilaria – Thymelaeaceae
Aralia – Araliaceae
Arbutus – Ericaceae
Archytaea – Theaceae
Ardisia – Myrsinaceae
Aromadendron, see Talauma
Artabotrys – Annonaceae
Arthrophyllum – Araliaceae
Artocarpus – Moraceae
Arytera – Sapindaceae

Aspidosperma – Apocynaceae
Asteropeia – Theaceae or
    Flacourtiaceae
Astronium – Anacardiaceae
Atherosperma – Monimiaceae
Aucoumea – Burseraceae
Aucuba – Cornaceae
Aulacocalyx – Rubiaceae
Auxemma – Boraginaceae
Averrhoa – Oxalidaceae
Avicennia – Verbenaceae
Axinandra – Lythraceae
Azadirachta – Meliaceae
Azara – Flacourtiaceae

Baccaurea – Euphorbiaceae
Backhousia – Myrtaceae
Bagassa – Moraceae
Baikiaea – Leguminosae
Balanites – Simaroubaceae
Balanocarpus – Dipterocarpa-
    ceae
Balfourodendron – Rutaceae
Baloghia – Euphorbiaceae
Banara – Flacourtiaceae
Banksia – Proteaceae
Baphia – Leguminosae
Barringtonia – Lecythidaceae
Barteria – Passifloraceae
Barylucuma – Sapotaceae
Bassia – Sapotaceae
Bauhinia – Leguminosae
Bedfordia – Compositae
Beilschmiedia – Lauraceae
Belangera – Cunoniaceae
Belencita – Capparidaceae
Bellota – Lauraceae
Bellucia – Melastomaceae
Bennettia – Flacourtiaceae
Bergsmia – Flacourtiaceae
Berlinia – Leguminosae
Bernouillia – Bombacaceae

---

[1] Anona and hence Anonaceae is preferred by some botanists to Annona and Annonaceae.

Berria, see Berrya
Berrya – Tiliaceae
Bersama – Melianthaceae
Bertholletia – Lecythidaceae
Betula – Betulaceae
Bischofia – Euphorbiaceae
Bixa – Bixaceae
Blepharocarya – Anacardiaceae
Blighia – Sapindaceae
Blumeodendron – Euphorbiaceae
Bombacopsis – Bombacaceae
Bombax – Bombacaceae
Boschia – Bombacaceae
Boscia – Capparidaceae
Bosquiea – Moraceae
Boswellia – Burseraceae
Bouea – Anacardiaceae
Bougainvillaea, see Buginvillaea
Bowdichia – Leguminosae
Brachylaena – Compositae
Brachystegia – Leguminosae
Bravaisia – Acanthaceae
Bridelia – Euphorbiaceae
Brieya – Annonaceae
Brosimopsis – Moraceae
Brosimum – Moraceae
Broussonetia – Moraceae
Bruguiera – Rhizophoraceae
Bruinsimia – Styracaceae
Brya – Leguminosae
Buchanania – Anacardiaceae
Buchenavia – Combretaceae
Buchholzia – Capparidaceae
Bucida – Combretaceae
Bucklandia – Hamamelidaceae
Buginvillaea – Nyctaginaceae
Bulnesia – Zygophyllaceae
Bumelia – Sapotaceae
Burkea – Leguminosae
Bursera – Burseraceae
Bussea – Leguminosae
Butryospermum – Sapotaceae
Buxus – Buxaceae
Byrsonima – Malpighiaceae

Cabralea – Meliaceae
Cadaba – Capparidaceae
Caesalpinia – Leguminosae
Calatola – Icacinaceae
Caldcluvia – Cunoniaceae
Calderonia – Rubiaceae
Callicoma – Cunoniaceae
Callistermon – Myrtaceae
Callisthene – Vochysiaceae
Calocarpum – Sapotaceae
Caloncoba – Flacourtiaceae
Calophyllum – Guttiferae
Calpocalyx – Leguminosae
Calycogonium – Melastomaceae
Calycophyllum – Rubiaceae
Campnosperma – Anacardiaceae
Canangium – Annonaceae
Canarium – Burseraceae
Canella – Canellaceae
Canthium – Rubiaceae
Cantleya – Icacinaceae
Capparis – Capparidaceae
Caraipa – Guttiferae or Theaceae
Carallia – Rhizophoraceae
Carapa — Meliaceae
Cardwellia – Proteaceae
Careya – Lecythidaceae
Cariniana – Lecythidaceae
Carnarvonia – Proteaceae
Carpinus – Betulaceae
Carpolobia – Polygalaceae
Carpotroche – Flacourtiaceae
Carya – Juglandaceae
Caryocar – Caryocaraceae
Casearia, see Gossypiospermum
Cassia – Leguminosae
Cassine – Celastraceae
Cassipourea – Rhizophoraceae
Castanea – Fagaceae
Castanopsis – Fagaceae
Castanospermum – Leguminosae
Castanospora – Sapindaceae
Castilla – Moraceae
Casuarina – Casuarinaceae

Catalpa – Bignoniaceae
Cathormion – Leguminosae
Cavanillesia – Bombacaceae
Ceanothus, see Ziziphus
Cecropia – Moraceae
Cedrela – Meliaceae
Ceiba – Bombacaceae
Celastrus – Celastraceae
Celtis – Ulmaceae
Centrolobium – Leguminosae
Cephalosphaera – Myristicaceae
Ceratopetalum – Cunoniaceae
Cerbera – Apocynaceae
Cereus – Cactaceae
Ceriops – Rhizophoraceae
Cespedesia – Ochnaceae
Chaetachne – Ulmaceae
Chaetocarpus – Euphorbiaceae
Champereia – Opiliaceae
Chaunochiton – Olacaceae
Cheilosa – Euphorbiaceae
Cheirodendron – Araliaceae
Chenolea – Chenopodiaceae
Chickrassia, see Chukrasia
Chidlowia – Leguminosae
Chilopsis – Bignoniaceae
Chionanthus, see Linociera
Chisocheton – Meliaceae
Chlorophora – Moraceae
Chloroxylon – Rutaceae or
    Meliaceae
Chorisia – Bombacaceae
Chromolucuma – Sapotaceae
Chrysobalanus – Rosaceae
Chrysophyllum – Sapotaceae
Chukrasia – Meliaceae
Chydenanthus – Lecythidaceae
Chytranthus – Sapindaceae
Chytroma – Lecythidaceae
Cinnamodendron – Canellaceae
Cinnamomum – Lauraceae
Cinnamosma – Canellaceae
Cistanthera – Tiliaceae
Citharexylum – Verbenaceae

Citrus – Rutaceae
Clarisia – Moraceae
Clausena – Rutaceae
Cleidion – Euphorbiaceae
Cleistanthus – Euphorbiaceae
Cleistopholis – Annonaceae
Clusia – Guttiferae
Coccoceras – Euphorbiaceae
Coccoloba – Polygonaceae
Coccolobis, see Coccoloba
Cochlospermum – Bixaceae
Coelocaryon – Myristicaceae
Coelostegia – Bombacaceae
Coffea – Rubiaceae
Cola – Sterculiaceae
Colletia – Rhamnaceae
Colubrina – Rhamnaceae
Columbia – Tiliaceae
Combretocarpus – Rhizophora-
    ceae
Combretodendron – Lecythi-
    daceae or Combretaceae
Combretum – Combretaceae
Commersonia – Sterculiaceae
Commiphora – Burseraceae
Comocladia – Anacardiaceae
Compsoneura – Myristicaceae
Condalia – Rhamnaceae
Connaropsis – Oxalidaceae
Conocarpus – Combretaceae
Conomorpha – Myrsinaceae
Conopharyngia – Apocynaceae
Copaifera – Leguminosae
Cordia – Boraginaceae
Cordyla – Leguminosae
Cornus – Cornaceae
Corylus – Corylaceae or Betula-
    ceae
Corynanthe – Rubiaceae
Cotinus – Anacardiaceae
Cotylelobium – Dipterocarpaceae
Couepia – Rosaceae
Coula – Olacaceae
Couma – Apocynaceae

Couratari – Lecythidaceae
Couroupita – Lecythidaceae
Coussapoa – Moraceae
Crataegus – Rosaceae
Crataeva – Capparidaceae
Cratoxylon – Guttiferae
Crescentia – Bignoniaceae
Crossopteryx – Rubiaceae
Crossostylis – Rhizophoraceae
Croton – Euphorbiaceae
Crudia – Leguminosae
Crypteronia – Crypteroniaceae
Cryptocarya – Lauraceae
Ctenolophon – Linaceae or Ola-
  caceae
Cudrania – Moraceae
Cunonia – Cunoniaceae
Curatella – Dilleniaceae
Curtisia – Cornaceae
Cussonia – Araliaceae
Cyathocalyx – Annonaceae
Cybianthus – Myrsinaceae
Cyclostemon – Euphorbiaceae
Cylicodiscus – Leguminosae
Cynometra – Leguminosae

Dacryodes – Burseraceae
Dalbergia – Leguminosae
Dalbergiella – Leguminosae
Daniella – Leguminosae
Daphnandra – Monimiaceae
Daphniphyllum – Euphorbiaceae
Daphnopsis – Thymelaeaceae
Dehaasia – Lauraceae
Deinbollia – Sapindaceae
Dennettia – Annonaceae
Deplanchea – Bignoniaceae
Derris – Leguminosae
Desbordesia – Simaroubaceae
Desmostachys – Icacinaceae
Desplatzia – Tiliaceae
Detarium – Leguminosae
Dialium – Leguminosae
Dialyanthera – Myristicaceae

Dichrostachys – Leguminosae
Dicorynia – Leguminosae
Dicranolepis – Thymelaeaceae
Dictyandra – Rubiaceae
Didelotia – Leguminosae
Didymopanax – Araliaceae
Dillenia – Dilleniaceae
Dilodendron – Sapindaceae
Dimorphandra – Leguminosae
Dimorphocalyx – Euphorbiaceae
Diospyros – Ebenaceae
Diphysa – Leguminosae
Diplodiscus – Tiliaceae
Diploglottis – Sapindaceae
Diplospora – Rubiaceae
Diplotropis – Leguminosae
Dipterocarpus – Dipterocar-
  paceae
Dipteryx – Leguminosae
Dirca – Thymelaeaceae
Discaria – Rhamnaceae
Discoglypremna – Euphorbia-
  ceae
Discophora – Icacinaceae
Dissomeria – Flacourtiaceae or
  Samydaceae
Distemonanthus – Leguminosae
Dodonaea – Sapindaceae
Doerpfeldia – Rhamnaceae
Dolichandrone – Bignoniaceae
Dombeya – Sterculiaceae
Doona – Dipterocarpaceae
Dorpyhora – Monimiaceae
Dracontomelum – Anacardiaceae
Drepananthus – Annonaceae
Drimycarpus – Anacardiaceae
Drimys – Magnoliaceae
Dryobalanops – Dipterocar-
  paceae
Drypetes – Euphorbiaceae
Duabanga – Sonneratiaceae
Duboscia – Tiliaceae
Duguetia – Annonaceae
Durio – Bombacaceae

Dyera – Apocynaceae
Dysoxylum – Meliaceae

Echinocarpus – Elaeocarpaceae
Echiochilon – Boraginaceae
Echirospermum – Leguminosae
Ehretia – Boraginaceae
Ekebergia – Meliaceae
Elaeocarpus – Elaeocarpaceae
Elaeodendron – Celastraceae
Elaeophorbia – Euphorbiaceae
Elateriospermum – Euphorbia-
　ceae
Emblica – Euphorbiaceae
Embothrium – Proteaceae
Enantia – Annonaceae
Endiandra – Lauraceae
Endospermum – Euphorbiaceae
Engelhardtia – Juglandaceae
Enicosanthum – Annonaceae
Entada – Leguminosae
Entandrophragma – Meliaceae
Enterolobium – Leguminosae
Eperua – Leguminosae
Eremophila – Myoporaceae
Eriobotrya – Rosaceae
Eriocoelum – Sapindaceae
Eriodendron, see Ceiba
Erioglossum – Sapindaceae
Erisma – Vochysiaceae
Ervatamia – Apocynaceae
Erythrina – Leguminosae
Erythrophloeum – Leguminosae
Erythropsis – Sterculiaceae
Erythroxylon, see Erythroxylum
Erythroxylum – Erythroxylaceae
Eschweilera – Lecythidaceae
Esenbeckia – Rutaceae
Eucalyptus – Myrtaceae
Eucryphia – Eucryphiaceae or
　Rosaceae
Eugenia – Myrtaceae
Euonymus – Celastraceae
Eupatorium – Compositae

Euphorbia – Euphorbiaceae
Euroschinus – Anacardiaceae
Eurya – Theaceae
Eusideroxylon – Lauraceae
Euxylophora – Rutaceae
Evodia – Rutaceae
Evonymus, see Euonymus
Exandra – Rubiaceae
Excoecaria – Euphorbiaceae
Exocarpus – Santalaceae
Eysenhardtia – Leguminosae

Fagara – Rutaceae
Fagraea – Loganiaceae
Fagus – Fagaceae
Faurea – Proteaceae
Fegimanra – Anacardiaceae
Ferolia – Rosaceae
Ferreirea – Leguminosae
Ficus – Moraceae
Fillaeopsis – Leguminosae
Firmiana – Sterculiaceae
Flacourtia – Flacourtiaceae
Flindersia – Rutaceae or Melia-
　ceae
Fluggea – Euphorbiaceae
Foetidia – Lecythidaceae
Fontanesia – Oleaceae
Forsythia – Oleaceae
Fraxinus – Oleaceae
Funifera – Thymelaeaceae
Funtumia – Apocynaceae
Fusanus – Santalaceae

Gaertnera – Loganiaceae
Gallesia – Phytolaccaceae
Ganophyllum – Sapindaceae
Ganua – Sapotaceae
Garcinia – Guttiferae
Gardenia – Rubiaceae
Garuga – Burseraceae
Geasonia – Rubiaceae
Geijera – Rutaceae
Geissois – Cunoniaceae

Genipa – Rubiaceae
Gilibertia – Araliaceae
Gleditschia, see Gleditsia
Gleditsia – Leguminosae
Glochidion – Euphorbiaceae
Gluema – Sapotaceae
Gluta – Anacardiaceae
Glycosmis – Rutaceae
Glycoxylon – Sapotaceae
Glyphaea – Tiliaceae
Gmelina – Verbenaceae
Gochnatia – Compositae
Goeldinia – Lecythidaceae
Gomphia, see Ouratea
Gonioma – Apocynaceae
Gonystylus – Gonystylaceae
Gordonia – Theaceae
Gossweilerodendron – Legu-
   minosae
Gossypiospermum – Flacourtia-
   ceae
Goupia – Goupiaceae
Grevillea – Proteaceae
Grewia – Tiliaceae
Grias – Lecythidaceae
Guaiacum – Zygophyllaceae
Guarea – Meliaceae
Guatteria – Annonaceae
Guazuma – Sterculiaceae
Guettarda – Rubiaceae
Guevina – Proteaceae
Guiera – Combretaceae
Guioa – Sapindaceae
Gustavia – Lecythidaceae
Gyminda – Celastraceae
Gymnacranthera – Myristica-
   ceae
Gymnanthes – Euphorbiaceae
Gymnocladus – Leguminosae
Gymnosporia – Celastraceae
Gynocardia – Flacourtiaceae
Gynotroches – Rhizophoraceae
Gyranthera – Bombacaceae
Gyrocarpus – Hernandiaceae

Haematoxylon – Leguminosae
Hakea – Proteaceae
Halfordia – Rutaceae
Haloxylon – Chenopodiaceae
Hamamelis – Hamamelidaceae
Hampea – Bombacaceae
Hancornia – Apocynaceae
Hannoa – Simaroubaceae
Haploclathra – Guttiferae
Haplormosia – Leguminosae
Hardwickia – Leguminosae
Harmandia – Olacaceae
Harpullia – Sapindaceae
Harungana – Hypericaceae
Hasseltia – Flacourtiaceae
Hedera – Araliaceae
Hedwigia – Burseraceae
Hedycarya – Monimiaceae
Heeria – Anacardiaceae
Heinsia – Rubiaceae
Heisteria – Olacaceae
Helicia – Proteaceae
Helicostylis – Moraceae
Helietta – Rutaceae
Heliocarpus – Tiliaceae
Heliotrium – Boraginaceae
Hennecartia – Monimiaceae
Henoonia – Sapotaceae
Henriettella – Melastomaceae
Heritiera – Sterculiaceae
Hernandia – Hernandiaceae
Heterodendron – Sapindaceae
Heterophragma – Bignoniaceae
Heterotrichum – Melastomaceae
Hevea – Euphorbiaceae
Hexalobus – Annonaceae
Hibiscus – Malvaceae
Hicoria, see Carya
Hieronymia – Euphorbiaceae
Hippomane – Euphorbiaceae
Hirtella – Rosaceae
Holarrhena – Apocynaceae
Holocalyx – Leguminosae
Holoptelea – Ulmaceae

Homalium – Samydaceae or
    Flacourtiaceae
Hopea – Dipterocarpaceae
Horsfieldia – Myristicaceae
Humiria – Humiriaceae
Hunteria – Apocynaceae
Hura – Euphorbiaceae
Hydnocarpus – Flacourtiaceae
Hymenaea – Leguminosae
Hymenocardia – Euphorbiaceae
Hymenodictyon – Rubiaceae
Hymenolobium – Leguminosae
Hymenostegia – Leguminosae
Hypodaphnis – Lauraceae

Ichthyomethia – Leguminosae
Ilex – Aquifoliaceae
Illicium – Winteraceae
Inga – Leguminosae
Intsia – Leguminosae
Irvingia – Simaroubaceae
Iryanthera – Myristicaceae
Isoberlinia – Leguminosae
Isolona – Annonaceae
Itea – Escalloniaceae
Iteadaphne – Lauraceae
Ixonanthes – Erythroxylaceae
Ixora – Rubiaceae

Jacaranda – Bignoniaceae
Jackia – Rubiaceae
Jacquinia – Myrsinaceae
Jasminum – Oleaceae
Jatropha – Euphorbiaceae
Joanesia – Euphorbiaceae
Jugastrum – Lecythidaceae
Juglans – Juglandaceae

Kandelia – Rhizophoraceae
Karwinski – Rhamnaceae
Kayea – Guttiferae
Khaya – Meliaceae
Kibara – Monimiaceae

2 B

Kibatalia, see Kickxia
Kibessia – Melastomaceae
Kickxia – Apocynaceae
Kigelia – Bignoniaceae
Kiggelaria – Flacourtiaceae
Killmevera – Guttiferae
Klainedoxa – Simaroubaceae
Kleinhovia – Sterculiaceae
Knema – Myristicaceae
Koodersiodendron – Anacardia-
    ceae
Koompassia – Leguminosae
Krugiodendron – Rhamnaceae
Kunstlerodendron – Euphorbia-
    ceae
Kurrimia – Celastraceae

Labatia – Sapotaceae
Labourdonnaisia – Sapotaceae
Laburnum – Leguminosae
Lachnopylis – Loganiaceae
Laetia – Flacourtiaceae
Lagerstroemia – Lythraceae
Lagetta – Thymelaeaceae
Laguncularia – Combretaceae
Lannea – Anacardiaceae
Lansium – Meliaceae
Laportea – Urticaceae
Lasiodiscus – Rhamnaceae
Laurelia – Monimiaceae
Lecaniodiscus – Sapindaceae
Lecythis – Lecythidaceae
Leea – Ampelidaceae
Lepisanthes – Sapindaceae
Leptactina – Rubiaceae
Leptanlus – Icacinaceae
Leptospermum – Myrtaceae
Leucadendron – Proteaceae
Licania – Rosaceae
Licaria – Lauraceae
Lichnophora – Compositae
Ligustrum – Oleaceae
Lindackeria – Flacourtiaceae
Lindera – Lauraceae

Linociera – Oleaceae
Liquidambar – Hamamelidaceae
Liriodendron – Magnoliaceae
Liriosma – Olacaceae
Lithospermum – Boraginaceae
Litsea – Lauraceae
Loesnera – Leguminosae
Lomatia – Proteaceae
Lonchocarpus – Leguminosae
Lophira – Ochnaceae
Lophopetalum – Celastraceae
Lovoa – Meliaceae
Loxopterygium – Anacardiaceae
Lucuma – Sapotaceae
Luehea – Tiliaceae
Lumnitzera – Combretaceae
Lunania – Flacourtiaceae
Lunasia – Rutaceae
Lysicarpus – Myrtaceae
Lysiloma – Leguminosae

Maba – Ebenaceae
Mabea – Euphorbiaceae
Macadamia – Proteaceae
Macaranga – Euphorbiaceae
Macarisia – Rhizophoraceae
Machaerium – Leguminosae
Machilus – Lauraceae
Maclura – Moraceae
Macrodendron – Cunoniaceae
Macrolobium – Leguminosae
Madhuca – Sapotaceae
Maerua – Capparidaceae
Maesa – Myrsinaceae
Maesobotrya – Euphorbiaceae
Maesopsis – Rhamnaceae
Magnolia – Magnoliaceae
Magonia – Sapindaceae
Maingaya – Hamamelidaceae
Malacantha – Sapotaceae
Mallotus – Euphorbiaceae
Malus – Rosaceae
Mammea – Guttiferae

Mangifera – Anacardiaceae
Manihot – Euphorbiaceae
Manilkara – Sapotaceae
Mannia – Simaroubaceae
Mansonia – Sterculiaceae
Mappa – Icacinaceae
Mareya – Euphorbiaceae
Markhamia – Bignoniaceae
Marquesia – Dipterocarpaceae
Mastixia – Cornaceae
Matisia – Bombacaceae
Mayna – Flacourtiaceae
Maytenus – Celastraceae
Melaleuca – Myrtaceae
Melanochyla – Anacardiaceae
Melanorrhoea – Anacardiaceae
Melanoxylon – Leguminosae
Melia – Meliaceae
Melicocca – Sapindaceae
Meliosma – Sabiaceae
Melochia – Sterculiaceae
Memecylon – Melastomaceae
Merrillia – Rutaceae
Mesua – Guttiferae
Metopium – Anacardiaceae
Mezzettia – Annonaceae
Michelia – Magnoliaceae
Miconia – Melastomaceae
Microcos – Tiliaceae
Microdesmis – Euphorbiaceae
Micromelum – Rutaceae
Micropholis – Sapotaceae
Miliusa – Annonaceae
Millettia – Leguminosae
Mimosa – Leguminosae
Mimusops, see Tieghemella
Minquartia – Olacaceae
Misanteca – Lauraceae
Mischocarpus – Sapindaceae
Mitragyna – Rubiaceae
Mitrephora – Annonaceae
Mollinedia – Monimiaceae
Monimia – Monimiaceae
Monodora – Annonaceae

Monopetalanthus – Bignoniaceae
Monotes – Dipterocarpaceae
Montanoa – Compositae
Moquinia – Compositae
Mora – Leguminosae
Morelia – Rubiaceae
Morinda – Rubiaceae
Moringa – Moringaceae
Morisonia – Capparidaceae
Moronobea – Guttiferae
Morquilea – Rosaceae
Morus – Moraceae
Mosquitoxyllum – Anacardiaceae
Mouriria – Melastomaceae
Murraya – Rutaceae
Musanga – Moraceae
Musgravea – Proteaceae
Mussaendopsis – Rubiaceae
Myoporum – Myoporaceae
Myrcia – Myrtaceae
Myrianthus – Moraceae
Myristica – Myristicaceae
Myrocarpus – Leguminosae
Myrospermum – Leguminosae
Myroxylon – Leguminosae
Myrsine – Myrsinaceae

Napeodendron – Sapindaceae
Napoleona – Lecythidaceae
Nauclea – Rubiaceae
Necespia – Euphorbiaceae
Nectandra, see Ocotea
Neesia – Bombacaceae
Neonauclea – Rubiaceae
Neoscortechinia – Euphorbiaceae
Neostenanthera – Annonaceae
Nephelium – Sapindaceae
Newbouldia – Bignoniaceae
Niebuhria – Capparidaceae
Norrisia – Loganiaceae
Nothofagus – Fagaceae
Nothophoebe – Lauraceae
Nuxia – Loganiaceae

Nuxia (most spp.), see Lachnopylis
Nyssa – Nyssaceae

Ochanostachys – Olacaceae
Ochna – Ochnaceae
Ochrocarpus – Guttiferae
Ochroma – Bombacaceae
Ocotea – Lauraceae
Octoknema – Octoknemataceae
Octomeles – Datiscaceae
Odina, see Lannea
Olax – Olacaceae
Oldfieldia – Euphorbiaceae
Olea – Oleaceae
Olearia – Compositae
Olinia – Oliniaceae
Olmedia – Moraceae
Omphalocarpum – Sapotaceae
Oncoba – Flacourtiaceae
Ongokea – Olacaceae
Oreodaphne – Lauraceae
Orites – Proteaceae
Ormosia – Leguminosae
Osmanthus – Oleaceae
Ostodes – Euphorbiaceae
Ostrya – Betulaceae or Corylaceae
Ostryoderris – Leguminosae
Otophora – Sapindaceae
Ougeinia – Leguminosae
Ouratea – Ochnaceae
Ovidia – Thymelaeaceae
Owenia – Meliaceae
Oxandra – Annonaceae
Oxystigma – Leguminosae
Oxytheca – Sapotaceae

Pachira – Bombacaceae
Pachyelsma – Leguminosae
Pachylobus – Burseraceae
Pachypodanthium – Annonaceae
Pachystela – Sapotaceae
Pajanelia – Bignoniaceae
Palaquium – Sapotaceae

Panax, see Tieghemopanax
Pancheria – Cunoniaceae
Panda – Pandaceae
Pangium – Flacourtiaceae
Paralabatia – Sapotaceae
Paramignya – Rutaceae
Parashorea – Dipterocarpaceae
Parastemon – Rosaceae
Paratecoma – Bignoniaceae
Paravallaris – Apocynaceae
Parinari – Rosaceae
Parinarium, see Parinari
Parishia – Anacardiaceae
Parkia – Leguminosae
Paropsia – Passifloraceae
Pasania – Fagaceae
Patagonula – Boraginaceae
Patrisia – Flacourtiaceae
Paulownia – Scrophulariaceae
Pausinystalia – Rubiaceae
Payena – Sapotaceae
Pellacalyx – Rhizophoraceae
Peltogyne – Leguminosae
Peltophorum – Leguminosae
Pentace – Tiliaceae
Pentaclethra – Leguminosae
Pentacme – Dipterocarpaceae
Pentadesma – Guttiferae
Pentapanax – Araliaceae
Pentaphylax – Theaceae
Pentaspadon – Anacardiaceae
Pera – Euphorbiaceae
Perebea – Moraceae
Pereskia – Cactaceae
Persea – Lauraceae
Pericopsis – Leguminosae
Perymenium – Compositae
Petitia – Verbenaceae
Peumus – Monimiaceae
Phebalium – Rutaceae
Phialodiscus – Sapindaceae
Phoebe – Lauraceae
Photinia – Rosaceae
Phyllanthus – Euphorbiaceae

Phyllostylon – Ulmaceae
Physocalymma – Lythraceae
Phytolacca – Phytolaccaceae
Picraena – Simaroubaceae
Picralima – Apocynaceae
Picrasma – Simaroubaceae
Pilocarpus – Rutaceae
Pimeleodendron – Euphorbia-
    ceae
Pimenta – Myrtaceae
Pinckneya – Rubiaceae
Piptadenia, see Piptadeniastrum
Piptadeniastrum – Lequminosae
Piptocarpha – Compositae
Piranhea – Euphorbiaceae
Piratinera, see Brosimum
Piscidia, see Ichthyomethia
Pisonia – Nyctaginaceae
Pistacia – Anacardiaceae
Pithecellobium – Leguminosae
Pithecolobium, see Pithecellobium
Pittosporum – Pittosporaceae
Placodiscus – Sapindaceae
Planchonella – Sapotaceae
Planchoria – Lecythidaceae
Planera – Ulmaceae
Platanus – Platanaceae
Plathymenia – Leguminosae
Platonia – Leguminosae
Platycarya – Juglandaceae
Platycyamus – Leguminosae
Platylophus – Cunoniaceae
Platymiscium – Leguminosae
Platymitra – Annonaceae
Platypodium – Leguminosae
Pleiocarpa – Apocynaceae
Pleiococca – Rutaceae
Pleiogynium – Anacardiaceae
Pleodendron – Canellaceae
Pleurostylia – Celastraceae
Plumeria – Apocynaceae
Poeciloneuron – Guttiferae
Poga – Rhizophoraceae
Polyalthia – Annonaceae

Polygala – Polygalaceae
Polyosma – Escalloniaceae
Polyscias – Araliaceae
Pometia – Sapindaceae
Pongamia – Leguminosae
Popowia – Annonaceae
Populus – Salicaceae
Poraqueiba – Icacinaceae
Porlieria – Zygophyllaceae
Pouteria – Sapotaceae
Pradosia – Sapotaceae
Prainea – Moraceae
Premna – Verbenaceae
Prioria – Leguminosae
Prockia – Flacourtiaceae
Prosopis – Leguminosae
Protca – Proteaceae
Protium – Burseraceae
Protomegabaria – Euphorbiaceae
Prunus – Rosaceae
Pseudocedrela – Meliaceae
Pseudomorus – Moraceae
Pseudosamanea – Leguminosae
Pseudospondias – Anacardiaceae
Psidium – Myrtaceae
Ptaeroxylon – Meliaceae
Pteleocarpa – Cardiopteridaceae
Pternandra – Melastomaceae
Pterocarpus – Leguminosae
Pterocymbium – Sterculiaceae
Pterogyne – Leguminosae
Pterospermum – Sterculiaceae
Pterygota – Sterculiaceae
Ptychopetalum – Olacaceae
Ptychopyxis – Euphorbiaceae
Punica – Punicaceae
Pycnanthus – Myristicaceae
Pygeum – Rosaceae
Pyrenaria – Theaceae
Pyrus – Rosaceae

Qualea – Vochysiaceae
Quararibea – Bombacaceae
Quassia – Simaroubaceae

Quebrachia, see Schinopsis
Quercus – Fagaceae
Quillaja – Rosaceae
Quintinia – Saxifragaceae

Randia – Rubiaceae
Rapanea – Myrsinaceae
Rauwolfia – Apocynaceae
Reevesia – Sterculiaceae
Reynosia – Rhamnaceae
Rhamnidium – Rhamnaceae
Rhamnus – Rhamnaceae
Rheedia – Guttiferae
Rhizophora – Rhizophoraceae
Rhodamnia – Myrtaceae
Rhodoleia – Hamamelidaceae
Rhodosphaera – Anacardiaceae
Rhus – Anacardiaceae
Ricinodendron – Euphorbiaceae
Ricinus – Euphorbiaceae
Rinorea – Violaceae
Robinia – Leguminosae
Rollinia – Annonaceae
Roupala – Proteaceae
Ruprechtia – Polygonaceae
Ryparosa – Flacourtiaceae

Sabia – Sabiaceae
Saccopetalum – Annonaceae
Sacoglottis – Humiriaceae
Sacrosperma – Sapotaceae
Sageraea – Annonaceae
Sageretia – Rhamnaceae
Salacia – Hippocrateaceae
Salix – Salicaceae
Salvadora – Salvadoraceae
Salvertia – Vochysiaceae
Sambucus – Caprifoliaceae
Sandoricum – Meliaceae
Santalum – Santalaceae
Santiria – Burseraceae
Sapindus – Sapindaceae
Sapium – Euphorbiaceae
Saraca – Leguminosae

Sarcocephalus, see Nauclea
Sarcomphalus – Rhamnaceae
Sassafras – Lauraceae
Saurauia – Saurauiaceae
Scalasia – Compositae
Scaphium – Sterculiaceae
Scaphopetalum – Sterculiaceae
Schaefferia – Celastraceae
Schefflera – Araliaceae
Schima – Theaceae
Schinopsis – Anacardiaceae
Schinus – Anacardiaceae
Schizolobium – Leguminosae
Schizomeria – Cunoniaceae
Schleichera – Sapindaceae
Schoepfia – Olacaceae
Schotia – Leguminosae
Schoutenia – Tiliaceae
Schrebera – Oleaceae
Schumanniophyton – Rubiaceae
Sciadodendron – Araliaceae
Sclerocarya – Anacardiaceae
Scolopia – Flacourtiaceae
Scorodocarpus – Olacaceae
Scottellia – Flacourtiaceae
Scutinanthe – Burseraceae
Scyphiphora – Rubiaceae
Scytopetalum – Scytopetalaceae
Sebastiana – Euphorbiaceae
Securidaca – Polygalaceae
Semecarpus – Anacardiaceae
Sersalisia – Sapotaceae
Shorea – Dipterocarpaceae
Sickingia – Rubiaceae
Sideroxylon – Sapotaceae
Silvia – Lauraceae
Simarouba – Simaroubaceae
Simaruba, see Simarouba
Sindora – Leguminosae
Siparuna – Monimiaceae
Siphonodon – Celastraceae
Skimmia – Rutaceae
Sloanea – Elaeocarpaceae
Sloetia – Moraceae

Smeathmannia – Passifloraceae
Sonneratia – Sonneratiaceae
Soyauxia – Passifloraceae
Spathodea – Bignoniaceae
Spiraeopsis – Cunoniaceae
Spirostachys – Euphorbiaceae
Spondianthus – Anacardiaceae
Spondias – Anacardiaceae
Staudtia – Myristicaceae
Stemonocoleus – Leguminosae
Stemonurus – Icacinaceae
Stenocarpus – Proteaceae
Sterculia – Sterculiaceae
Stereospermum – Bignoniaceae
Sterigmapetalum – Rhizophoraceae
Stewartia – Theaceae
Strephonema – Combretaceae
Strombosia – Olacaceae
Strychnos – Loganiaceae
Stylocoryna – Rubiaceae
Styrax – Styracaceae
Suriana – Surianaceae or Simaroubaceae
Swartzia – Leguminosae
Sweetia – Leguminosae
Swietenia – Meliaceae
Swintonia – Anacardiaceae
Sycopsis – Hamamelidaceae
Sygyiopsis – Sapotaceae
Symphonia – Guttiferae
Symplocos – Symplocaceae
Syncarpia – Myrtaceae
Synoum – Meliaceae
Synsepalum – Sapotaceae
Syringa – Oleaceae
Syzygium – Myrtaceae

Tabebuia – Bignoniaceae
Tabernaemontana – Apocynaceae
Talauma – Magnoliaceae
Talbotiella – Leguminosae
Tamarindus – Leguminosae

Tamarix – Tamaricaceae
Tapiria – Anacardiaceae
Taraktogenos, see Hydnocarpus
Tarrietia – Sterculiaceae
Taxotrophis – Moraceae
Teclea – Rutaceae
Tecoma – Bignoniaceae
Tectona – Verbenaceae
Terminalia – Combretaceae
Ternstroemia – Theaceae
Tessaria – Compositae
Tetracentron – Magnoliaceae
Tetragastris – Burseraceae
Tetrameles – Datiscaceae
Tetramerista – Marcgraviaceae
Tetrapleura – Leguminosae
Theobroma – Sterculiaceae
Thespesia – Malvaceae
Tibouchina – Melastomaceae
Tieghamopanax – Ancardiaceae
Tieghemella – Sapotaceae
Tilia – Tiliaceae
Timonius – Rubiaceae
Tipuana – Leguminosae
Toona, see Cedrela
Torresea – Leguminosae
Torricellia – Cornaceae
Tournefortia – Boraginaceae
Toxylon, see Maclura
Trachylobium – Leguminosae
Treculia – Moraceae
Trema – Ulmaceae
Trewia – Euphorbiaceae
Trichadenia – Flacourtiaceae
Trichanthera – Acanthaceae
Trichilia – Meliaceae
Trichoscypha – Anacardiaceae
Trichospermum – Tiliaceae
Trigoniastrum – Trigoniaceae
Trigonopleura – Euphorbiaceae
Triomma – Burseraceae
Triplaris – Polygonaceae
Triplochiton – Sterculiaceae
Tristania – Myrtaceae

Tristira – Sapindaceae
Trochodendron – Magnoliaceae
Turraea – Meliaceae
Turraeanthus – Meliaceae
Tylostemon – Lauraceae

Uapaca – Euphorbiaceae
Ulmus – Ulmaceae
Umbellularia – Lauraceae
Unonopsis – Annonaceae
Upuna – Dipterocarpaceae
Urandra – Icacinaceae
Urophyllum – Rubiaceae
Uvaria – Annonaceae
Uvariastrum – Annonaceae

Vantanea – Humiriaceae
Vateriopsis – Dipterocarpaceae–
Vatica – Dipterocarpaceae
Vernonia – Compositae
Viburnum – Caprifoliaceae
Villaresia – Icacinaceae
Virola – Myristicaceae
Vismia – Hypericaceae
Vitex – Verbenaceae
Vochysia – Vochysiaceae
Vouacapoua – Leguminosae

Wallenia – Myrsinaceae
Warburgia – Canellaceae
Weihea – Rhizophoraceae
Weinmannia – Cunoniaceae
Wetria – Euphorbiaceae
Wormia – Dilleniaceae
Wrightia – Apocynaceae

Xanthophyllum – Polygalaceae
Xanthostemon – Myrtaceae
Xerospermum – Sapindaceae
Ximenia – Olacaceae
Xylia – Leguminosae
Xylocarpus – Meliaceae
Xylomelum – Proteaceae
Xylopia – Annonaceae

Xylosma – Flacourtiaceae
Xymalos – Monimiaceae

Zanha – Burseraceae
Zanthoxylum – Rutaceae

Zelkova – Ulmaceae
Zizyphus – Rhamnaceae
Zollernia – Leguminosae
Zuelania – Flacourtiaceae
Zygogynum – Magnoliaceae

APPENDIX III

TABLE OF CONVERSION FACTORS

Imperial Measure	Metric Units	Multiply by
inches	to millimetres	25·4
inches	,, centimetres	2·54
feet	,, millimetres	304·8
feet	,, centimetres	30·48
feet	,, metres	0·3047
square inches	,, square centimetres	6·45
square feet	,, square metres	0·093
cubic inches	,, cubic centimetres	16·39
cubic feet	,, litres	28·3
pounds (avoir)	,, kilogrammes	0·4536
gallons	,, litres	4·546
lb. per sq. in.	,, gm. per sq. cm.	70·3
lb. per sq. ft.	,, kgm per sq. metre	4·883

Selected Bibliography

The literature has grown to very large dimensions and only some of the more important books and papers have been listed. Serious students will find fuller bibliographies on the different subjects in the publications listed.

CHAPTER 1

Publications relating to standard trade names

Great Britain: British Standards Institution: BS 881 and 589: 1955. (A revision of this standard is in preparation.)

Australia: Standards Association of Australia: A.S.S. O.2: 1940, 'Nomenclature of Australian timbers'.

India: *Official list of trade names of Indian timbers* (Revised 3rd Edition). Indian Forest Records (New Series). Utilization, Vol. I, No. 7, 1938.

Wood Anatomy

CLARKE, S. H.: *Fine structure of the plant cell wall.* Nature, No. 3603, pp. 899-904, November 1938.

EAMES, A. J., and MACDANIELS, L. H.: *An introduction to plant anatomy.* McGraw-Hill Book Co., Inc., New York and London.

WISE, L. E.: *Wood chemistry.* Reinhold Publishing Cor., New York, 1944.

CHAPTER 2

BROWN, H. P.: *An elementary manual of wood technology.* Gov. of India, Central Publication Branch, Calcutta, 1925.

PHILLIPS, E. W. J.: *The identification of softwoods by their microscopic structure.* Forest Products Research Laboratory Bull. 22. H.M. Stationery Office, London, 1948.

CHAPTER 3

BROWN, H. P.: *An elementary manual of wood technology.* Gov. of India, Central Publication Branch, Calcutta, 1925.

RECORD, S. J.: *Identification of the timbers of temperate North America.* John Wiley & Co., New York, 1934.

Other publications covering Chapters 1 to 3 and not specifically referred to elsewhere in this Bibliography

FORSAITH, C. C.: *The technology of New York State timbers.* Technical Bulletin No. 18. New York State College of Forestry, Syracuse, 1926.

HALE, J. D.: *The identification of woods commonly used in Canada.* Bulletin 81. Canadian Forest Service, Ottawa, 1932.

HENDERSON, F. Y.: *Timber; its properties, pests and preservation* (2nd Edition). Crosby Lockwood & Son Ltd., London, 1944.

CHAPTER 4

Cross, diagonal, and spiral grain in timber. Commonwealth of Australia, Council for Sci. Ind. Res. Division of Forest Products, Trade Circular No. 13, 1933.

Selecting ash by inspection selection (supersedes leaflet No. 37). Princes Risborough Laboratory, Technical Note No. 54.

CLARKE, S. H.: *Home-grown timbers. Their anatomical structure and its relation to physical properties: Elm.* For. Prod. Res. Lab. Bull. 7. H.M. Stationery Office, London, 1930.

CLARKE, S. H.: *The influence of cell wall composition on the physical properties of beech wood.* Forestry, Vol. 10, pp. 143-148, 1936.

CLARKE, S. H.: *The distribution, structure and properties of tension wood in beech.* Forestry, Vol. 11, pp. 85-91, 1937.

ORSLER, R. J.: *The effects of irritant timbers.* Woodworking Industry, Vol. 26 (5), 28–9, 1969, Timberlab Paper No. 11.

PHILLIPS, E. W. J.: *The occurrence of compression wood in African pencil cedar.* Empire Forestry Journal, Vol. XVI, pp. 54-57, 1937.

PILLOW, M. Y., and LUXFORD, R. F.: *The structure, occurrence and properties of compression wood.* Technical Bulletin No. 546, U.S. Dept. Agric., 1937.

CHAPTERS 5 AND 6

General

The identification of timbers (supersedes leaflet No. 34). Princes Risborough Laboratory, Technical Note No. 56.

Identification of hardwoods. A lens key. Forest Products Research Laboratory Bull. 25. H.M. Stationery Office, London, 1960.

An atlas of end-grain photomicrographs for the identification of hardwoods. Forest Products Research Laboratory Bull. 26. H.M. Stationery Office, London, 1953.

Identification of hardwoods. A microscopic key. Forest Products Research Laboratory Bull. 46. H.M. Stationery Office, London, 1971.

CLARKE, S. H.: *A multiple-entry perforated card key with special reference to the identification of hardwoods.* New Phytologist, Vol. XXXVII, pp. 369-374, 1938.

RENDLE, B. J., and CLARKE, S. H.: *The problem of variation in the structure of wood.* Tropical Woods, No. 38, pp. 1-8, 1934.

RENDLE, B. J., and CLARKE, S. H.: *The diagnostic value of measurements in wood anatomy.* Tropical Woods, No. 40, pp. 27-37, 1934.

Descriptions of timbers of different regions

Handbook of Hardwoods. H.M. Stationery Office, London, 1956.

Handbook of Softwoods. H.M. Stationery Office, London, 1960.

BROWN, H. P., PANSHIN, A. J., and FORSAITH, C. C.: The text-book of wood technology, vol. I, *Structure, identification, defects, and uses of the commercial woods of the United States.* McGraw-Hill Book Co., New York, 2nd Edition, 1965.

CHALK, L., and RENDLE, B. J.: *British hardwoods.* Forest Products Research Laboratory Bulletin 3. H.M. Stationery Office, London, 1929.

DADSWELL, H. E., and BURNELL, M.: *Identification of the coloured woods of the genus Eucalyptus.* Council for Sci. and Ind. Res. Commonwealth of Australia. Bulletin No. 67, 1932.

DADSWELL, H. E., BURNELL, M., and ECKERSLEY, A. M.: *Identification of the light-coloured woods of the genus Eucalyptus.* Council for Sci. and Ind. Res. Commonwealth of Australia. Bulletin No. 78, 1934.

DADSWELL, H. E., and ECKERSLEY, A. M.: *Identification of the principal commercial Australian timbers other than Eucalyptus.* Council for Sci. and Ind. Res. Commonwealth of Australia. Bulletin No. 90, 1935.

DESCH, H. E.: *Manual of Malayan timbers.* Vols. I and II. Malayan Forest Records No. 15.

McELHANNEY, T. A., and associates: *Canadian woods, their properties and uses.* Dept. of Interior, Ottawa, 1935.

MOLL, J. W., and JANSSONIUS, H. H.: *Mikrographie des Holzes der auf Java vorkommenden Baumarten.* Vols. I to VI. E. J. Brill, Leiden.

PEARSON, R. S., and BROWN, H. P.: *Commercial timbers of India.* Vols. I and II. Gov. of India, Central Publication Branch, Calcutta, 1932.

RECORD, S. J., and HESS, R. W.: *Timbers of the New World.* Yale School of Forestry, 1943.

RECORD, S. J., and MELL, C. D.: *Timbers of Tropical America.* Yale University Press, 1924.

REYES, L. J.: *Philippine woods,* Tech. Bull. No. 7, Common. Philipp. Dept. Agric. Comm., 1938.

CHAPTER 7

Moisture content determination by the oven-drying method. Forest Products Research Laboratory, Technical Note No. 35, May 1969.

Report of the Forest Products Research Board, 1930. H.M. Stationery Office, London, 1931.

The moisture content of wood with special reference to furniture manufacture. Forest Products Research Laboratory, Bull. 5. H.M. Stationery Office, London, 1929.

Report of a seminar on moisture content determination of wood. Princes Risborough Laboratory, Timberlab Paper No. 24.

The moisture content of timber in new buildings. Record No. 5. H.M. Stationery Office, London.

Moisture of construction as a cause of decay in suspended ground floors. Princes Risborough Laboratory, Technical Note No. 14, Feb. 1966.

Loss of moisture from painted wood. Princes Risborough Laboratory, Technical Note No. 34, April 1969.

The movement of timbers (supersedes leaflet No. 47). Princes Risborough Laboratory, Technical Note No. 38, May 1969.

The moisture content of timber in use (supersedes leaflet No. 9). Princes Risborough Laboratory, Technical Note No. 46, October 1970.

BARKAS, W. W.: *Recent work on moisture in wood.* Forest Products Research Laboratory, Special Report No. 4. H.M. Stationery Office, London, 1938.

BATESON, R. G.: *Timber drying* (2nd Edition). Crosby Lockwood & Sons Ltd., 1946.

BÜSGEN, M.: *The structure and life of forest trees.* English translation by T. Thompson. Chapman & Hall, London, 1929.

CLARKE, S. H.: *The differential shrinkage of wood.* Forestry, Vol. 4, pp. 93-104, 1930.

CLARKE, S. H.: *Fine structure of the plant cell wall.* Nature, No. 3603, pp. 899-904, 1938.

CLARKE, S. H.: *Recent work on the growth, structure, and properties of wood.* Forest Products Research Laboratory, Special Report No. 5. H.M. Stationery Office, London, 1939.

DUNLAP, M. E.: *Electrical moisture meters for wood.* F.P.L., Madison, U.S.A. R1146 (Revised), 1944.

FREY-WYSSLING, A.: *Der Aufbau der pflanzlichen Zellwande.* Protoplasma No. 25, pp. 261-300, 1936.

JACCARD, P., and FREY-WYSSLING, A.: *Résistance et structure microscopique des bois.* Ber. eidgenöss. Mat. Prüf. Anst. No. 119, 1938.

KNIGHT, R. A. G.: *The determination of the moisture content of timber.* F.P.R.L. Bulletin No. 14. H.M. Stationery Office, London.

KOEHLER, A.: *The structure, properties and uses of wood.* McGraw-Hill Book Co., New York, 1924.

PILLOW, M. Y., and LUXFORD, R. F.: *The structure, occurrence and properties of compression wood.* Technical Bulletin No. 546, U.S. Dept. Agric., 1937.

RENDLE, B. J., and FRANKLIN, G. L.: Project 18, Progress Report 10, 1938 (unpublished report F.P.R.L., Princes Risborough).

RITTER, G. J., and MITCHELL, R. L.: Paper Trade Journal, Vol. 108, No. 6, p. 33, 1939.

TIEMANN, H. D.: *Wood technology.* Pitman Publishing Corporation, New York and Chicago, 3rd Edition, 1951.

CHAPTER 8

CLARKE, S. H.: *Recent work on the growth, structure, and properties of wood.* Forest Products Research Laboratory, Special Report No. 5. H.M. Stationery Office, 1939.

NEWLIN, J. A., and WILSON, T. R. C.: *The relation of the shrinkage and strength properties of wood to its specific gravity.* Bulletin 676. U.S. Dept. Agric., 1919.

CHAPTER 9

The testing of timbers at the Forest Products Research Laboratory. Record No. 1. H.M. Stationery Office, London, 1928.

American Society for Testing Materials. *Standard methods of testing small clear specimens of timber.* A.S.T.M. Designation D143-27 & A.S.T.M. Standards, Pt. 2, Non-metallic materials, pp. 408-444, 1933.

Grade stresses for structural timbers. (For Visually and mechanically graded material.) 3rd Edition, Metric Units Bulletin No. 47, H.M. Stationery Office, London, 1968.

Strength tests of structural timbers. Pt. 1, *General principles, with data on redwood from Gefle and Archangel.* Record No. 2. H.M. Stationery Office, London.

Strength tests of structural timbers. Pt. 2, *General procedure of selecting and testing joists, with data on British Columbian Douglas fir.* Record No. 8. H.M. Stationery Office, London.

Strength tests of structural timbers. Pt. 3, *The development of safe loads and stresses, with data on Baltic redwood and Eastern Canadian spruce.* Record No. 15. H.M. Stationery Office, London.

Strength tests of structural timbers. Pt. 4, *The development of the 800 lb. f. grade for redwood.* Record No. 28. H.M. Stationery Office, London.

The strength properties of European redwood and whitewood. Princes Risborough Laboratory, Special Report No. 24, 1967.

The strength of timber (supersedes leaflet No. 55). Princes Risborough Laboratory, Technical Note No. 10, May 1969.

Stress graded timber. Princes Risborough Laboratory, Technical Note No. 20, Jan. 1967.

Stress grading of timber. Princes Risborough Laboratory, Technical Note No. 25, May 1968.

Stress grading machines. Princes Risborough Laboratory, Technical Note No. 26, May 1968.

The strength properties of timber. Forest Products Research Laboratory Bull. 50, 2nd Edition. H.M. Stationery Office, London, 1969.

Grade stresses for European redwood and whitewood. Princes Risborough Laboratory Bull. 52, H.M. Stationery Office, 1969.

CHAPLIN, C. J.: *Mechanical and physical properties of timber. Tests of small, clear specimens.* Forest Products Research Project, I. H.M. Stationery Office, London, 1928.

CHAPLIN, C. J.: *A development of the compression tests in timber research.* Forest Products Research Laboratory, Princes Risborough, unpublished report.

CLARKE, S. H.: *On estimating the mechanical strength of the wood of ash.* Forestry, Vol. 7, pp. 26-31, 1933.

CLARKE, S. H.: *Recent work on the growth, structure, and properties of wood.* Forest Products Research Laboratory, Special Report No. 5. H.M. Stationery Office, London.

CLARKE, S. H.: *A comparison of certain properties of temperate and tropical timbers.* Tropical Woods, No. 52, pp. 1-11, 1937.

COVINGTON, S. A.: *Interim values for machine stress-graded Canadian Western hemlock.* Princes Risborough Laboratory, Timberlab Paper No. 37.

CURRY, W. T.: *Progress in stress-grading of boards.* Timber Trades Journ. 6.11. 65.

CURRY, W. T.: *Mechanical stress grading of timber.* Timberlab Paper No. 18.

CURRY, W. T.: *Interim stress values for machine stress-graded European redwood and whitewood.* Timberlab Paper No. 21.

GARRATT, G. A.: *The mechanical properties of wood.* John Wiley & Sons, New York, 1931.

HEARMON, R. F. S., and RIXON, B. E.: *Recommended limiting spans for stress-graded European redwood and whitewood.* Timberlab Paper No. 30.

KOEHLER, A.: *Causes of brashness in wood.* Technical Bulletin No. 342, U.S. Dept. Agric., 1933.

MARKWARDT, L. J.: *Comparative strength properties of woods grown in the United States.* Technical Bulletin No. 158, U.S. Dept. Agric., 1930.

SUNLEY, J. G.: *Review of non-destructive testing of timber.* Timberlab Paper No. 19.

SUNLEY, J. G., and CURRY, W. T.: *A machine to stress-grade timber.* Timber Trades Journ. 26.5.62.

SUNLEY, J. G., and HUDSON, W. M.: *Machine-grading of timber.* Forest Products Journal, April, 1964.

WILSON, T. R. C.: *Guide to the grading of structural timbers and the determination of working stresses.* Miscellaneous Publications No. 185, U.S. Dept. Agric., 1934.

Publications containing data from strength tests of timbers or allowable working stresses

Australia: *Handbook of structural timber design.* Division of Forest Products. Technical Paper No. 32, 1939.

British Standard Code of Practice 112 (1967). British Standards Institution, London.

Canada: *The mechanical properties of Canadian woods, together with their related physical properties.* Bulletin 82. Canadian Forest Service, Ottawa, 1933.

Recommendations for trussed rafters for domestic roofs. Princes Risborough Laboratory, Timberlab Paper No. 29.

The mechanical properties of wood, by G. H. ROCHESTER. Timber of Canada, Toronto, Canada, pp. 45-63, 1946.

The trussed rafter. Princes Risborough Laboratory, Technical Note No. 21, December 1966.

GRAINGER, G. D.: *The design and construction of trussed rafters (Wood,* Vol. 33 (5), 27–30, 1968). Princes Risborough Laboratory, Timberlab Paper No. 6.

Handbook of Hardwoods. H.M. Stationery Office, London, 1956.

Handbook of Softwoods. H.M. Stationery Office, London, 1960.

India: *The physical and mechanical properties of woods grown in India.* Indian Forest Records, Vol. XVIII, Pt. X, 1933.

Malaysia: *Malayan timbers tested in a green condition.* Malayan Forest Service, Trade Leaflet No. 5.

MAYO, A. P.: *Recommended spans for link and fan trussed rafters.* Princes Risborough Laboratory, Timberlab Paper No. 32.

U.S.A.: *Strength and related properties of woods grown in the United States.* Technical Bulletin No. 479, U.S. Dept. Agric., 1935.

CHAPTER 10

Gas producers for motor vehicles and their operation with forest fuels. Technical Communication. Imperial Forestry Bureau, Oxford, 1942.

BARKAS, W. W., HEARMON, R. F. S., and PRATT, C. H.: *Electrical resistance of wood.* Nature, No. 151, p. 83, 1943.

LUNDBERG, H. A.: *The calorific value of a cubic metre of wood.* Svensk. Papptidn., No. 46, pp. 293-300, 1943.

CHAPTER 11

The air-seasoning of sawn timber (supersedes leaflet No. 21). Princes Risborough Laboratory, Technical Note No. 36, May 1969.
Humidity controllers for kilns. Wood, No. 7, p. 216, 1942.
Some observations on timber-drying kilns and their proper use. F.P.R.L. Leaflet No. 36
General observations on the design of timber-drying kilns. F.P.R.L. Leaflet No. 30, 1943.
Timber seasoning. F.P.R.L. Record No. 4. H.M. Stationery Office, London.
Types of timber kilns. F.P.R.L. Record No. 13. H.M. Stationery Office, London.
Methods of kiln operation. F.P.R.L. Record No. 19. H.M. Stationery Office, London.
Kiln-drying schedules. F.P.R.L. Record No. 26. H.M. Stationery Office, London.
Kiln-drying schedules (supersedes leaflet No. 42). Princes Risborough Laboratory, Technical Note No. 37, May 1969.
Kiln operator's handbook. A guide to the kiln drying of timber. H.M. Stationery Office, 1961.
Forest Products Research Laboratory Leaflet No. 10, 1945 (double-stack kiln).
Forest Products Research Laboratory Leaflet No. 18, 1945 (single-stack kiln).
A primer on the chemical seasoning of Douglas fir. Forest Products Laboratory, Madison, Mimeographed Report No. R1278, 1938.
BATESON, R. G.: *Timber drying* (2nd Edition). Crosby Lockwood & Son Ltd., 1946.
PECK, E. C.: *The hygroscopic and antishrink values of chemicals in relation to chemical seasoning of wood.* F.P.L., Madison, Mimeographed Report No. R1270, 1941.
PRATT, G. H., and SKINNER, N. P.: *Timber drying in the U.K.—A survey of the plant and operating methods used in kiln drying.* Princes Risborough Laboratory, Timberlab Paper No. 41.
STEVENS, W. C.: *Practical kiln drying.* F.P.R.L. Special Report No. 3. H.M. Stationery Office, London.
STEVENS, W. C., HARRIS, P., and PRATT, G. H.: *A furnace heated timber-drying kiln.* Wood, No. 9, pp. 211-214, 1944.
STILLWELL, S. T. C.: *The principles of kiln-seasoning of timber.* F.P.R.L. Special Report No. 2. H.M. Stationery Office, London.
TIEMANN, H. D.: *Lessons in kiln-drying.* Southern Lumberman. Nashville, Tenn., 1938.
TORGESON, O. W.: *Furnace-type lumber dry kiln.* F.P.L., Madison, Mimeographed Report No. R1474, 1945.
WYNANDS, R. H.: *Estimation of kiln drying time for timber.* Princes Risborough Laboratory, Timberlab Paper No. 48.

CHAPTER 12

TIEMANN, D. H.: *Wood technology.* (Chapter on collapse.) Pitman Publishing Corporation, New York and Chicago, 1942.

CHAPTERS 13 AND 16

Blue-stain prevention in the United Kingdom. Princes Risborough Laboratory, Technical Note No. 2, January 1968.

Dry rot in wood. Forest Products Research Laboratory Bulletin 1. H.M. Stationery Office, London.

Decay in buildings (supersedes leaflet No. 6). Princes Risborough Laboratory, Technical Note No. 44, December 1969.

Decay in skirting boards in new houses. Princes Risborough Laboratory, Technical Note No. 13, February 1966.

Fungus growths in buildings following wetting from burst pipes. Princes Risborough Laboratory, Technical Note No. 15, June 1969.

Prevention of decay of wood in boats. Princes Risborough Laboratory Bull. No. 31, 1961.

Prevention of timber decay in industrialised building. Princes Risborough Laboratory, Technical Note No. 1, Revised February 1971.

Timber decay and its control (supersedes leaflet No. 39). Princes Risborough Laboratory, Technical Note No. 53, December 1971.

Sap-stain in timber—its cause, recognition, etc. (supersedes leaflet No. 12). Princes Risborough Laboratory, Technical Note No. 50, June 1971.

Dry rot investigations in an experimental house. F.P.R.L. Record No. 14. H.M. Stationery Office, London.

CARTWRIGHT, K. ST. G., and FINDLAY, W. P. K.: *Decay of timber and its prevention.* 2nd Edition, 1958. H.M. Stationery Office, London.

FINDLAY, W. P. K.: *Dry rot and other timber troubles.* Hutchinson Scientific and Technical Publications, London, 1953.

FINDLAY, W. P. K. *Timber pests and diseases.* Pergamon Press, Oxford, 1967.

Report of a symposium on painting wood treated with a water repellent preservative. Princes Risborough Laboratory, Timberlab Paper No. 36.

The resistance of timbers to impregnation with creosote. Bull. No. 54, H.M. Stationery Office, 1971.

Preservative treatments for external joinery timber. Princes Risborough Laboratory, Technical Note No. 24, Revised April 1971.

SAVORY, J. G.: *Dry rot—causes and remedies* (*Illus. Corp. Bldr.*, Vol. 162 (4882), 31–36, 1971). Princes Risborough Laboratory, Timberlab Paper No. 44.

SAVORY, J. G.: *Prevention of blue-stain in packaged Baltic redwood.* Princes Risborough Laboratory, Timberlab Paper No. 47.

SCHEFFER, T. C., and LINDGREN, R. M.: *Stains of sapwood and sapwood products and their control.* Technical Bulletin 714, U.S. Dept. Agric., 1940.

CHAPTERS 14 AND 16

Defects caused by ambrosia (pinhole borer) beetles: origin and recognition. Princes Risborough Laboratory, Technical Note No. 55 (supersedes leaflet No. 50).

Insect and marine borer damage to timber and woodwork. H.M. Stationery Office, London, 1967.

Recognition of decay and insect damage in timbers for aircraft and other purposes. Forest Products Research Laboratory. H.M. Stationery Office, London.

Termites and the protection of timber. Princes Risborough Laboratory, Leaflet No. 38 (Revised 1968).

The kiln sterilisation of Lyctus infested timber. F.P.R.L. Technical Note No. 43.

The house longhorn beetle (supersedes leaflet No. 14). F.P.R.L. Technical Note No. 39 (Revised March 1969).

Lyctus powder-post beetles. F.P.R.L. Leaflet No. 3 (Revised 1969).

The death-watch beetle. F.P.R.L. Leaflet No. 45 (Revised 1967).

The common furniture beetle. Princes Risborough Laboratory. Technical Note No. 45, October 1970 (supersedes leaflet No. 4).

The common furniture beetle. Princes Risborough Laboratory, Technical Note No. 47, November 1970 (supersedes leaflet No. 8).

Ernobius mollis. A bark borer of softwoods. F.P.R.L. Leaflet No. 54, 1963.

The protection of buildings and timber against termites. Forest Products Research Laboratory Bull. 24. H.M. Stationery Office, London, 1951.

Marine borers and methods of preserving timber against their attacks. F.P.R.L. Leaflet No. 46. H.M. Stationery Office, London.

FISHER, R. C.: *Beetles injurious to timber and furniture.* Forest Products Research Laboratory Bull. 19. H.M. Stationery Office, London, 1945.

KOFOID, C. A., and others: *Termites and termite control.* Berkeley University of California Press, 1934.

CHAPTER 15

Experiments on the preservation of mine timbers. F.P.R.L. Record No. 3. H.M. Stationery Office, London.

High density wood. Bull. resin. Prod. Chem. Co. No. 7, 1944.

Insecticidal smokes for control of wood-boring insects. Princes Risborough Laboratory, Technical Note No. 7 (Revised January 1968).

Methods of applying wood preservatives. Pt. 1: *Non-pressure methods.* F.P.R.L. Record No. 9. H.M. Stationery Office, London.

Non-pressure methods of applying wood preservatives. Princes Risborough Laboratory, Record No. 31, 1961.

Preservation of building timbers by boron diffusion treatment. Princes Risborough Laboratory, Technical Note No. 41, October 1969.

The fireproofing of timber. Technical Leaflet No. 4. Timber Research and Development Association.

The hot-and-cold open tank process of impregnating timber. Princes Risborough Laboratory, Technical Note No. 42, November 1969 (supersedes leaflet No. 11).

The preservation of farm timbers. Princes Risborough Laboratory, Technical Note No. 23 (2nd Edition, October 1970).

The preservation of Western Red Cedar shingles. Princes Risborough Laboratory, Technical Note No. 3, November 1968.

The preservative treatment of timber by brushing, spraying and immersion. Princes Risborough Laboratory, Leaflet No. 53, 1962.

Timbers for flooring. Forest Products Research Laboratory, Bull. 40. H.M. Stationery Office, London, 1957.

Wood preservatives. F.P.R. Record No. 17. H.M. Stationery Office, London.

BS 144: 1954: *Coal tar creosote for the preservation of timber.* Add. March, 1963.

BS 838: 1961: *Methods of test for the toxicity of wood preservatives to fungi.* Add. February, 1965.

BS 913: 1954: *Pressure creosoting of timber.*

BS 1282: 1959: *Classification of wood preservatives and their methods of application.* Add. August, 1961.

BS 3051: 1958: *Coal tar oil types of wood preservatives (other than creosote to BS 144)*. Add. March, 1963.

BS 3452: 1962: *Copper/chrome water-borne wood preservatives and their application.*

BS 3453: 1962: *Fluoride/arsenate/chromate/dinitrophenol water-borne wood preservatives and their application.*

BS 3651: 1962: *Method of test for toxicity of wood preservatives to the wood-boring insects* Anobium punctatum *and* Hylotrupes bajulus *by larval transfer.*

BS 3652: 1963: *Method of test for toxicity of wood preservatives to the wood-boring insect* Anobium punctatum *by egg-laying and larval survival.*

BS 3653: 1963: *Method of test for toxicity of surface-applied wood preservatives to the powder-post beetle* Lyctus brunneus.

BS Code of Practice 98: *Preservative treatments for constructional timber.*

BANKS, W. B.: *A standard test to measure the effectiveness of water repellent preservative solutions.* Princes Risborough Laboratory, Timberlab Paper No. 40.

COCKCROFT, R.: *Timber preservatives and methods of treatment.* Princes Risborough Laboratory, Timberlab Paper No. 46.

FINDLAY, W. P. K.: *Preservation of timber.* A. & C. Black, London, 1962.

HUNT, G. N., and GARRATT, G. A.: *Wood preservation.* McGraw-Hill Book Co., New York, 2nd Edition, 1953.

MOORE, G. E.: *Improved wood.* Forest Products Laboratories of Canada, Dominion Forest Service, 1945.

SMITH, D. N.: *Field trials on coal-tar creosote and copper/chrome/arsenic preservatives. Results from a new method of assessment.* (*Holzforschung*, Vol. 23 (6), 185–91, 1969). Princes Risborough Laboratory, Timberlab Paper No. 31.

SMITH, D. N., and WILLIAMS, A. I.: *Wood preservation by the boron diffusion process—the effect of moisture content on diffusion time* (*Journal of the Institute of Wood Science*, Vol. 4 (4) 3–10, 1969).

TACK, C. H.: *The economics of timber preservation in house-building* (*Building*, 19 September, 217 (6592), 38/92–38/95, 1969. Princes Risborough Laboratory, Technical Note No. 17.

WHITE, M. G.: *The inspection and treatment of houses for wood-boring insects* (*Journal of the Institute of Municipal Engineers*, Vol. 95 (7), 212–15, 1968). Princes Risborough Laboratory, Timberlab Paper No. 33.

WHITE, M. G.: *The sterilisation of exported packaging timber (to meet Quarantine Regulations).* Princes Risborough Laboratory, Timberlab Paper No. 49.

WHITE, M. G.: *Timber in freight container construction (with particular reference to international preservation requirements.* Princes Risborough Laboratory, Timberlab Paper No. 38.

CHAPTER 17

The more important published grading rules are as under

Grading rules and standard sizes for Empire hardwoods (square-edged boards and planks). 2nd Edition. Imperial Institute Advisory Committee on Timbers, 1937.

'Stress Grading Rules': BS 940, Pts. 1 and 2. British Standards Institution, London (Replaced by BS 1860: Part 1, 1959: and BSCP 112).

Structural use of timber: BS Code of Practice 112. Structural timber, measurement of characteristics affecting strength. Part 1, Softwood: BS 1860: 1959. British Standards Institution, London. Note: BS 1860 will shortly be republished in a revised form.

The Malayan grading rules for sawn hardwood timber. Forest Department, Malaysia, 1960.

Finland: Committee of Forest Products Association of Finland, 1936.

Eastern Canadian spruce; Maritime Lumber Bureau, Amherst, Nova Scotia.

Douglas fir (B.C.): Export grading rules and schedule of Douglas fir – Lumber. The Physics Lumber Inspection Bureau, Inc., Seattle, Washington, U.S.A.

American softwoods: Guide to the grading of structural timbers and the determination of working stresses. U.S. Dept. Agric., February 1934. Misc. Publications 185.

American lumber associations with their own rules

Douglas fir, Western red cedar, Western hemlock, Sitka spruce: West Coast Lumbermen's Association, 364 Stuart Building, Seattle 1, Washington, U.S.A.

Sequoia (Californian redwood): Californian Redwood Association, 405 Montgomery Street, San Francisco 4, California.

Southern cypress: Southern Cypress Manufacturers Association, 723 Barnett National Bank Buildings, Jacksonville 2, Florida.

Longleaf and shortleaf southern pine (the pitch pine group): Southern Pine Association, Canal Buildings, New Orleans 4, U.S.A.

North Carolina pine: North Carolina Pine Association, Norfolk, Virginia.

Eastern hemlock, tamarack: Northern Hemlock and Hardwood Manufacturers Association, Oshkosh, Wisconsin.

Sugar pine, California white pine, white fir, incense cedar, Douglas fir: Californian White and Sugar Pine Manufacturers Association, San Francisco, California.

Pondosa pine, Idaho white pine, larch, Douglas fir, white fir, cedar, spruce: Western Pine Manufacturers Association, Yeon Buildings, Portland 4, Oregon.

Northern white pine, Norway pine, Eastern spruce, tamarack: Northern Pine Manufacturers Association, 911 North Larrabee Street, Chicago 10, Illinois.

Hardwoods: National Hardwood Lumber Association Rules. National Hardwood Lumber Association, 59 East van Buren Street, Chicago 5, Illinois.

CHAPTER 18

Efficiency of adhesives for wood. Bull. No. 38, 4th Edition, H.M. Stationery Office, 1968.

Glues for wood. Princes Risborough Laboratory, Technical Note No. 9, October 1967.

Gluing preservative-treated wood. Princes Risborough Laboratory, Technical Note No. 31, May 1968.

Requirements and properties of adhesives for wood. Bull. No. 20, 5th Edition. H.M. Stationery Office, 1971.

Structural adhesives—the effect of the resorcinol shortage. Princes Risborough Laboratory, Timberlab Paper No. 35.

The gluing of wooden components. Princes Risborough Laboratory, Technical Note No. 4, October, 1967.

PERKINS, N. S., LANDSEM, P., and TRAYER, G. W.: *Modern connectors for timber construction.* U.S. Dept. Commerce, Natl. Wood Utilization, and U.S. Dept. Agric. Forest Serv., 1933.

PERRY, T. D.: *Modern plywood.* Pitman Publishing Corporation, New York and Chicago, 1942.

REECE, P. O.: *Recent experiences in the design of timber structures.* Journal Royal Institute of British Architects, Vol. 51, Third Series, No. 5, pp. 118-126, March 1944.

REECE, P. O.: *An introduction to the design of timber structures.* F. and M. Spon, 19.

SCHOLTEN, J. A.: *Timber connector joints, their strength and design.* Technical Bulletin No. 865. U.S. Dept. Agric., 1944.

WILSON, T. R. C.: *The glued laminated wooden arch.* Technical Bulletin No. 691. U.S. Dept. Agric., 1939.

WOOD, A. D., and LINN, T. G.: *Plywoods.* W. & A. K. Johnston Ltd., Edinburgh, 1946.

General

Wood handbook. Forest Products Laboratory, U.S. Dept. Agric., 1935.

Trade Literature: Timber Engineering Co., 1319 Eighteenth Street, N.W., Washington, D.C.

CHAPTER 19

An investigation of some properties of tempered hardboard. Timberlab Paper No. 25, 1970.

High density wood. Bull. resin Prod. Chem. Co., No. 7, 1944.

Jablo Airscrew Blades, their manufacture from the Log to the Finished Component by F. Hills & Sons Ltd. Aircraft Production, April 1942.

Properties of Particleboard in Building. A guide to its manufacture and use. Timber Research & Development Association, Hughenden Valley, High Wycombe, Bucks.

AINGE, ROY: *A progress report from C.P.A. Timber and Plywood* Particleboard Supplement, November 15, 1972.

BEECH, J. C.: *Performance of wood particleboard under adverse conditions. Timber and Plywood* Supplement, November 15, 1972.

BENNETT, G. A.: *An investigation of some properties of medium hardboard.* Timberlab Paper No. 4, 1969.

BENNETT, G. A.: *Finding out more about fibre building board.* Timberlab Paper No. 42, 1971.

BENNETT, G. A., and HUDSON, R. W.: *Wood-based panel products as predictable 'engineering' materials.* New Technology, October 1971.

BROWN, W. H.: *Testing facilities available to particleboard. Timber and Plywood* Particleboard Supplement, November 15, 1972.

FARMER, R. H.: *Chemistry in the utilization of wood.* Pergamon, 1967.

FIDOR: Fibre Building Board Development Organisation Limited, 6/7 Buckingham Street, London, WC2N 6BZ, publishes technical leaflets on all types of fibre building boards.

HARWELL INDUSTRIAL SERVICES: *Industrial radiation processing. Gamma-rays. Electron beams materials. Services.* Applied Radiation Chemistry Group, A.E.R.E., Harwell,

HAYWARD, C. H.: *Manufacture and Uses of Improved Wood. Wood*, May 1939.

HERN, B. H. K.: *The market potential for wood chipboard. Timber and Plywood* Particleboard Supplement, November 15, 1972.

HUDSON, R. W.: *Timberlab research on exterior grade particleboard. Timber and Plywood* Particleboard Supplement, November 24, 1971.

HUDSON, R. W.: *Improving chipboard. Timberlab News*, No. 7, December 1970.

HUDSON, R. W., and GRANT, C.: *Fungicides and fire-retardants in UF bonded chipboard. Results of tests to determine effect and feasibility. Timber Trades Journal*, June 26, 1971.

JABLONSKY, BRUNO: *The Case for Wood. Mechanical and Other Properties Improved by Impregnation with Synthetic Resins. Flight*, December 16, 1937.

LAIDLAW, R. A., and HUDSON, R. W.: *Chipboard developments in Europe.* Timberlab Paper No. 20, 1969.

LYNAM, F. C.: *The international picture. Timber and Plywood* Particleboard Supplement, November 15, 1972.

MOORE, F. C.: *Improved wood.* Forest Products Laboratories of Canada, Dominion Forest Service, 1945.

PERRY, T. D.: *Modern plywood.* Pitman, New York and Chicago, 1942.

SAVORY, J. G.: *Fungus Resistance of Standard Hardboard.* Timberlab Paper No. 15, 1969.

SAVORY, J. G.: *Testing the Fungus Resistance of Board Materials.* Timberlab Paper No. 26, 1970.

STAMM, A. J.: *Wood and cellulose science, dimensional stabilization.* The Ronald Press Company, New York, 1964.

STAMM, A. J. and HARRIS, E. E.: *Chemical processing of wood.* Chemical Publishing Co., 1953.

WOOD, A. D. and LINN, T. G.: *Plywood.* Johnston, Edinburgh, 1946.

Index

DUDLEY

COLLEGE
OF
EDUCATION